THE FANTASTIC INVENTIONS OF NIKOLA TESLA

BY
NIKOLA TESLA
&
DAVID H. CHILDRESS

ADVENTURES UNLIMITED
STELLE, ILLINOIS

N. TESLA.

APPARATUS FOR TRANSMITTING ELECTRICAL ENERGY.

APPLICATION FILED JAN. 18, 1902. RENEWED MAY 4, 1907.

1,119,732. Patented Dec. 1, 1914.

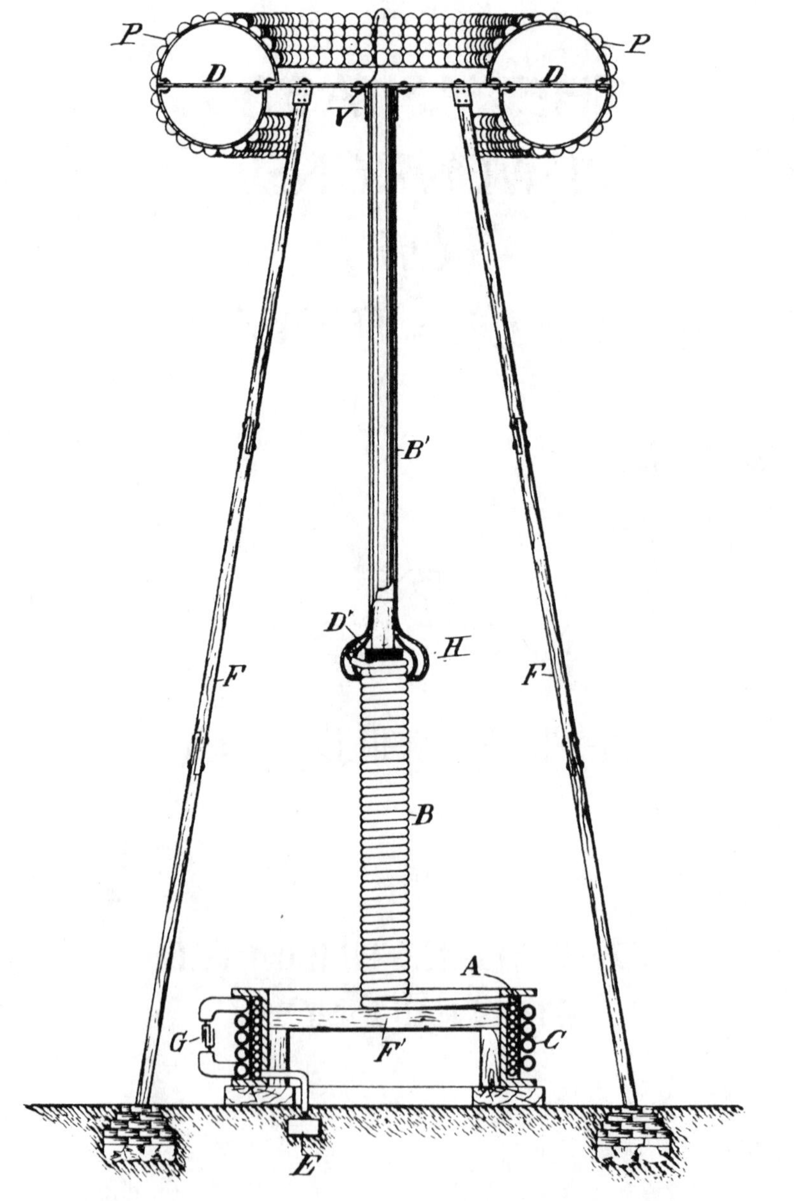

WITNESSES:

M. Lawson Dyer

Benjamin Miller.

Nikola Tesla, INVENTOR,

BY Kerr, Page & Cooper, his ATTORNEYS.

Nikola Tesla

This book is dedicated to Nikola Tesla and to the scientists and engineers who continue to forge ahead with an open mind into Tesla Technology.

Special thanks to Oliver Nichelson, John Ratzlaff, Mark Seifer, Leland Anderson, Mark Carlotto, Metascience Foundation, Moray B. King, Col. Tom Bearden, Kelley Net, The International Tesla Society, Steve Elswick, Toby Grotz, the Unarius Academy of Science, the Stellar Research Institute, Electric Spacecraft Journal, The Tesla Book Co., and all who publish on Nikola Tesla and his work.

The Fantastic Inventions of Nikola Tesla

ISBN 10: 0-932813-19-4
ISBN 13: 978-0-932813-19-0

Published by:
Adventures Unlimited Press
One Adventure Place
Kempton, Illinois 60946 USA
auphq@frontiernet.net

www.adventuresunlimitedpress.com

12 11 10 9 8 7 6 5 4 3 2

Other Books in this Series:
THE TESLA PAPERS
ETHER TECHNOLOGY
SECRETS OF THE UNIFIED FIELD
TAPPING THE ZERO POINT ENERGY
QUEST FOR ZERO POINT ENERGY
SECRETS OF THE UNIFIED FIELD
THE ANTI-GRAVITY HANDBOOK
ANTI-GRAVITY & THE WORLD GRID
ANTI-GRAVITY & THE UNIFIED FIELD
THE FREE ENERGY DEVICE HANDBOOK
THE TIME TRAVEL HANDBOOK
ATLANTIS & THE POWER SYSTEM OF THE GODS

PG-277

THE FANTASTIC INVENTIONS OF NIKOLA TESLA

TABLE OF CONTENTS

Chapter 1

ORIGINAL 1890'S BIOGRAPHICAL SKETCH

While a large portion of the European family has been surging westward during the last three or four hundred years, settling the vast continents of America, another, but smaller, portion has been doing frontier work in the Old World, protecting the rear by beating back the "unspeakable Turk" and reclaiming gradually the fair lands that endure the curse of Mohammedan rule. For a long time the Slav people—who, after the battle of Kosovopjolje, in which the Turks defeated the Servians, retired to the confines of the present Montenegro, Dalmatia, Herzegovina and Bosnia, and "Borderland" of Austria—knew what it was to deal, as our Western pioneers did, with foes ceaselessly fretting against their frontier ; and the races of these countries, through their strenuous struggle against the armies of the Crescent, have developed notable qualities of bravery and sagacity, while maintaining a patriotism and independence unsurpassed in any other nation.

It was in this interesting border region, and from among these valiant Eastern folk, that Nikola Tesla was born in the year 1857, and the fact that he, to-day, finds himself in America and one of our foremost electricians, is striking evidence of the extraordinary attractiveness alike of electrical pursuits and of the country where electricity enjoys its widest application.

Mr. Tesla's native place was Smiljan, Lika, where his father was an eloquent clergyman of the Greek Church, in which, by the way, his family is still prominently represented. His mother enjoyed great fame throughout the country side for her skill and originality in needlework, and doubtless transmitted her ingenuity to Nikola; though it naturally took another and more masculine direction.

The boy was early put to his books, and upon his father's removal to Gospic he spent four years in the public school, and later, three years in the Real School, as it is called. His escapades were such as most quickwitted boys go through, although he varied the programme on one occasion by getting imprisoned in a remote mountain chapel rarely visited for service; and on another occasion by falling headlong into a huge kettle of boiling milk, just drawn from the paternal herds. A third curious episode was that connected with his efforts to fly when, attempting to navigate the air with the aid of an old umbrella, he had, as might be expected, a very bad fall, and was laid up for six weeks.

About this period he began to take delight in arithmetic and physics. One queer notion he had was to work out everything by three or the power of three. He was now sent to an aunt at Cartstatt, Croatia, to finish his studies in what is known as the Higher Real School. It was there that, coming from the rural fastnesses, he saw a steam engine for the first time with a pleasure that he remembers to this day. At Cartstatt he was so diligent as to compress the four years' course into three, and graduated in 1873. Returning home during an epidemic of cholera, he was stricken down by the disease and suffered so seriously from the consequences that his studies were interrupted for fully two years. But the time was not wasted, for he had become passionately fond of experimenting, and as much as his means and leisure permitted devoted his energies to electrical study and investigation. Up to this period it had been his father's intention to make a priest of him, and the idea hung over the young physicist like a very sword of Damocles. Finally he prevailed upon his worthy but reluctant sire to send him to Gratz in Austria to finish his studies at the Polytechnic School, and to prepare for work as professor of mathematics and physics. At Gratz he saw and operated a Gramme machine for the first time, and was so struck with the objections to the use of commutators and brushes that he made up his mind there and then to remedy that defect in dynamo-electric machines. In the second year of his course he abandoned the intention of becoming a teacher and took up the engineering curriculum. After three years of absence he returned home, sadly, to see his father die; but, having resolved to settle down in Austria, and recognizing the value of linguistic acquirements, he went to Prague and then to Buda-Pesth with the view of mastering the languages he deemed necessary. Up to this time he had never realized the enormous sacrifices that his parents had made in promoting his education, but he now began to feel the pinch and to grow unfamiliar with the image of Francis Joseph I. There was considerable lag between his dispatches and the corresponding remittance from home; and when the mathematical expression for the value of the lag assumed the shape of an eight laid flat on its back, Mr. Tesla became a very fair example of high thinking and plain living, but he made up his mind to the struggle and determined to go through depending solely on his own resources. Not desiring the fame of a faster, he cast about for a livelihood, and through the help of friends he secured a berth as assistant in the engineering department of the government telegraphs. The salary was five dollars a week. This brought him into direct contact with practical electrical work and ideas, but it is needless to say that his means did not admit of much experimenting. By the time he had extracted several hundred thousand square and cube roots for the public benefit, the limitations, financial and otherwise, of the position had become painfully apparent, and he concluded that the best thing to do was to make a valuable invention. He proceeded at once to make inventions, but their value was visible only to the eye of faith, and they brought no grist to the mill. Just at this time the telephone made its appearance in Hungary, and the success of that great invention determined his career, hopeless as the profession had thus far seemed to him. He associated himself at once with telephonic work, and made various telephonic inventions, including an operative repeater; but it did not take him long to discover that, being so remote from the scenes of electrical activity, he was apt to spend time on aims and results already reached by others, and to lose touch. Longing for new opportunities and anxious for the development of which he felt himself possible, if once he could place himself within the genial and direct influences of the gulf streams of electrical thought, he broke away from the ties and traditions of the past, and in 1881 made his way to Paris. Arriving in that city, the ardent young Likan obtained employment as an electrical engineer with one of the largest electric lighting companies. The next year he went to Strasburg to install a plant, and on returning to Paris sought to carry out a number of ideas that had now ripened into inventions. About this time, however, the remarkable progress of America in electrical industry attracted his attention, and once again staking everything on a single throw, he crossed the Atlantic.

Mr. Tesla buckled down to work as soon as he landed on these shores, put his best thought and skill into it, and soon saw openings for his talent. In a short while a proposition was made to him to start his own company, and, accepting the terms, he at once worked up a practical system of arc lighting, as well as a potential method of dynamo regulation, which in one form is now known as the "third brush regulation." He also devised a thermo-magnetic motor and other kindred devices, about which little was published, owing to legal complications. Early in 1887 the Tesla Electric Company of New York was formed. and not long after that Mr. Tesla produced his admirable and epoch-marking motors for multiphase alternating currents, in which, going back to his ideas of long ago, he evolved machines having neither commutator nor brushes. It will be remembered that about the time that Mr. Tesla brought out his motors, and read his thoughtful paper before the American Institute of Electrical Engineers, Professor Ferraris, in Europe, published his discovery of principles analogous to those ennunciated by Mr. Tesla. There is no doubt, however, that Mr. Tesla was an independent inventor of this rotary field motor, for although anticipated in dates by Ferraris, he could not have known about Ferraris' work as it had not been published. Professor Ferraris stated himself, with becoming modesty, that he did not think Tesla could have known of his (Ferraris') experiments at that time, and adds that he thinks Tesla was an independent and original inventor of this principle. With such an acknowledgment from Ferraris there can be little doubt about Tesla's originality in this matter.

Mr. Tesla's work in this field was wonderfully timely, and its worth was promptly appreciated in various quarters. The Tesla patents were acquired by the Westinghouse Electric Company, who undertook to develop his motor and to apply it to work of different kinds. Its use in mining, and its employment in printing, ventilation, etc., was described and illustrated in *The Electrical World* some years ago. The immense stimulus that the announcement of Mr. Tesla's work gave to the study of alternating current motors would, in itself, be enough to stamp him as a leader.

Mr. Tesla is only 35 years of age. He is tall and spare, with a clean-cut, thin, refined face, and eyes that recall all the stories one has read of keenness of vision and phenomenal ability to see through things. He is an omnivorous reader, who never forgets; and he possesses the peculiar facility in languages that enables the least educated native of eastern Europe to talk and write in at least half a dozen tongues. A more congenial companion cannot be desired for the hours when one "pours out heart affluence in discursive talk," and when the conversation, dealing at first with things near at hand and next to us, reaches out and rises to the greater questions of life, duty and destiny.

In the year 1890 he severed his connection with the Westinghouse Company, since which time he has devoted himself entirely to the study of alternating currents of high frequencies and very high potentials, with which study he is at present engaged. No comment is necessary on his interesting achievements in this field; the famous London lecture published in this volume is a proof in itself. His first lecture on his researches in this new branch of electricity, which he may be said to have created, was delivered before the American Institute of Electrical Engineers on May 20, 1891, and remains one of the most interesting papers read before that society. It will be found reprinted in full in *The Electrical World,* July 11, 1891. Its publication excited such interest abroad that he received numerous requests from English and French electrical engineers and scientists to repeat it in those countries, the result of which has been the interesting lecture published in this volume.

The present lecture presupposes a knowledge of the former, but it may be read and understood by any one even though he has not read the earlier one. It forms a sort of continuation of the latter, and includes chiefly the results of his researches since that time.

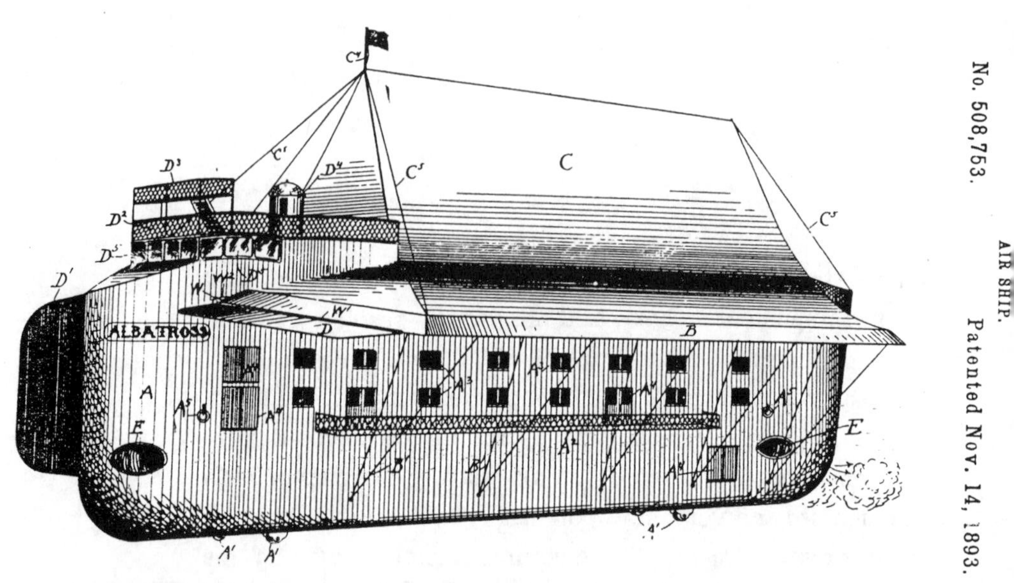

An 1893 design for an electric spacecraft. Tesla was keenly aware of other inventions and patents.

A 1924 design for a centrifugal spacecraft launcher by Mark P. Madden published in *Science and Invention.*magazine. The design also included transmitting power to the craft, which dragged a long aerial behind it, an idea obviously taken from Tesla.

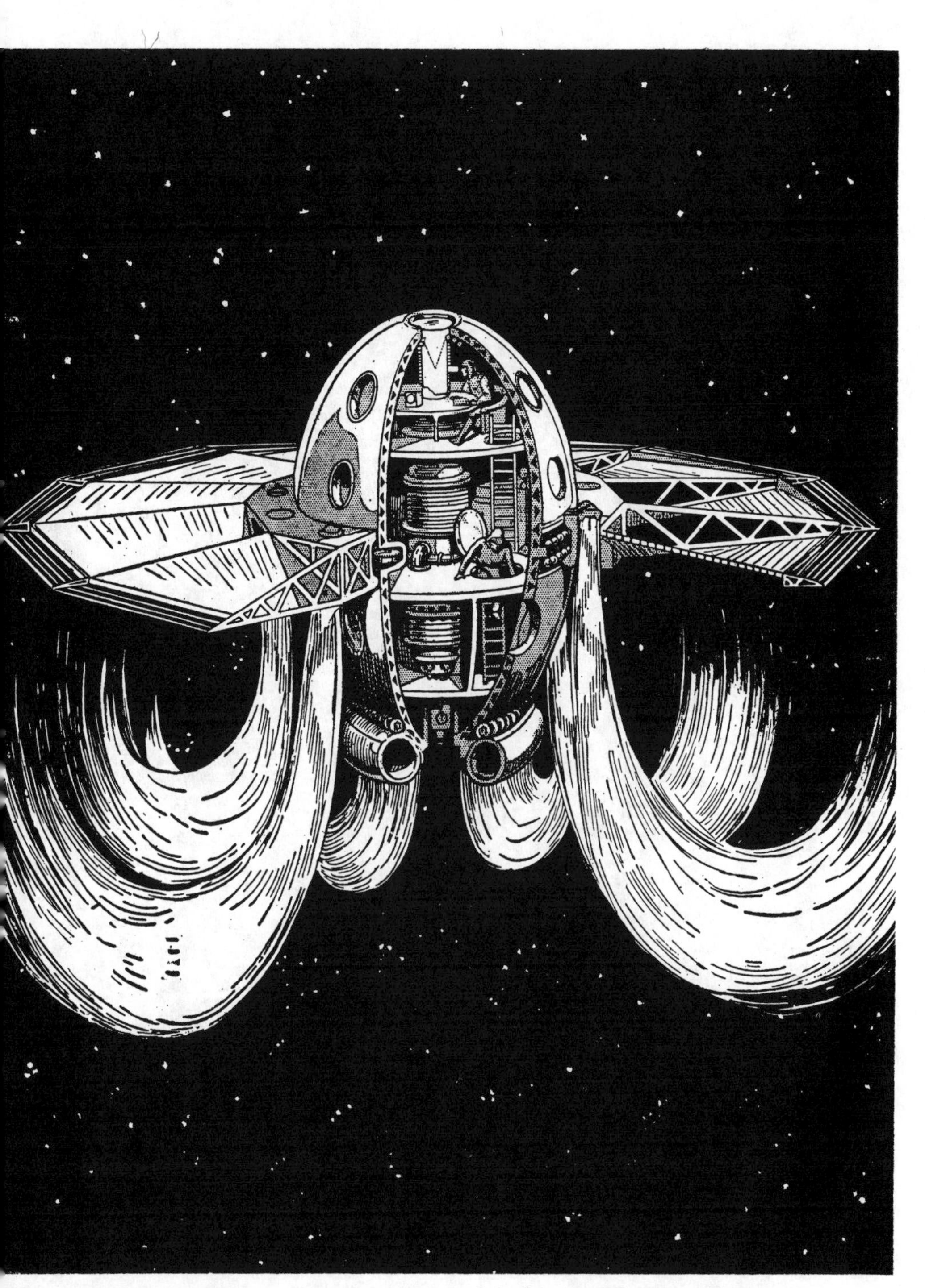

This spaceship design of an "electron wind" craft is from 1927! Inventor Franz A. Ulinski described a number of spacecraft in the mid-1920s, including a spherical "interplanetary ship," and a similar "cosmic ship."

Electrical oscillator activity ten million Horsepower

Power transmission without wires

New York, January 1, 1904

I wish to announce that in connection wtth the commercial introduction of my inventions I shall render professional services in the general capacity of consulting electrician and engineer.

The near future, I expect with confidence, will be a witness of revolutionary departures in the production, transformation and transmission of energy, transportation, lighting, manufacture of chemical compounds, telegraphy, telephony and other arts and industries.

In my opinion, these advances are certain to follow from the universal adoption of high-potential and high-frequency currents and novel regenerative processes of refrigeration to very low temperatures.

Much of the old apparatus will have to be improved, and much of the new developed, and I believe that while furthering my own inventions, I shall be more helpful in this evolution by placing at the disposal of others the knowledge and experience I have gained.

Special attention will be given by me to the solution of problems requiring both expert information and inventive resource—work coming within the sphere of my constant training and predilection.

I shall undertake the experimental investigation and perfection of ideas, methods and appliances, the devising of useful expedients and, in particular, the design and construction of machinery for the attainment of desired results.

Any task submitted to and accepted by me, will be carried out thoroughly and conscientiously.

Laboratory, Long Island, N. Y.
Residence, Waldorf, New York City.

Nikola Tesla

Burning atmospheric nitrogen by high frequency discharge twelve million volts

Secrecy non-interference and multiplicity of messages

Chapter 2

THE FIRST PATENTS (1886–1888)

(No Model.) 2 Sheets—Sheet 1.

N. TESLA.

ELECTRIC ARC LAMP.

No. 335,786. Patented Feb. 9, 1886.

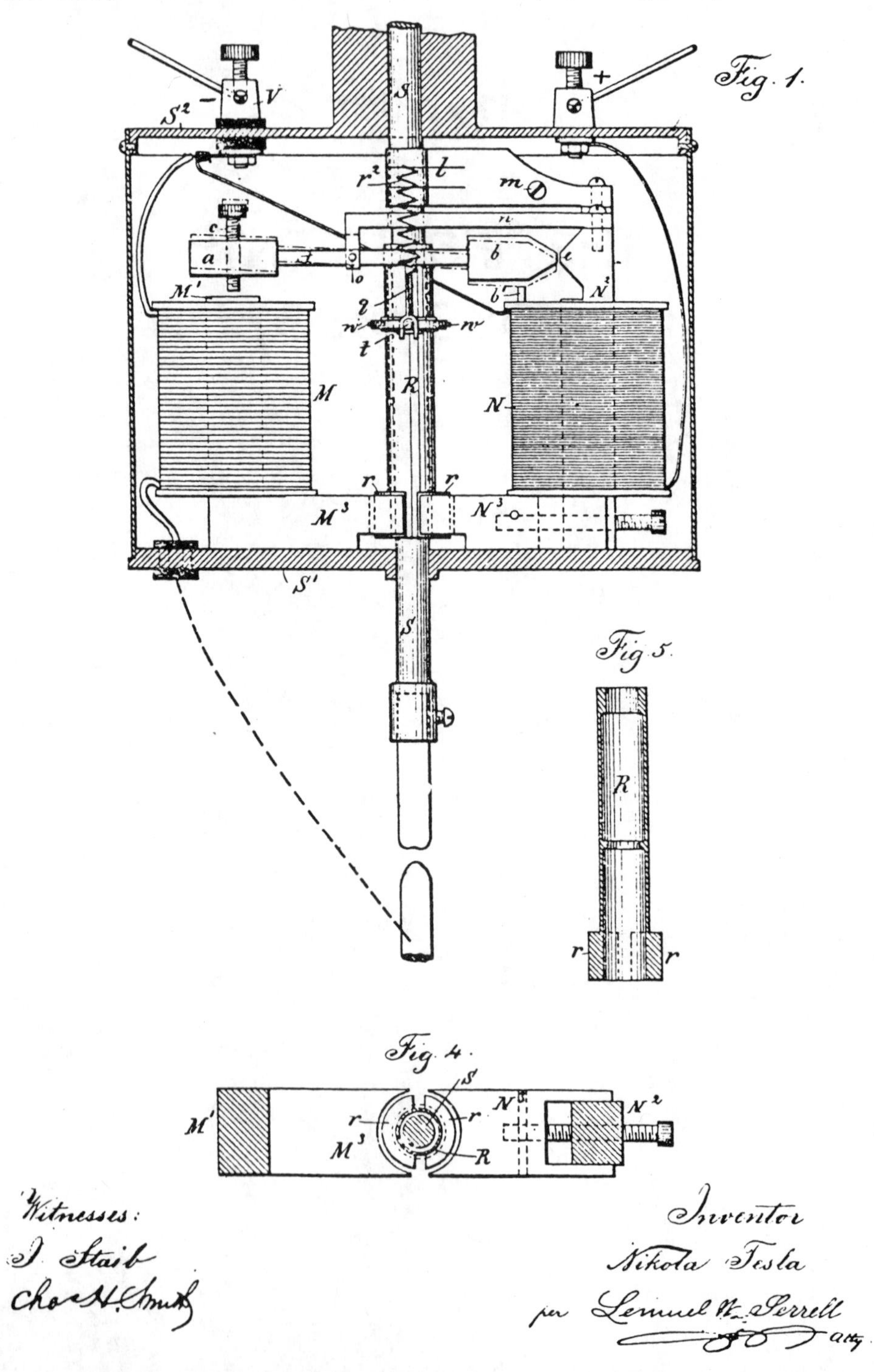

Tesla's first patent, applied for in 1885, and given to him on Feb. 9, 1886.

(No Model.)

N. TESLA.

COMMUTATOR FOR DYNAMO ELECTRIC MACHINES.

No. 334,823. Patented Jan. 26, 1886.

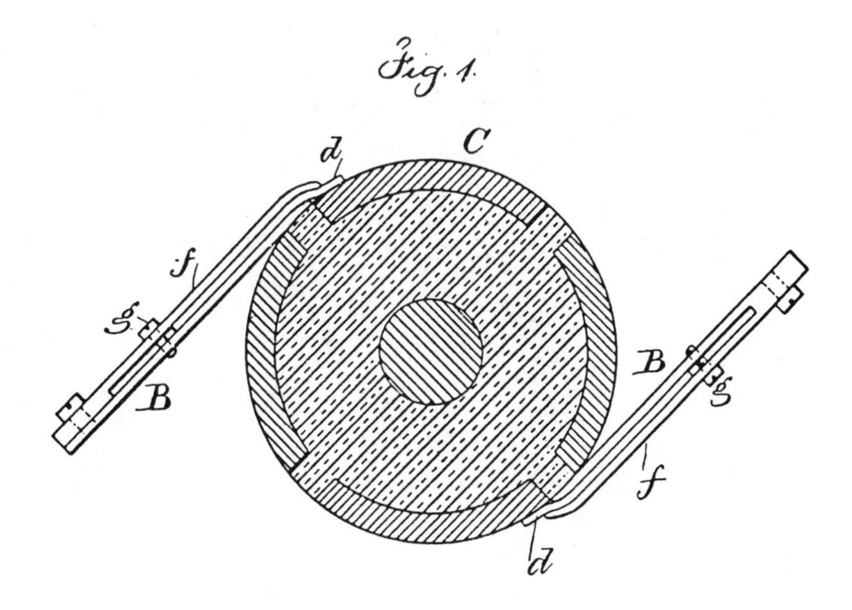

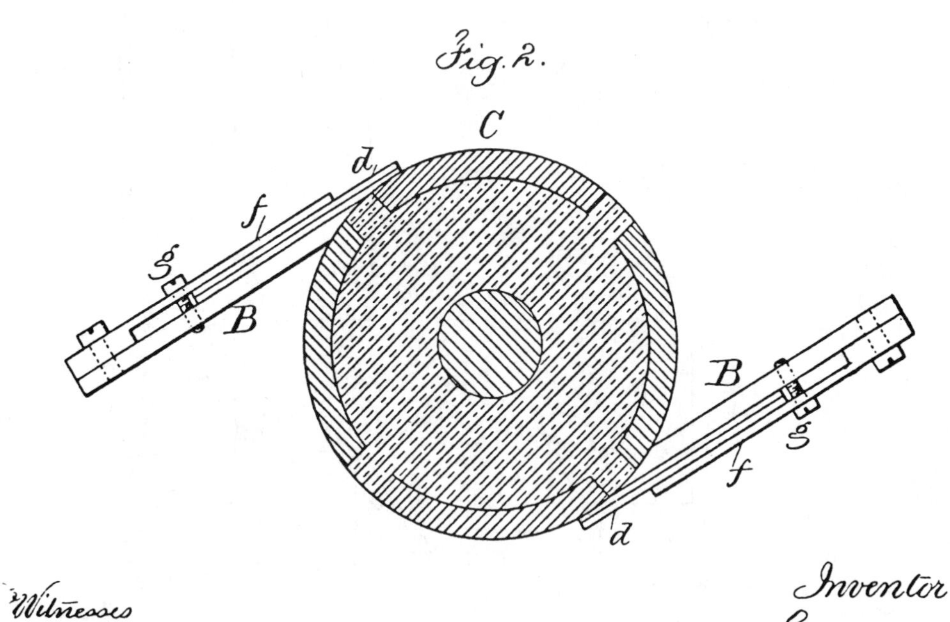

Witnesses

Chas. H. Smith

J. Staib

Inventor

Nikola Tesla.

per Lemuel W. Serrell

att'y.

(No Model.) 2 Sheets—Sheet 1.

N. TESLA.

REGULATOR FOR DYNAMO ELECTRIC MACHINES.

No. 336,961. Patented Mar. 2, 1886.

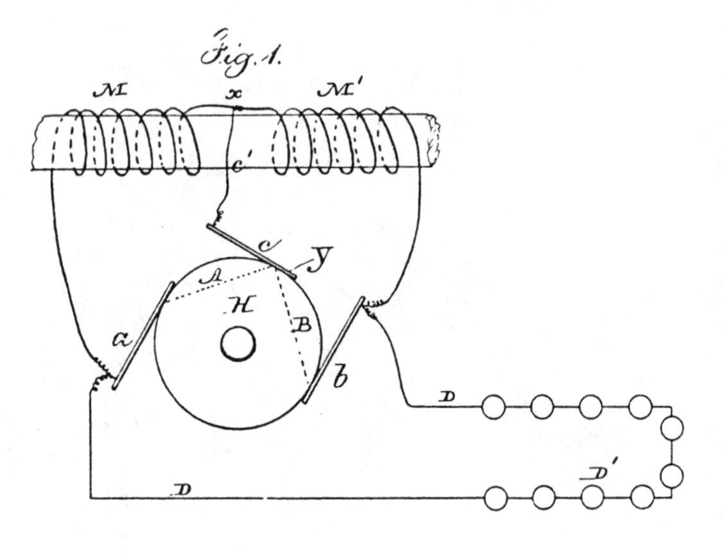

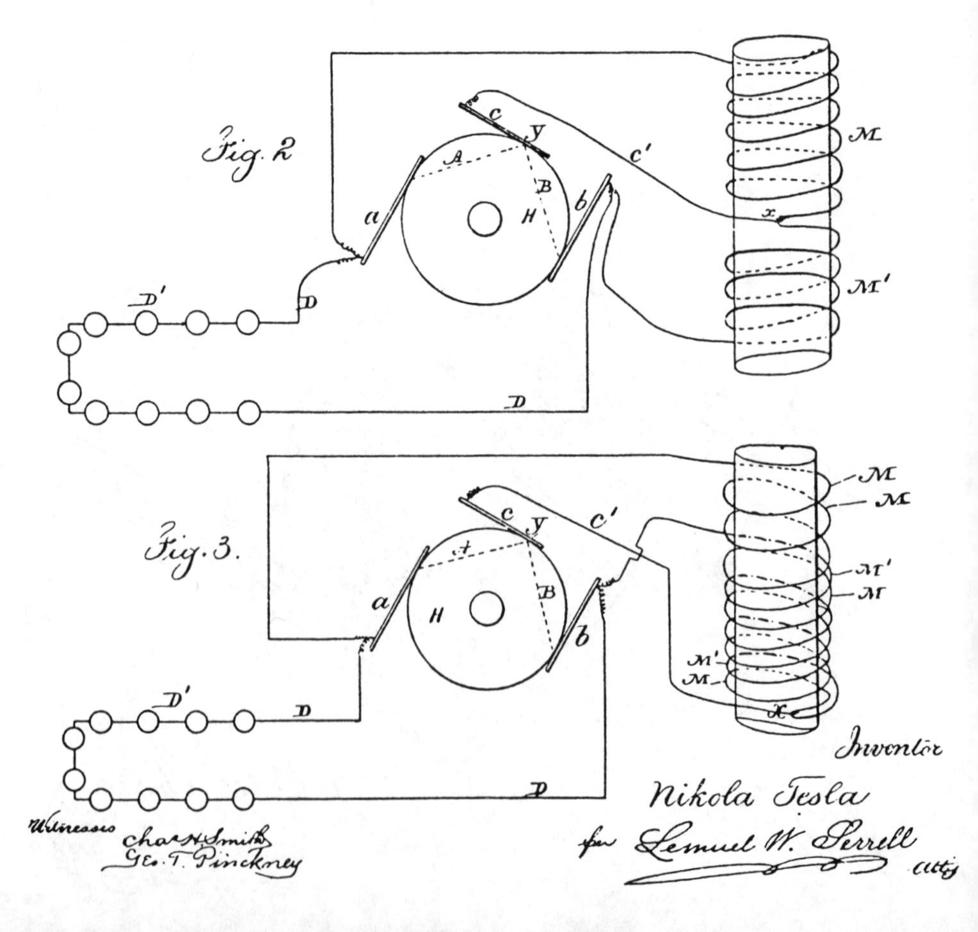

(No Model.) 3 Sheets—Sheet 1.

N. TESLA.

DYNAMO ELECTRIC MACHINE.

No. 359,748. Patented Mar. 22, 1887.

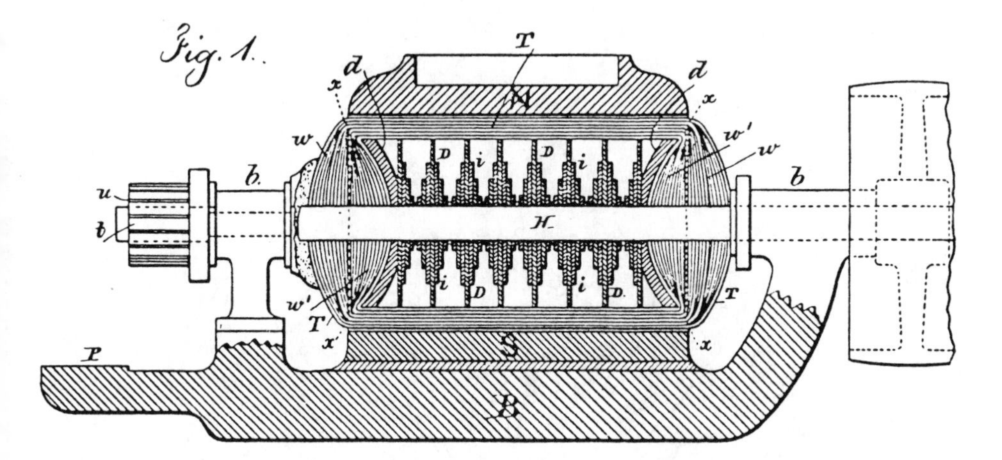

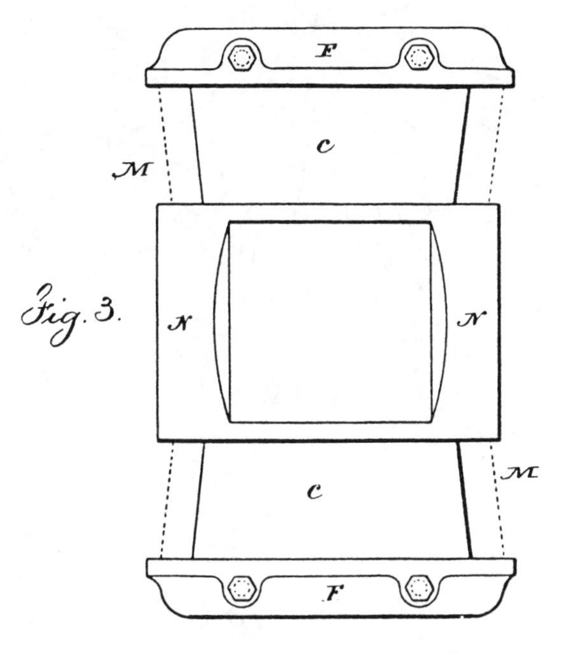

Witnesses

Chas H. Smith

J. Stail

Inventor

Nikola Tesla

per Lemuel W. Serrell

att

(No Model.)

3 Sheets—Sheet 3.

N. TESLA.

DYNAMO ELECTRIC MACHINE.

No. 359,748.

Patented Mar. 22, 1887.

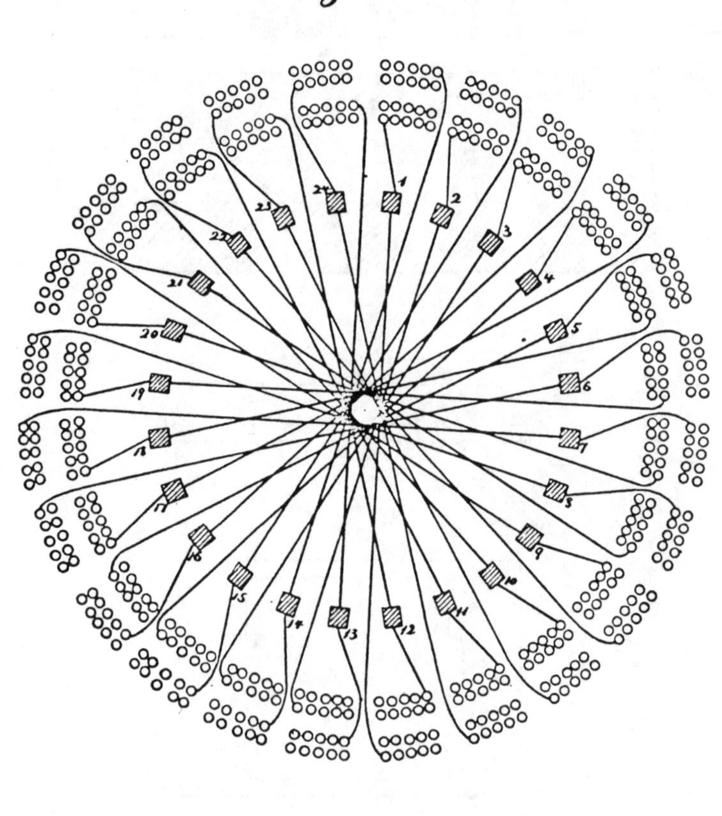

Inventor

Witnesses

Chas H. Smith
Geo. T. Pinckney

Nikola Tesla

for Lemuel W. Serrell

atty

(No Model.) 2 Sheets—Sheet 1.

N. TESLA.

COMMUTATOR FOR DYNAMO ELECTRIC MACHINES.

No. 382,845. Patented May 15, 1888.

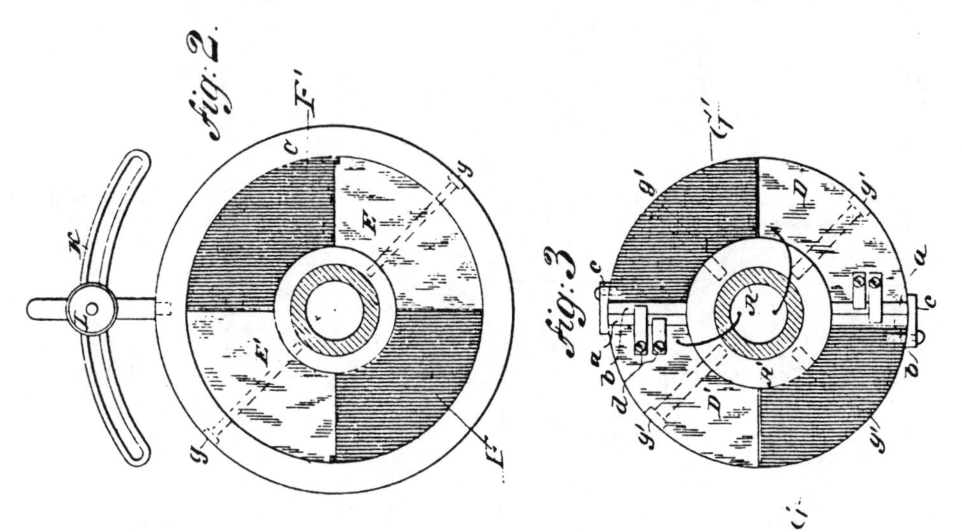

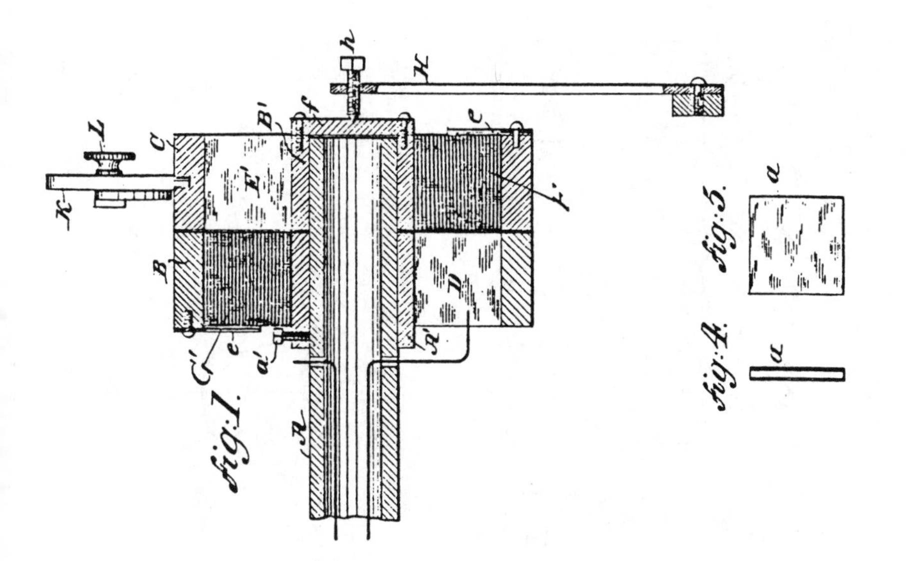

WITNESSES:
Robt. F. Gaylord
Robt. P. Harlow

INVENTOR.
Nikola Tesla.
BY
Duncan, Curtis & Page
ATTORNEYS.

(No Model.) 4 Sheets—Sheet 1.

N. TESLA.

ELECTRO MAGNETIC MOTOR.

No. 381,968. Patented May 1, 1888.

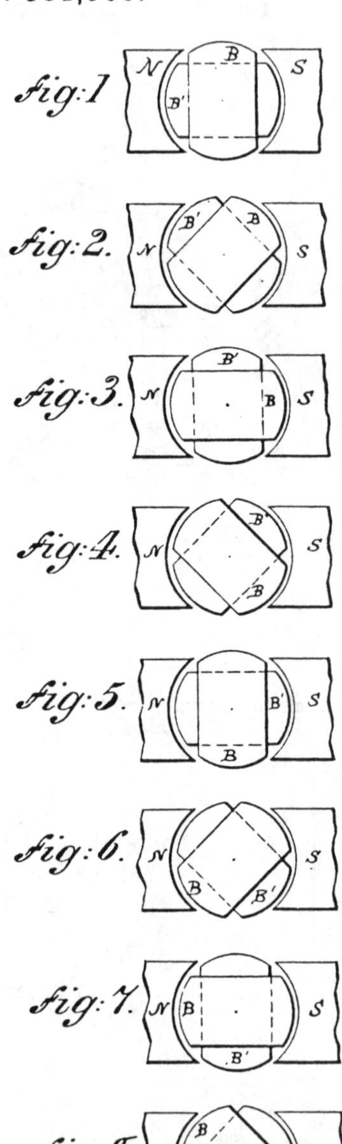

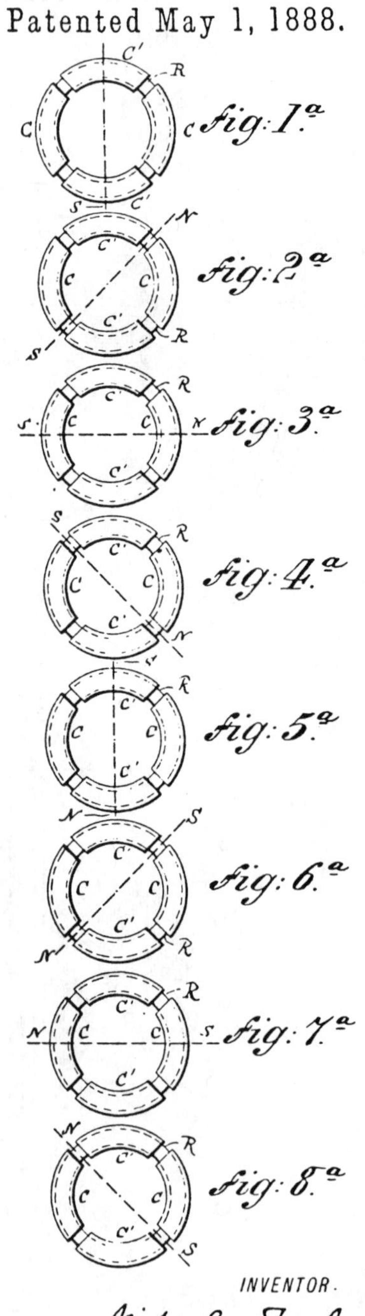

WITNESSES:

Frank E. Hartley

Frank B. Murphy

INVENTOR.

Nikola Tesla

BY

Duncan, Curtis &

ATTORNEYS.

(No Model.)

4 Sheets—Sheet 3.

N. TESLA.

ELECTRO MAGNETIC MOTOR.

No. 381,968.

Patented May 1, 1888.

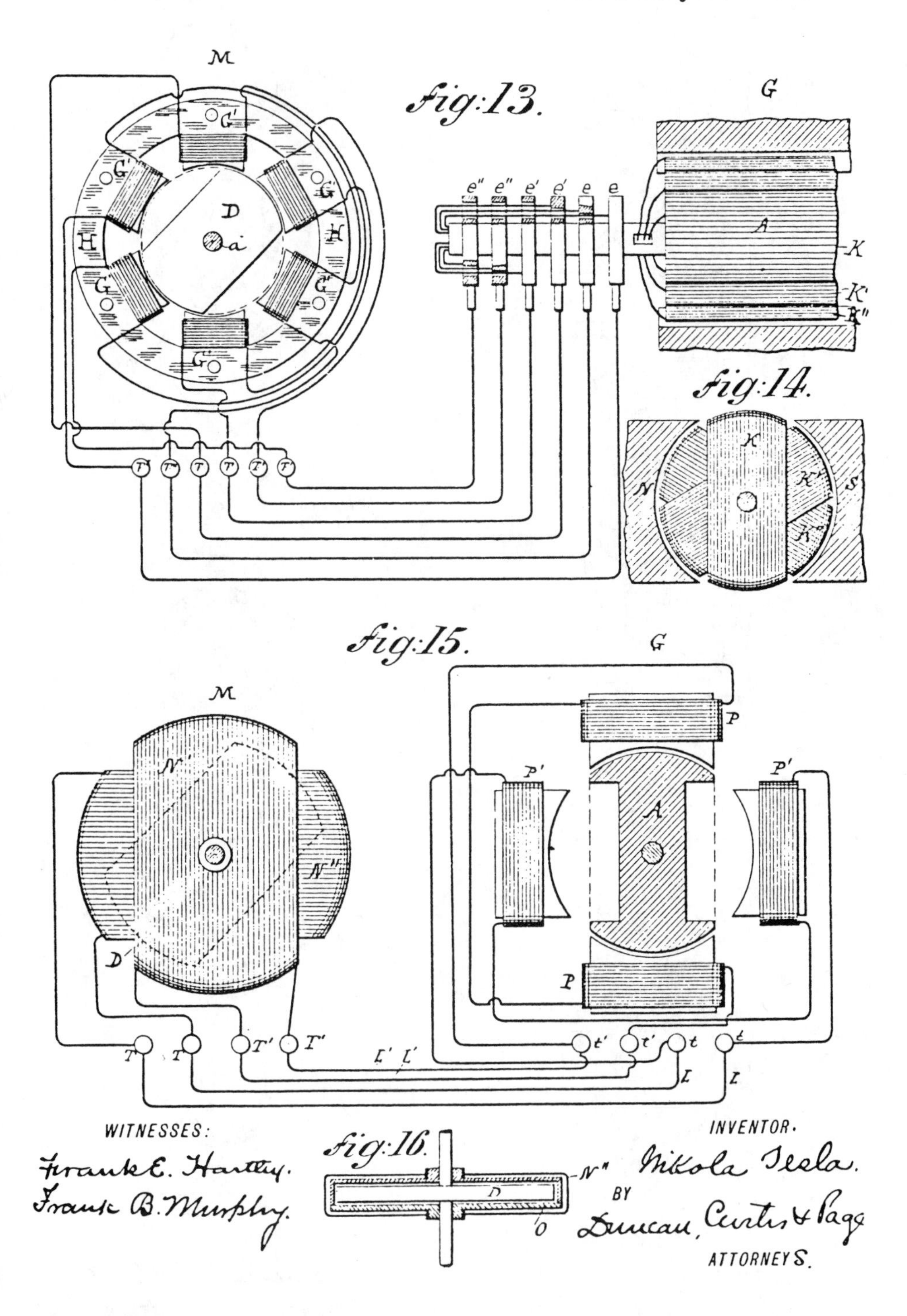

WITNESSES:

Frank E. Hartley.

Frank B. Murphy.

INVENTOR.

Nikola Tesla.

BY

Duncan, Curtis & Page

ATTORNEYS.

(No Model.) 4 Sheets—Sheet 1.

N. TESLA.

ELECTRICAL TRANSMISSION OF POWER.

No. 382,280. Patented May 1, 1888.

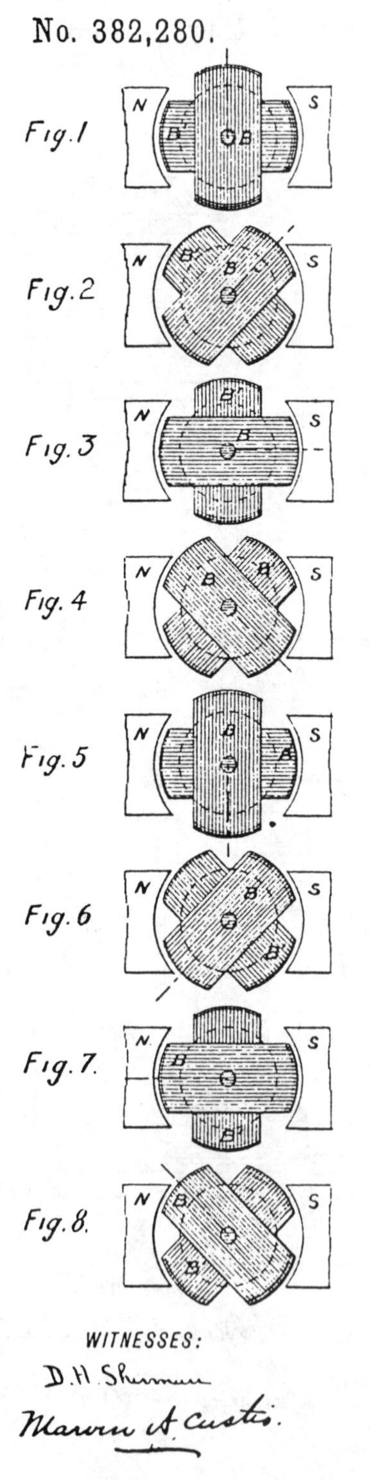

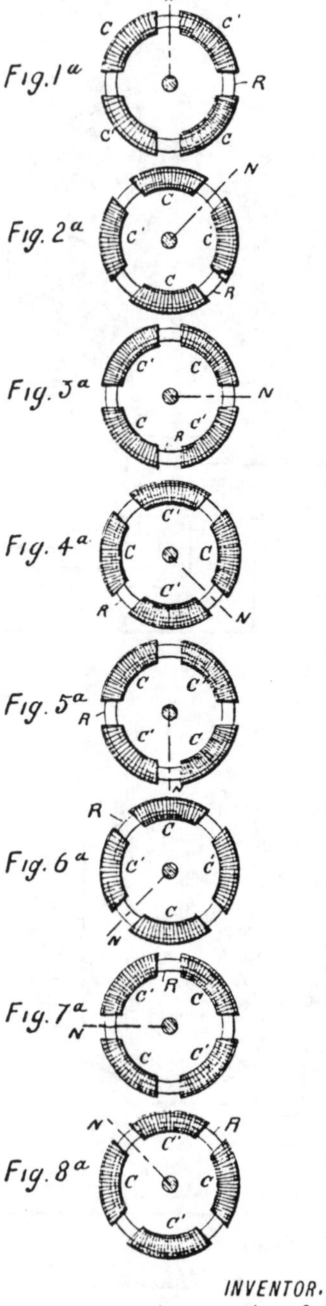

WITNESSES:

D. H. Sherman

Marvin A. Custis.

INVENTOR.

Nikola Tesla.

BY

Duncan, Curtis & Page.

ATTORNEYS.

(No Model.)

4 Sheets—Sheet 2.

N. TESLA.

ELECTRICAL TRANSMISSION OF POWER.

No. 382,280. Patented May 1, 1888.

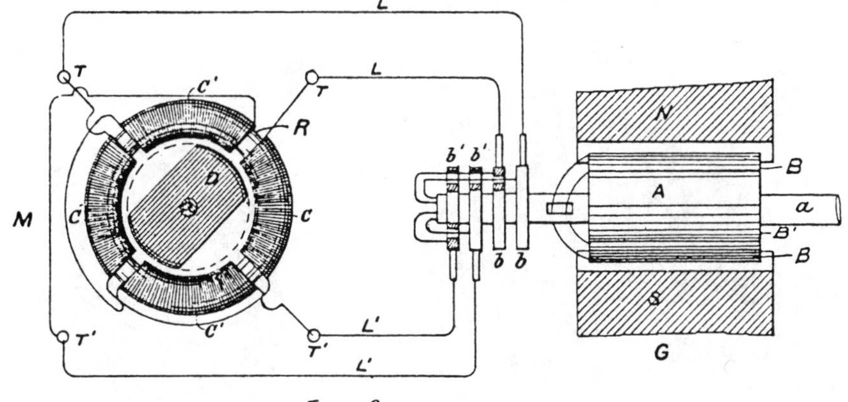

Fig. 9

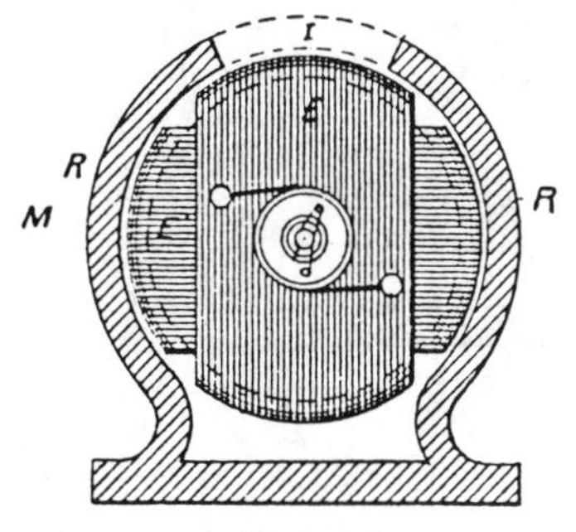

Fig. 10

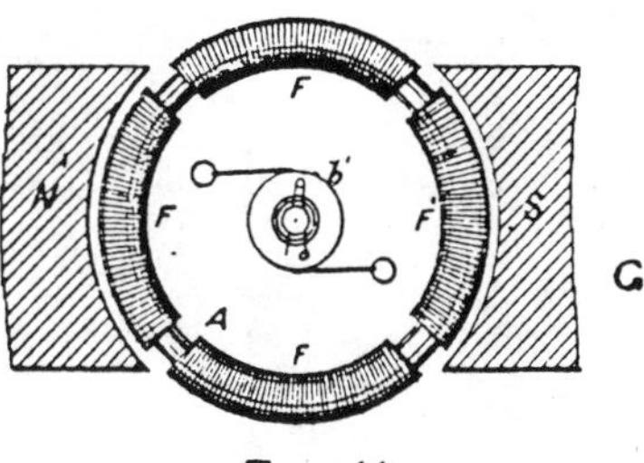

Fig. 11

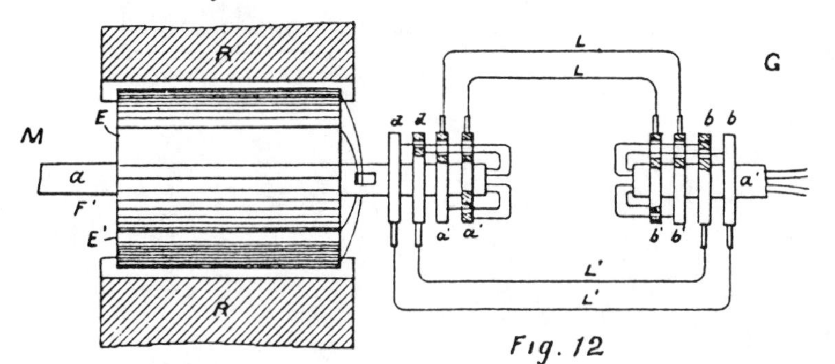

Fig. 12

WITNESSES:

D. H. Sherman.

Marvin A. Curtis.

INVENTOR.

Nikola Tesla.

BY

Duncan, Curtis & Page

ATTORNEYS.

(No Model.) 4 Sheets—Sheet 4.

N. TESLA.

ELECTRICAL TRANSMISSION OF POWER.

No. 382,280. Patented May 1, 1888.

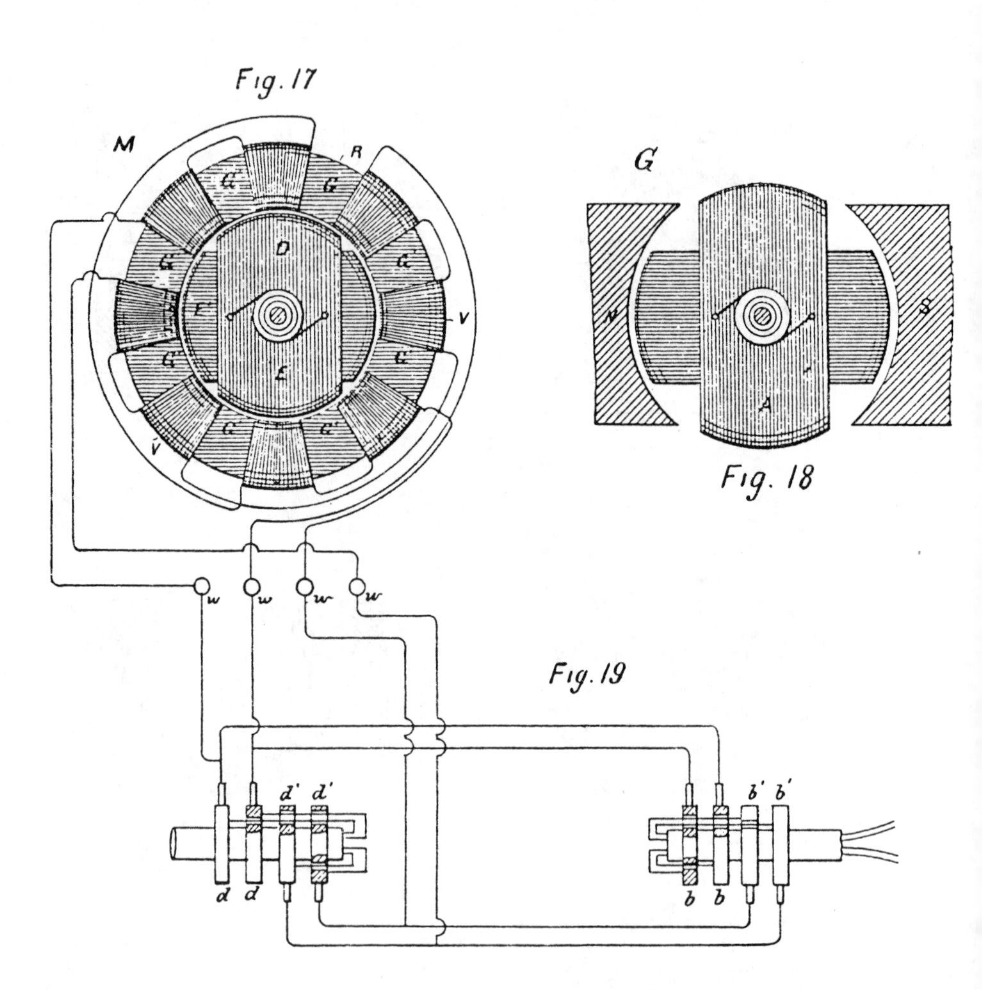

WITNESSES:

D. H. Sherman.

Marvin A. Custis.

INVENTOR.

Nikola Tesla.

BY

Duncan, Curtis & Page

ATTORNEYS.

(No Model.) 2 Sheets—Sheet 1.

N. TESLA.

ELECTRO MAGNETIC MOTOR.

No. 382,279. Patented May 1, 1888.

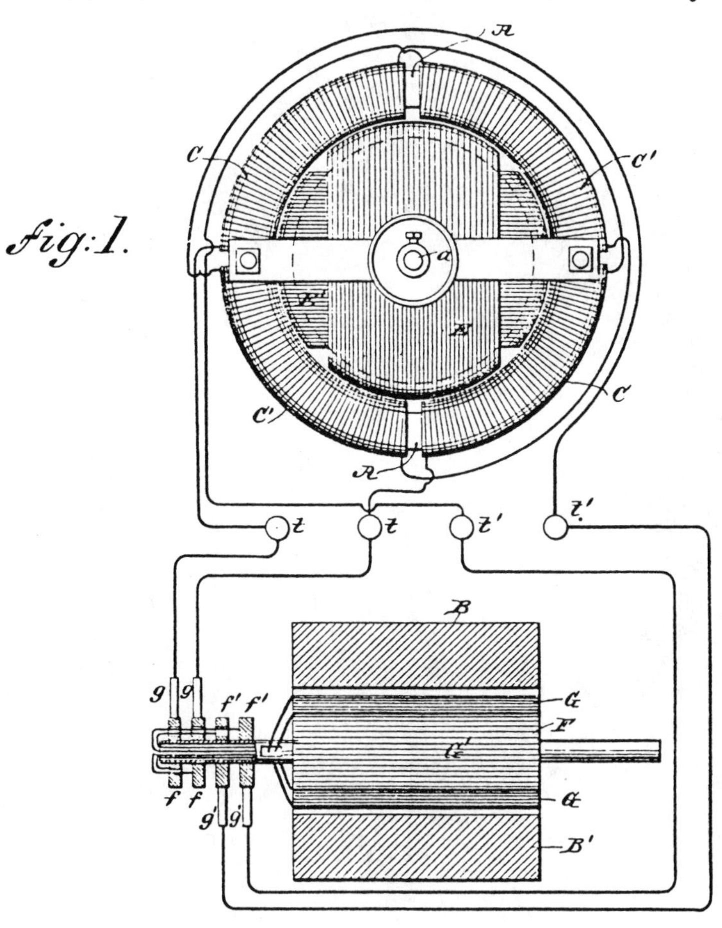

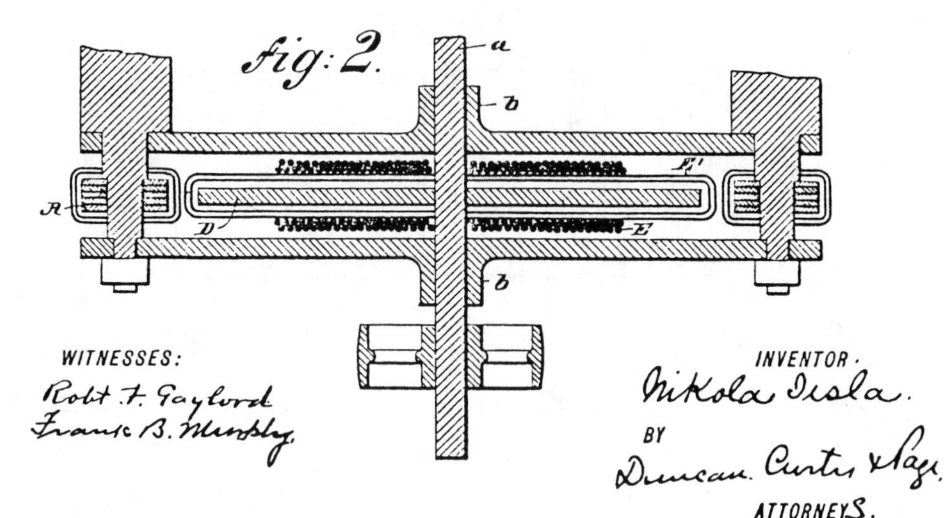

WITNESSES:

Robt F. Gaylord
Frank B. Murphy

INVENTOR.

Nikola Tesla.

BY

Duncan, Curtis & Page.

ATTORNEYS.

(No Model.) 2 Sheets—Sheet 1

N. TESLA.

SYSTEM OF ELECTRICAL DISTRIBUTION.

No. 381,970. Patented May 1, 1888.

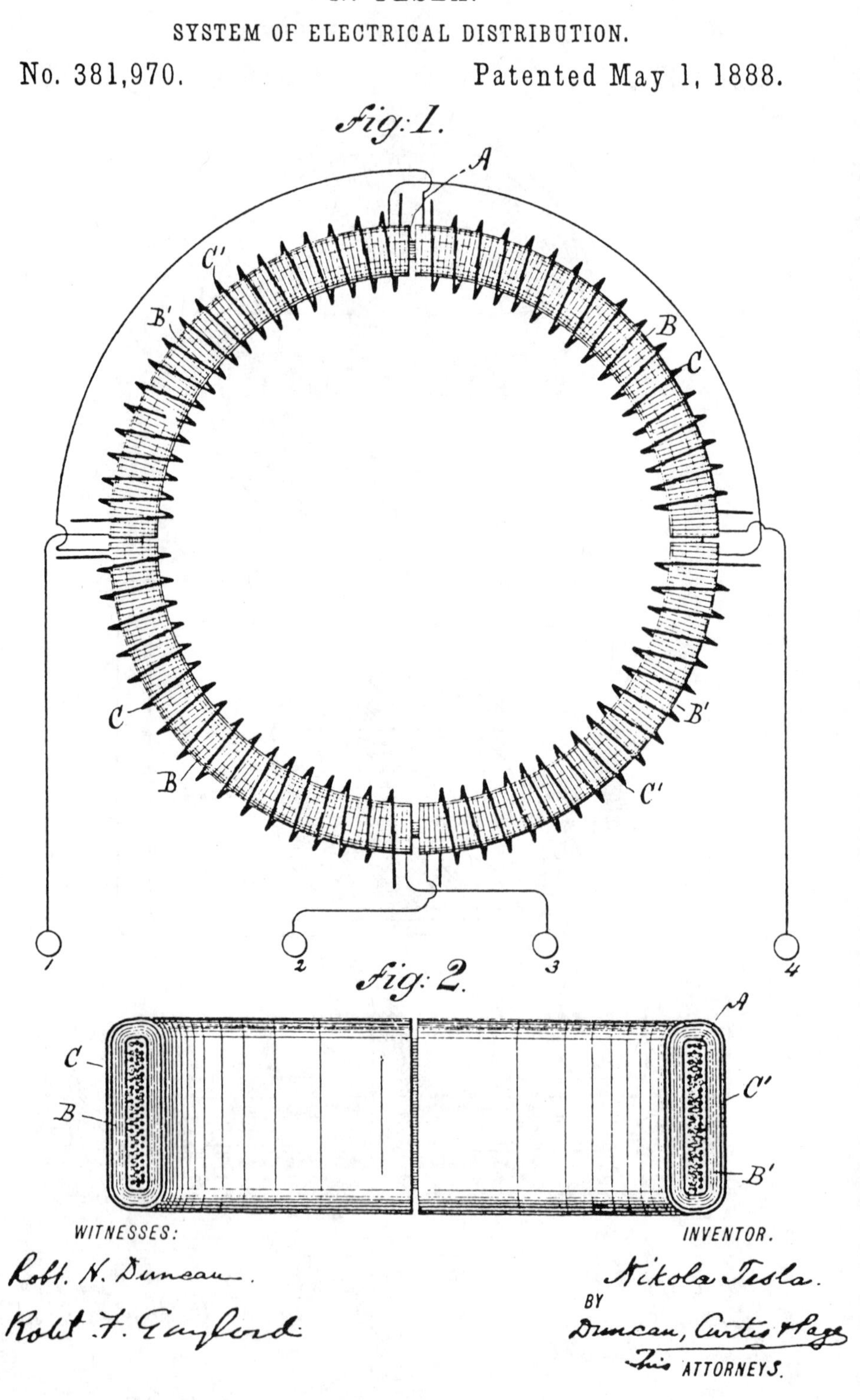

(No Model.)

2 Sheets—Sheet 2.

N. TESLA.

SYSTEM OF ELECTRICAL DISTRIBUTION.

No. 381,970.

Patented May 1, 1888.

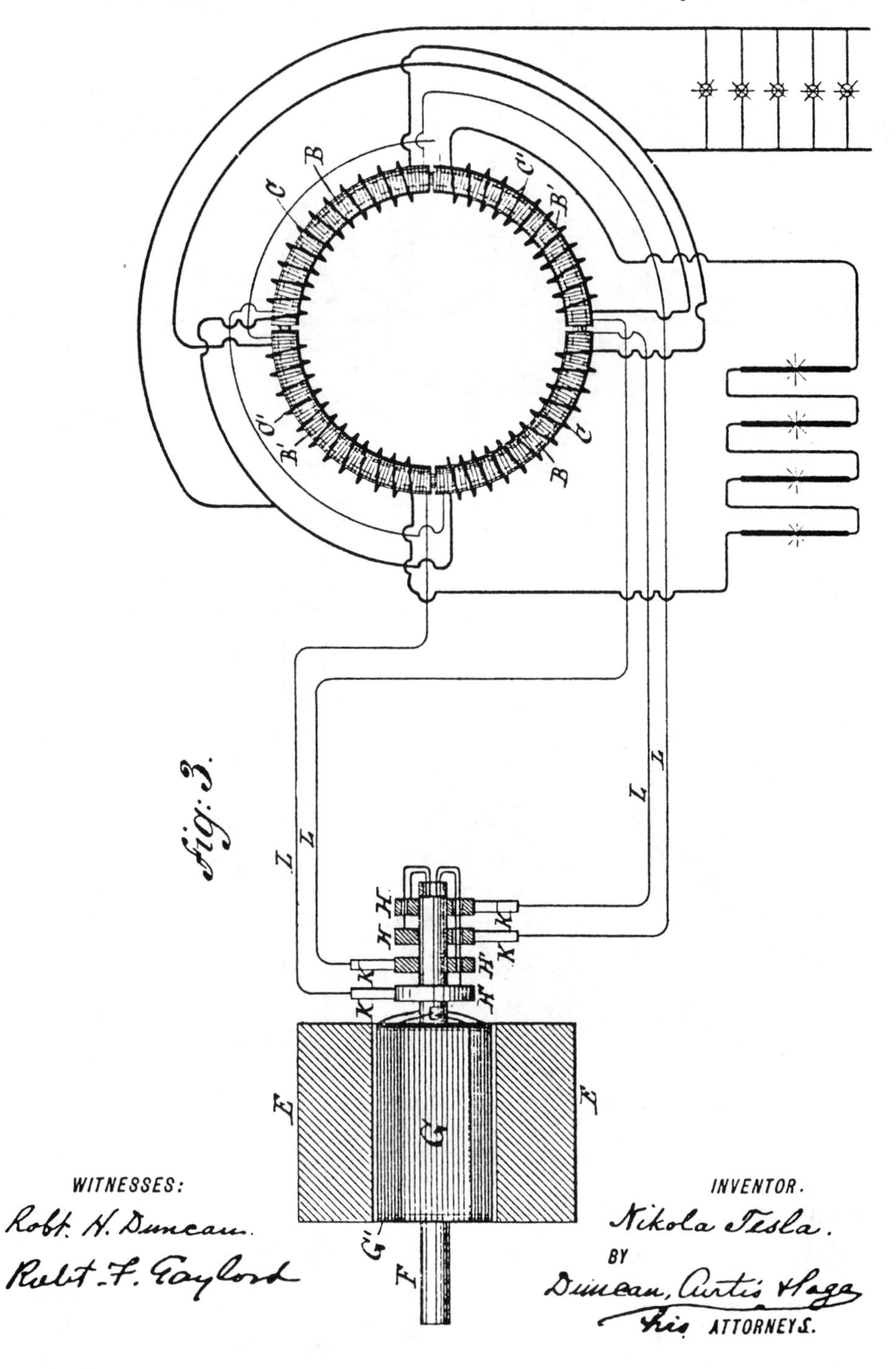

WITNESSES:

Robt. H. Duncan.

Robt. F. Gaylord

INVENTOR.

Nikola Tesla.

BY

Duncan, Curtis & Page

his ATTORNEYS.

(No Model.) 3 Sheets—Sheet 1.

N. TESLA.

SYSTEM OF ELECTRICAL DISTRIBUTION.

No. 390,413. Patented Oct. 2, 1888.

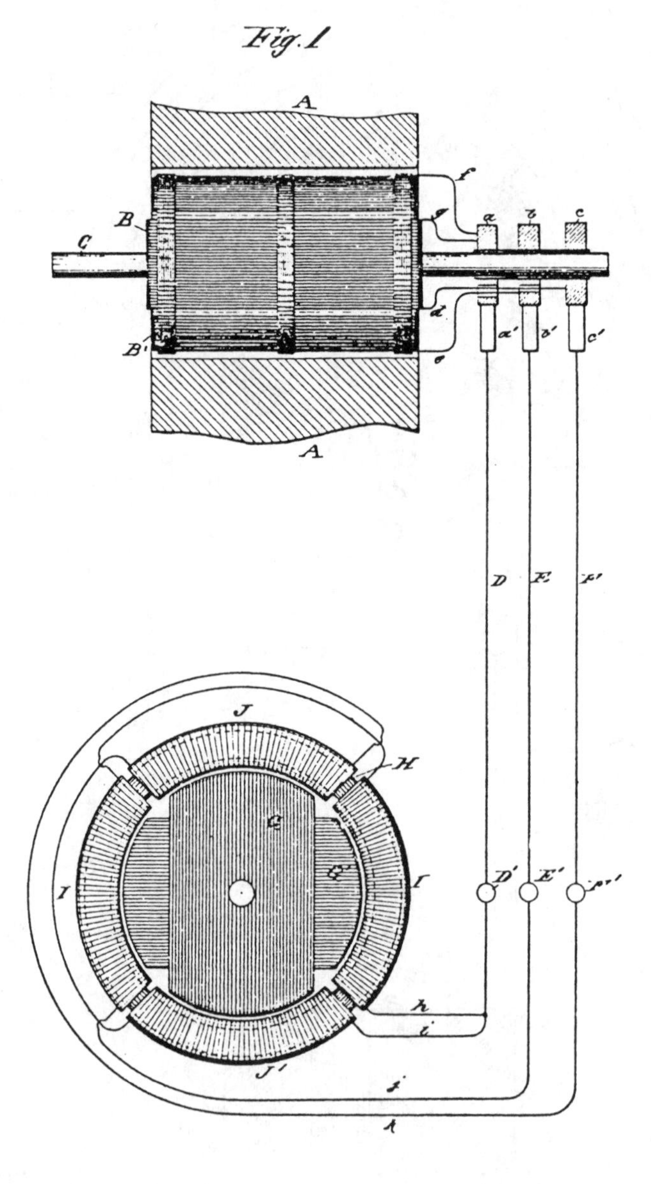

WITNESSES:

Raphaël Netter

Frank B. Murphy

INVENTOR

Nikola Tesla

BY

Duncan, Curtis & Page

ATTORNEY

(No Model.) 3 Sheets—Sheet 2.

N. TESLA.

SYSTEM OF ELECTRICAL DISTRIBUTION.

No. 390,413. Patented Oct. 2, 1888.

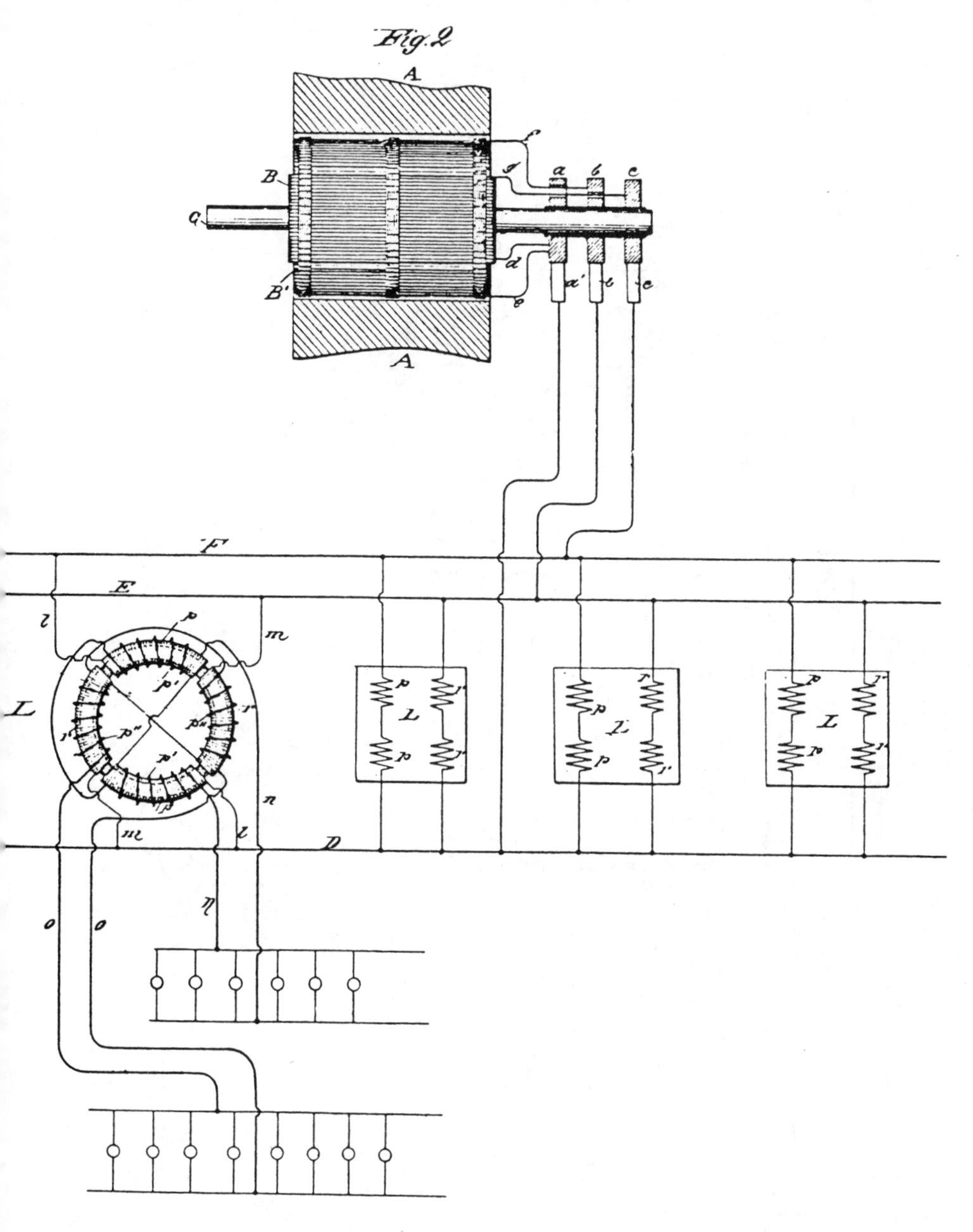

WITNESSES:

Raphaël Netter

Frank B. Murphy

INVENTOR

Nikola Tesla.

BY

Duncan, Curtis & Page

ATTORNEYS.

(No Model.)

2 Sheets—Sheet 1.

N. TESLA.

DYNAMO ELECTRIC MACHINE.

No. 390,414.

Patented Oct. 2, 1888.

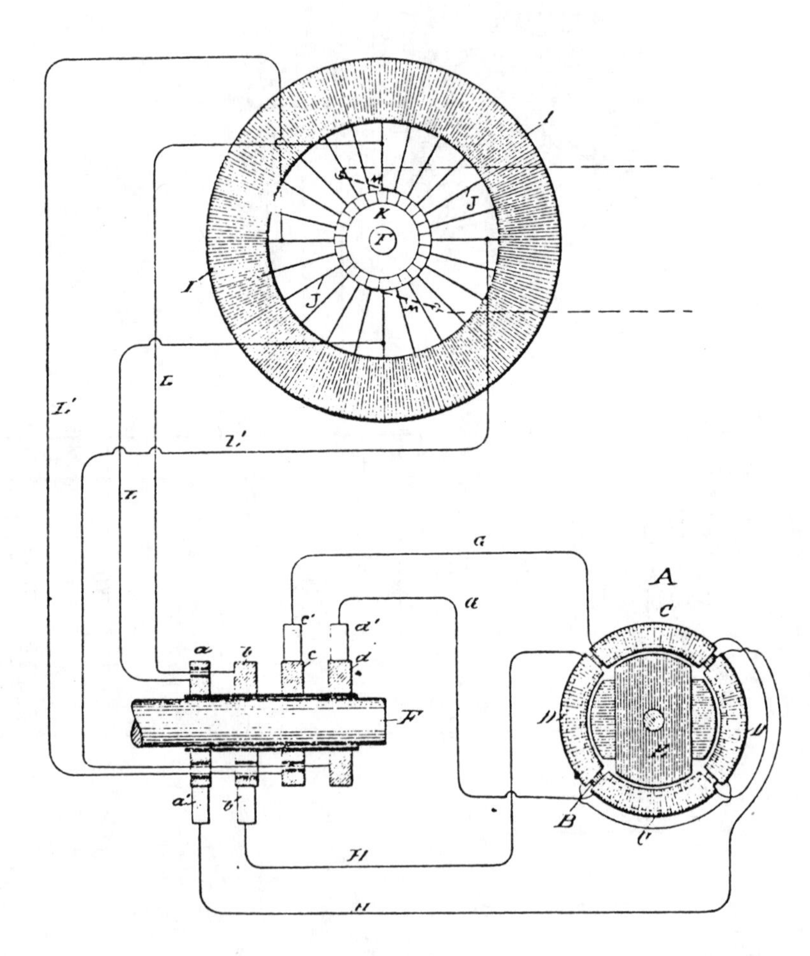

WITNESSES:

Raphaël Netter

Frank E. Hartley

INVENTOR

Nikola Tesla

BY

Duncan, Curtis & Page

ATTORNEYS

(No Model.) 2 Sheets—Sheet 1.

N. TESLA.

REGULATOR FOR ALTERNATE CURRENT MOTORS.

No. 390,820. Patented Oct. 9, 1888.

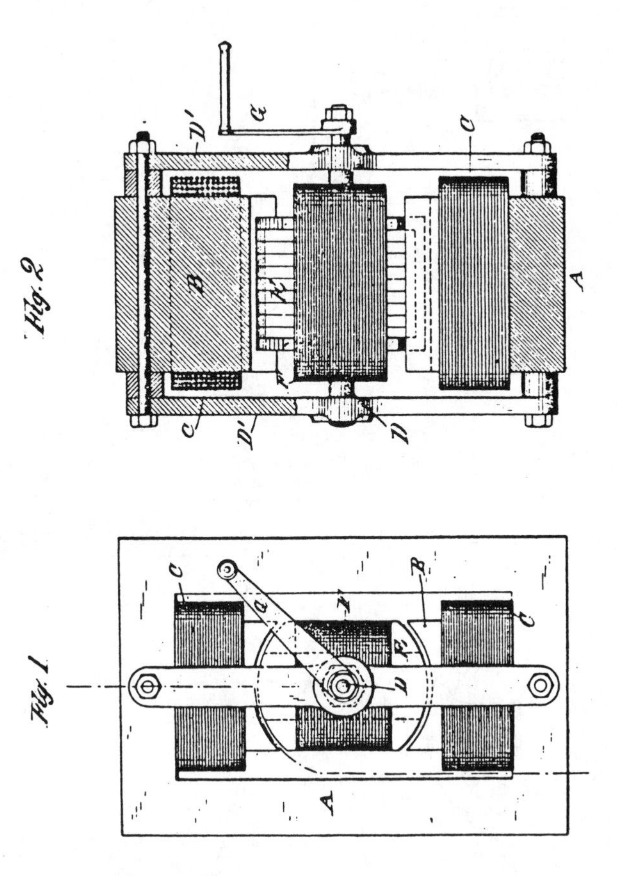

WITNESSES:

Raphaël Netter.

Robt. F. Gaylord

INVENTOR.

Nikola Tesla

BY

Duncan, Curtis & Page.

ATTORNEYS

(No Model.)

N. TESLA.

DYNAMO ELECTRIC MACHINE.

No. 390,721. Patented Oct. 9, 1888.

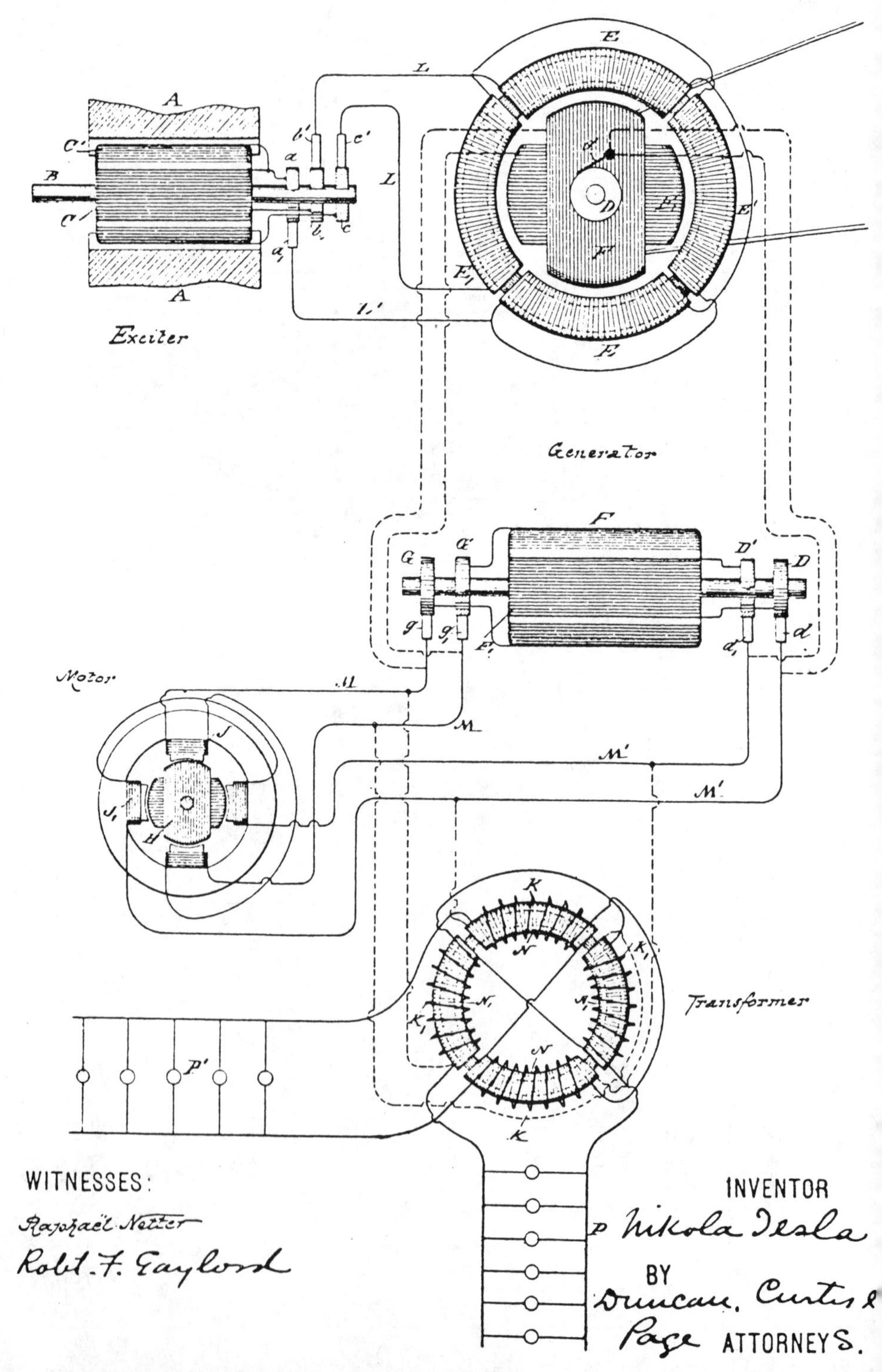

(No Model.)

N. TESLA.

DYNAMO ELECTRIC MACHINE OR MOTOR.

No. 390,415. Patented Oct. 2, 1888.

Fig. 1

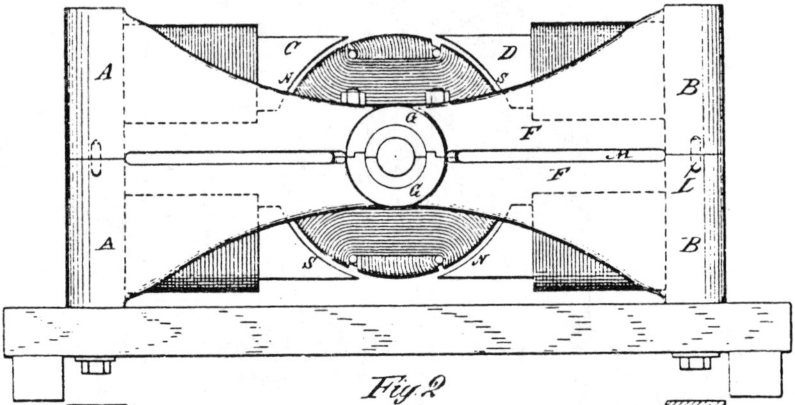

Fig. 2

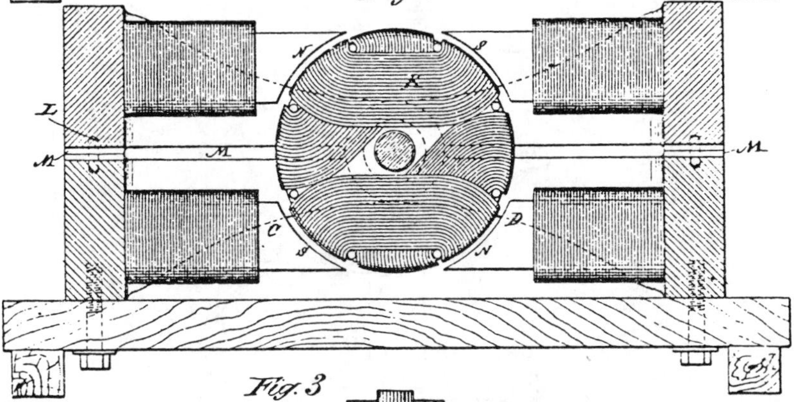

Fig. 3

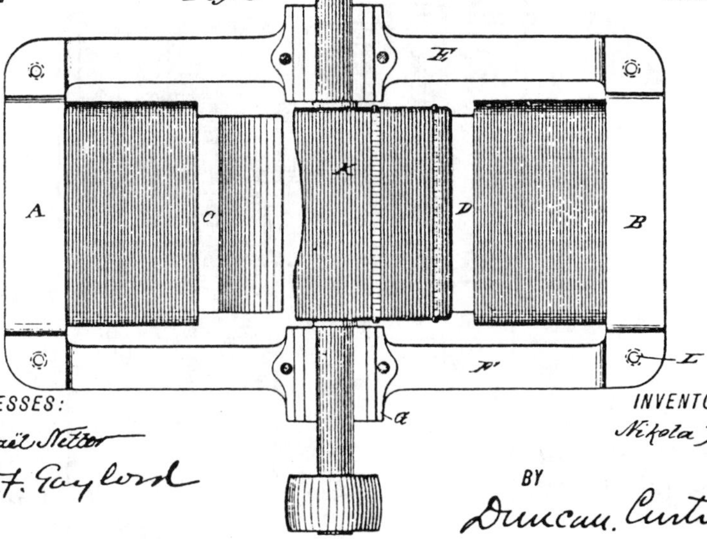

WITNESSES:

Raphaël Netter

Robt. F. Gaylord

INVENTOR

Nikola Tesla

BY

Duncan, Curtis & Page

ATTORNEYS

(No Model.)

2 Sheets—Sheet 1.

N. TESLA.

ELECTRO MAGNETIC MOTOR.

No. 381,969. Patented May 1, 1888.

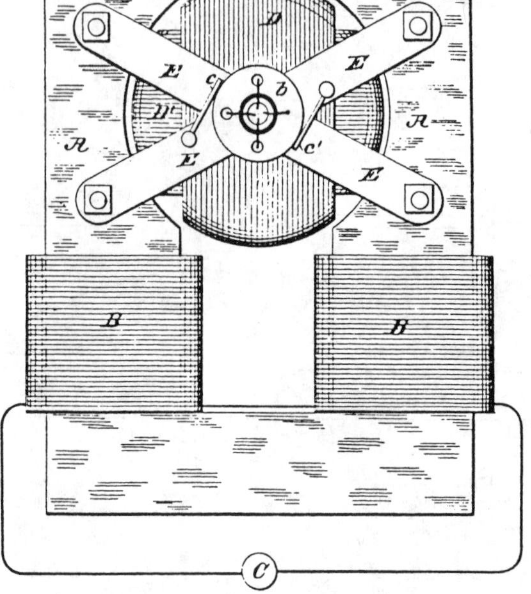

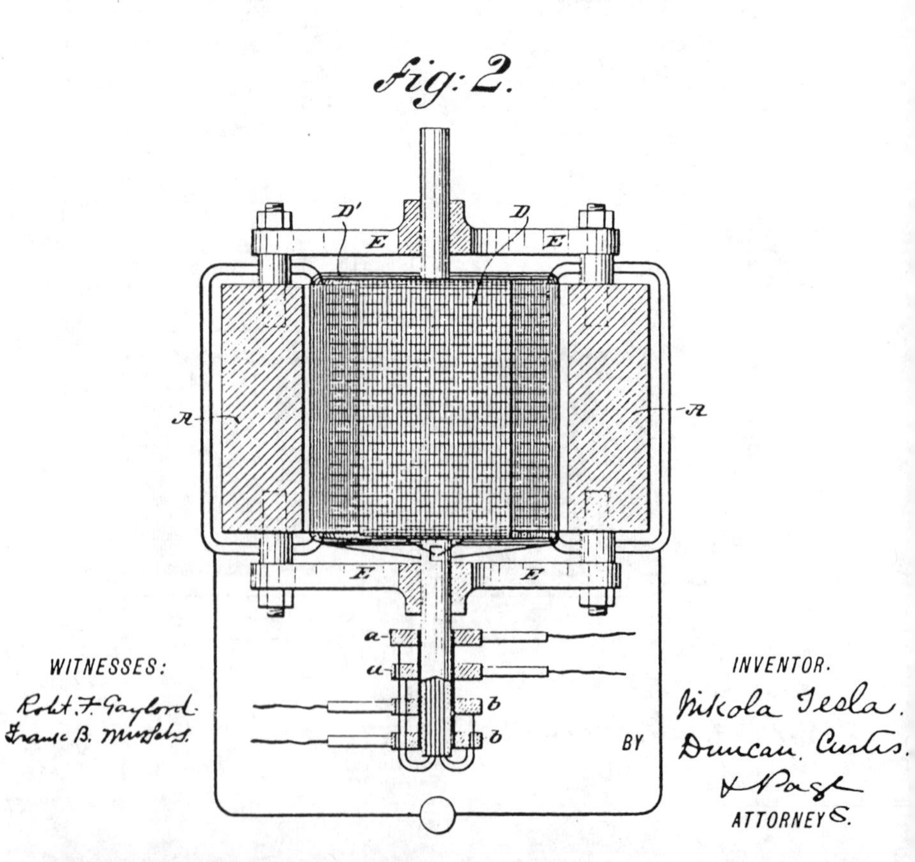

WITNESSES:

Robt. F. Gaylord
Frank B. Murphy

INVENTOR.

Nikola Tesla.

BY Duncan, Curtis & Page

ATTORNEYS.

Chapter 3

EXPERIMENTS WITH ALTERNATE CURRENTS OF HIGH POTENTIAL & HIGH FREQUENCY

I cannot find words to express how deeply I feel the honor of addressing some of the foremost thinkers of the present time, and so many able scientific men, engineers and electricians, of the country greatest in scientific achievements.

The results which I have the honor to present before such a gathering I cannot call my own. There are among you not a few who can lay better claim than myself on any feature of merit which this work may contain. I need not mention many names which are world-known—names of those among you who are recognized as the leaders in this enchanting science ; but one, at least, I must mention—a name which could not be omitted in a demonstration of this kind. It is a name associated with the most beautiful invention ever made : it is Crookes!

When I was at college, a good time ago, I read, in a translation (for then I was not familiar with your magnificent language), the description of his experiments on radiant matter. I read it only once in my life—that time—yet every detail about that charming work I can remember this day. Few are the books, let me say, which can make such an impression upon the mind of a student.

But if, on the present occasion, I mention this name as one of many your institution can boast of, it is because I have more than one reason to do so. For what I have to tell you and to show you this evening concerns, in a large measure, that same vague world which Professor Crookes has so ably explored ; and, more than this, when I trace back the mental process which led me to these advances—which even by myself cannot be considered trifling, since they are so appreciated by you—I believe that their real origin, that which started me to work in this direction, and brought me to them, after a long period of constant thought, was that fascinating little book which I read many years ago.

And now that I have made a feeble effort to express my homage and acknowledge my indebtedness to him and others among you, I will make a second effort, which I hope you will not find so feeble as the first, to entertain you.

Give me leave to introduce the subject in a few words.

A short time ago I had the honor to bring before our American Institute of Electrical Engineers* some results then arrived at by me in a novel line of work. I need not assure you that the many evidences which I have received that English scientific men and engineers were interested

---

*For Mr. Tesla's American lecture on this subject see THE ELECTRICAL WORLD of July 11, 1891, and for a report of his French lecture see THE ELECTRICAL WORLD of March 26, 1892.

in this work have been for me a great reward and encouragement. I will not dwell upon the experiments already described, except with the view of completing, or more clearly expressing, some ideas advanced by me before, and also with the view of rendering the study here presented self-contained, and my remarks on the subject of this evening's lecture consistent.

This investigation, then, it goes without saying, deals with alternating currents, and, to be more precise, with alternating currents of high potential and high frequency. Just in how much a very high frequency is essential for the production of the results presented is a question which, even with my present experience, would embarrass me to answer. Some of the experiments may be performed with low frequencies; but very high frequencies are desirable, not only on account of the many effects secured by their use, but also as a convenient means of obtaining, in the induction apparatus employed, the high potentials, which in their turn are necessary to the demonstration of most of the experiments here contemplated.

Of the various branches of electrical investigation, perhaps the most interesting and immediately the most promising is that dealing with alternating currents. The progress in this branch of applied science has been so great in recent years that it justifies the most sanguine hopes. Hardly have we become familiar with one fact, when novel experiences are met with and new avenues of research are opened. Even at this hour possibilities not dreamed of before are, by the use of these currents, partly realized. As in nature all is ebb and tide, all is wave motion, so it seems that in all branches of industry alternating currents—electric wave motion—will have the sway.

One reason, perhaps, why this branch of science is being so rapidly developed is to be found in the interest which is attached to its experimental study. We wind a simple ring of iron with coils ; we establish the connections to the generator, and with wonder and delight we note the effects of strange forces which we bring into play, which allow us to transform, to transmit and direct energy at will. We arrange the circuits properly, and we see the mass of iron and wires behave as though it were endowed with life, spinning a heavy armature, through invisible connections, with great speed and power—with the energy possibly conveyed from a great distance. We observe how the energy of an alternating current traversing the wire manifests itself—not so much in the wire as in the surrounding space —in the most surprising manner, taking the forms of heat, light, mechanical energy, and, most surprising of all, even chemical affinity. All these observations fascinate us, and fill us with an intense desire to know more about the nature of these phenomena. Each day we go to our work in the hope of discovering,—in the hope that some one, no matter who, may find a solution of one of the pending great problems,—and each succeeding day we return to our task with renewed ardor ; and even if we *are* unsuccessful, our work has not been in vain, for in these strivings, in these efforts, we have found hours of untold pleasure, and we have directed our energies to the benefit of mankind.

We may take—at random, if you choose—any of the many experiments which may be performed with alternating currents; a few of which only, and by no means the most striking, form the subject of this evening's demonstration; they are all equally interesting, equally inciting to thought.

Here is a simple glass tube from which the air has been partially exhausted. I take hold of it; I bring my body in contact with a wire conveying alternating currents of high potential, and the tube in my hand is brilliantly lighted. In whatever position I may put it, wherever I may move it in space, as far as I can reach, its soft, pleasing light persists with undiminished brightness.

Here is an exhausted bulb suspended from a single wire. Standing on an insulated support, I grasp it, and a platinum button mounted in it is brought to vivid incandescence.

Here, attached to a leading wire, is another bulb, which, as I touch its metallic socket, is filled with magnificent colors of phosphorescent light.

Here still another, which by my fingers' touch casts a shadow—the Crookes shadow, of the stem inside of it.

Here, again, insulated as I stand on this platform, I bring my body in contact with one of the terminals of the secondary of this induction coil—with the end of a wire many miles long—and you see streams of light break forth from its distant end, which is set in violent vibration.

Here, once more, I attach these two plates of wire gauze to the terminals of the coil, I set them a distance apart, and I set the coil to work. You may see a small spark pass between the plates. I insert a thick plate of one of the best dielectrics between them, and instead of rendering altogether impossible, as we are used to expect, I *aid* the passage of the discharge, which, as I insert the plate, merely changes in appearance and assumes the form of luminous streams.

Is there, I ask, can there be, a more interesting study than that of alternating currents?

In all these investigations, in all these experiments, which are so very, very interesting, for many years past—ever since the greatest experimenter who lectured in this hall discovered its principle—we have had a steady companion, an appliance familiar to every one, a plaything once, a thing of momentous importance now—the induction coil. There is no dearer appliance to the electrician. From the ablest among you, I dare say, down to the inexperienced student, to your lecturer, we all have passed many delightful hours in experimenting with the induction coil. We have watched its play, and thought and pondered over the beautiful phenomena which it disclosed to our ravished eyes. So well known is this apparatus, so familiar are these phenomena to every one, that my courage nearly fails me when I think that I have ventured to address so able an audience, that I have ventured to entertain you with that same old subject. Here in reality is the same apparatus, and here are the same phenomena, only the apparatus is operated somewhat differently, the phenomena are presented in a different aspect. Some of the results we find as expected, others surprise us, but all captivate our attention, for in scientific investigation each novel result achieved may be the centre of a new departure, each novel fact learned may lead to important developments.

Usually in operating an induction coil we have set up a vibration of moderate frequency in the primary, either by means of an interrupter or break, or by the use of an alternator. Earlier English investigators, to mention only Spottiswoode and J. E. H. Gordon, have used a rapid break in connection with the coil. Our knowledge and experience of to-day enables us to see clearly why these coils under the conditions of the tests did not disclose any, remarkable phenomena, and why able experimenters failed to preceive many of the curious effects which have since been observed.

In the experiments such as performed this evening, we operate the coil either from a specially constructed alternator capable of giving many thousands of reversals of current per second, or, by disruptively discharging a condenser through the primary, we set up a vibration in the secondary circuit of a frequency of many hundred thousand or millions per second, if we so desire; and in using either of these means we enter a field as yet unexplored.

It is impossible to pursue an investigation in any novel line without finally making some interesting observation or learning some useful fact. That this statement is applicable to the subject of this lecture the many curious and unexpected phenomena which we observe afford a convincing proof. By way of illustration, take for instance the most obvious phenomena, those of the discharge of the induction coil.

Here is a coil which is operated by currents vibrating with extreme rapidity, obtained by disruptively discharging a Leyden jar. It would not surprise a student were the lecturer to say that the secondary of this coil consists of a small length of comparatively stout wire; it would not surprise him were the lecturer to state that, in spite of this, the coil is capable of giving any potential which the best insulation of the turns is able to withstand; but although he may be prepared, and even be indifferent as to the anticipated result, yet the aspect of the discharge of the coil will surprise and interest him. Every one is familiar with the discharge of an ordinary coil; it need not be reproduced here. But, by way of contrast, here is a form of discharge of a coil, the primary current of which is vibrating several hundred thousand times per second. The discharge of an ordinary coil appears as a simple line or band of light. The discharge of this coil appears in the form of powerful brushes and luminous streams issuing from all points of the two straight wires attached to the terminals of the secondary. (Fig. 1.)

Now compare this phenomenon which you have just witnessed with the discharge of a Holtz or Wimshurst machine—that other interesting appliance so dear to the experimenter. What a difference there is between these phenomena! And yet, had I made the necessary arrangements—which could have been made easily, were it not that they would interfere with other experiments—I could have produced with this coil sparks which, had I the coil hidden from your view and only two knobs exposed, even the keenest observer among you would find it difficult, if not impossible, to distinguish from those of an influence or friction machine. This may be done in many ways—for instance, by operating the induction coil which charges the condenser from an alternating-current machine of very low frequency, and preferably adjusting the discharge circuit so that there are no oscillations set up in it. We then ob

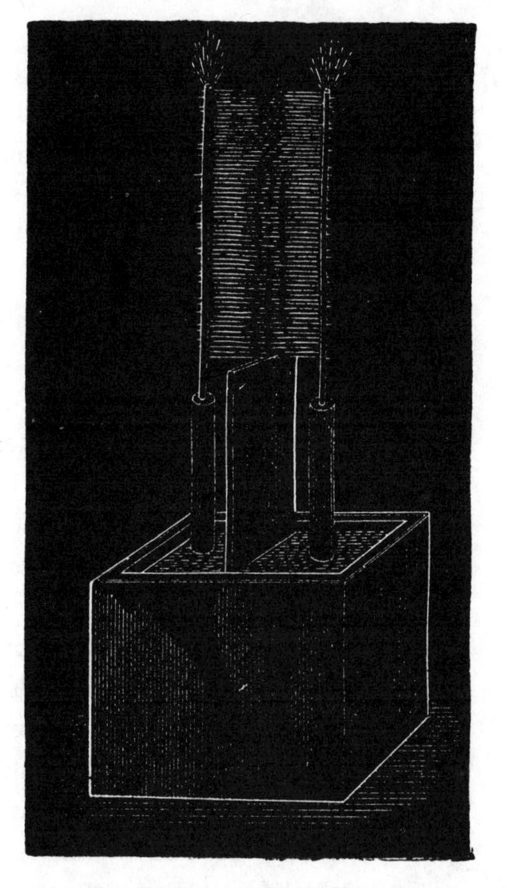

FIG. 1.—DISCHARGE BETWEEN TWO WIRES WITH FREQUENCIES OF A FEW HUNDRED THOUSAND PER SECOND.

tain in the secondary circuit, if the knobs are of the required size and properly set, a more or less rapid succession of sparks of great intensity and small quantity, which possess the same brilliancy, and are accompanied by the same sharp crackling sound, as those obtained from a friction or influence machine.

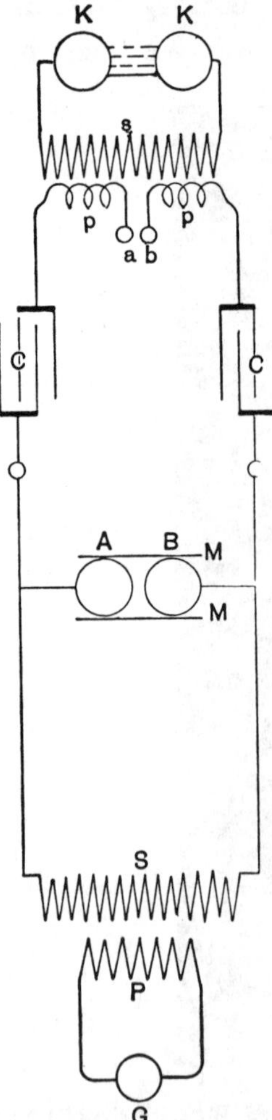

FIG. 2.—IMITATING THE SPARK OF A HOLTZ MACHINE.

Another way is to pass through two primary circuits, having a common secondary, two currents of a slightly different period, which produce in the secondary circuit sparks occurring at comparatively long intervals. But, even with the means at hand this evening, I may succeed in imitating the spark of a Holtz machine. For this purpose I establish between the terminals of the coil which charges the condenser a long, unsteady arc, which is periodically interrupted by the upward current of air produced by it. To increase the current of air I place on each side of the arc, and close to it, a large plate of mica. The condenser charged from this coil discharges into the primary circuit of a second coil through a small air gap, which is necessary to produce a sudden rush of current through the primary. The scheme of connections in the present experiment is indicated in Fig. 2.

G is an ordinarily constructed alternator, supplying the primary *P* of an induction coil, the secondary *S* of which charges the condensers or jars *C C*. The terminals of the secondary are connected to the inside coatings of the jars, the outer coatings being connected to the ends of the primary *p p* of a second induction coil. This primary *p p* has a small air gap *a b*.

The secondary *s* of this coil is provided with knobs or spheres *K K* of the proper size and set at a distance suitable for the experiment.

A long arc is established between the terminals *A B* of the first induction coil. *M M* are the mica plates.

Each time the arc is broken between *A* and *B* the jars are quickly charged and discharged through the primary *p p*, producing a snapping spark between the knobs *K K*. Upon the arc forming between *A* and *B* the potential falls, and the jars cannot be charged to such high potential as to break through the air gap *a b* until the arc is again broken by the draught.

In this manner sudden impulses, at long intervals, are produced in the primary *p p*, which in the secondary *s* give a corresponding number of impulses of great intensity. If the secondary knobs or spheres, *K K*, are of the proper size, the sparks show much resemblance to those of a Holtz machine.

But these two effects, which to the eye appear so very different, are only two of the many discharge phenomena. We only need to change the conditions of the test, and again we make other observations of interest.

When, instead of operating the induction coil as in the last two experiments, we operate it from a high frequency alternator, as in the next experiment, a systematic study of the phenomena is rendered much more easy. In such case, in varying the strength and frequency of the currents through the primary, we may observe five distinct forms of discharge, which I have described in my former paper on the subject* before the American Institute of Electrical Engineers, May 20, 1891.

It would take too much time, and it would lead us too far from the subject presented this evening, to reproduce all these forms, but it seems to me desirable to show you one of them. It is a brush discharge, which is interesting in more than one respect. Viewed from a near position it resembles much a jet of gas escaping under great pressure. We know that the phenomenon is due to the agitation of the molecules near the terminal, and we anticipate that some heat must be developed by the impact of the molecules against the terminal or against each other. Indeed, we find that the brush is hot, and only a little thought leads us to the conclusion that, could we but reach sufficiently high frequencies, we could produce a brush which would give intense light and heat, and which would resemble in every particular an ordinary flame, save, perhaps, that both phenomena might not be due to the same agent—save, perhaps, that chemical affinity might not be *electrical* in its nature.

As the production of heat and light is here due to the impact of the molecules, or atoms of air, or something else besides, and, as we can augment the energy simply by raising the potential, we might, even with frequencies ob-

---

*See THE ELECTRICAL WORLD, July 11, 1891.

tained from a dynamo machine, intensify the action to such a degree as to bring the terminal to melting heat. But with such low frequencies we would have to deal always with something of the nature of an electric current. If I approach a conducting object to the brush, a thin little spark passes, yet, even with the frequencies used this evening, the tendency to spark is not very great. So, for instance, if I hold a metallic sphere at some distance above the terminal you may see the whole space between the terminal and sphere illuminated by the streams without the spark passing; and with the much higher frequencies obtainable by the disruptive discharge of a condenser, were it not for the sudden impulses, which are comparatively few in number, sparking would not occur even at very small distances. However, with incomparably higher frequencies, which we may yet find means to produce efficiently, and provided that electric impulses of such high frequencies could be transmitted through a conductor, the electrical characteristics of the brush discharge would completely vanish – no spark would pass, no shock would be felt—yet we would still have to deal with an *electric* phenomenon, but in the broad, modern interpretation of the word. In my first paper before referred to I have pointed out the curious properties of the brush, and described the best manner of producing it, but I have thought it worth while to endeavor to express myself more clearly in regard to this phenomenon, because of its absorbing interest.

When a coil is operated with currents of very high frequency, beautiful brush effects may be produced, even if the coil be of comparatively small dimensions. The experimenter may vary them in many ways, and, if it were nothing else, they afford a pleasing sight. What adds to their interest is that they may be produced with one single terminal as well as with two—in fact, often better with one than with two.

But of all the discharge phenomena observed, the most pleasing to the eye, and the most instructive, are those observed with a coil which is operated by means of the disruptive discharge of a condenser. The power of the brushes, the abundance of the sparks, when the conditions are patiently adjusted, is often amazing. With even a very small coil, if it be so well insulated as to stand a difference of potential of several thousand volts per turn, the sparks may be so abundant that the whole coil may appear a complete mass of fire.

Curiously enough the sparks, when the terminals of the coil are set at a considerable distance, seem to dart in every possible direction as though the terminals were perfectly independent of each other. As the sparks would soon destroy the insulation it is necessary to prevent them. This is best done by immersing the coil in a good liquid insulator, such as boiled-out oil. Immersion in a liquid may be considered almost an absolute necessity for the continued and successful working of such a coil.

It is of course out of the question, in an experimental lecture, with only a few minutes at disposal for the performance of each experiment, to show these discharge phenomena to advantage, as to produce each phenomenon at its best a very careful adjustment is required. But even if imperfectly produced, as they are likely to be this evening, they are sufficiently striking to interest an intelligent audience.

Before showing some of these curious effects I must, for the sake of completeness, give a short description of the

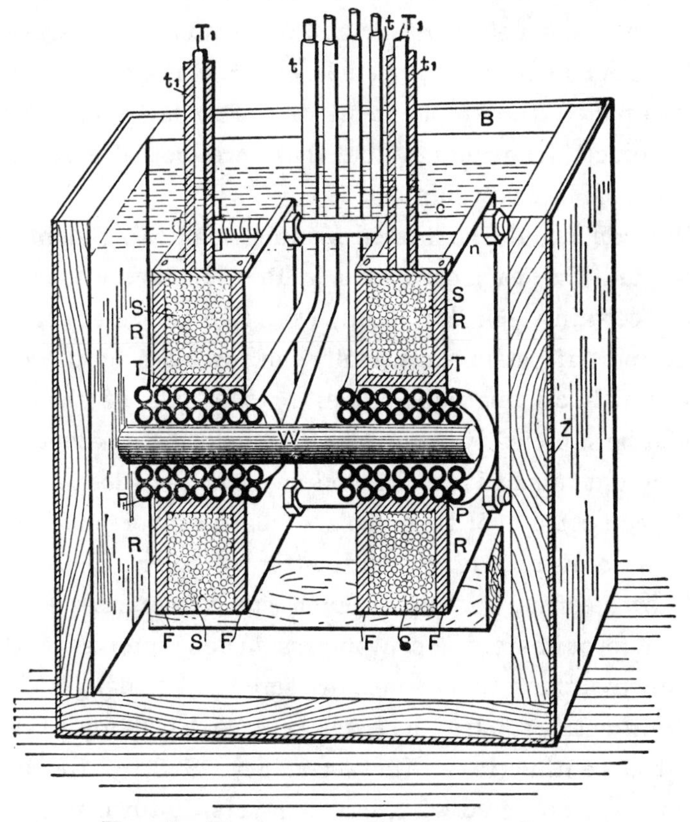

Fig. 3.—Disruptive Discharge Coil.

coil and other apparatus used in the experiments with the disruptive discharge this evening.

It is contained in a box *B* (Fig. 3) of thick boards of hard wood, covered on the outside with zinc sheet *Z*, which is carefully soldered all around. It might be advisable, in a strictly scientific investigation, when accuracy is of great importance, to do away with the metal cover, as it might introduce many errors, principally on account of its complex action upon the coil, as a condenser of very small capacity and as an electrostatic and electromagnetic screen. When the coil is used for such experiments as are here contemplated, the employment of the metal cover offers some practical advantages, but these are not of sufficient importance to be dwelt upon.

The coil should be placed symmetrically to the metal cover, and the space between should, of course, not be too small, certainly not less than, say, five centimetres, but much more if possible; especially the two sides of the zinc box, which are at right angles to the axis of the coil, should be sufficiently remote from the latter, as otherwise they might impair its action and be a source of loss.

The coil consists of two spools of hard rubber *R R*, held apart at a distance of 10 centimetres by bolts *c* and nuts *n*, likewise of hard rubber. Each spool comprises a tube *T* of approximately 8 centimetres inside diameter, and 3 millimetres thick, upon which are screwed two flanges *F F*, 24 centimetres square, the space between the flanges being about 3 centimetres. The secondary, *S S*, of the best gutta percha-covered wire, has 26 layers, 10 turns in each, giving for each half a total of 260 turns. The two halves are wound oppositely and connected in series, the connection between both being made over the primary. This disposition, besides being convenient, has the advantage that when the coil is well balanced—that is, when both of its terminals T_1 T_1 are connected to bodies or devices of equal capacity—there is not much danger of breaking through to the primary, and the insulation between the primary and the secondary need not be thick. In using the coil it is advisable to attach to *both* terminals devices of nearly equal capacity, as, when the capacity of the terminals is not equal, sparks will be apt to pass to the primary. To avoid this, the middle point of the secondary may be connected to the primary, but this is not always practicable.

The primary P P is wound in two parts, and oppositely, upon a wooden spool W, and the four ends are led out of the oil through hard rubber tubes t t. The ends of the secondary T_1 T_1 are also led out of the oil through rubber tubes t_1 t_1 of great thickness. The primary and secondary layers are insulated by cotton cloth, the thickness of the insulation, of course, bearing some proportion to the difference of potential between the turns of the different layers. Each half of the primary has four layers, 24 turns in each, this giving a total of 96 turns. When both the parts are connected in series, this gives a ratio of conversion of about 1 : 2.7, and with the primaries in multiple, 1 : 5.4; but in operating with very rapidly alternating currents this ratio does not convey even an approximate idea of the ratio of the E. M. Fs. in the primary and secondary circuits. The coil is held in position in the oil on wooden supports, there being about 5 centimetres thickness of oil all round. Where the oil is not specially needed, the space is filled with pieces of wood, and for this purpose principally the wooden box B surrounding the whole is used.

The construction here shown is, of course, not the best on general principles, but I believe it is a good and convenient one for the production of effects in which an excessive potential and a very small current are needed.

In connection with the coil I use either the ordinary form of discharger or a modified form. In the former I have introduced two changes which secure some advantages, and which are obvious. If they are mentioned, it is only in the hope that some experimenter may find them of use

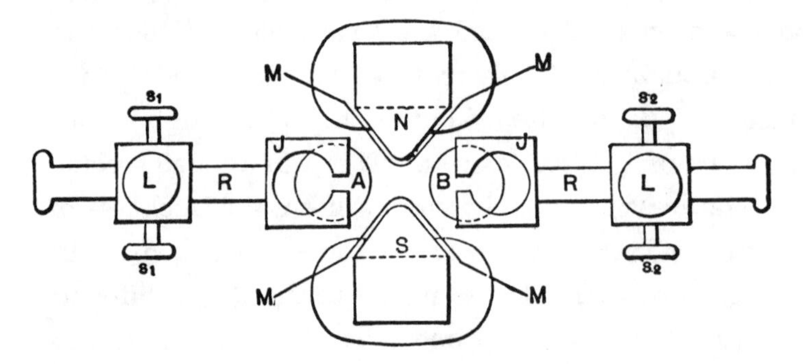

FIG. 4.—ARRANGEMENT OF IMPROVED DISCHARGER AND MAGNET.

One of the changes is that the adjustable knobs *A* and *B* (Fig. 4), of the discharger are held in jaws of brass, *J J*, by spring pressure, this allowing of turning them successively into different positions, and so doing away with the tedious process of frequent polishing up.

The other change consists in the employment of a strong electromagnet *N S*, which is placed with its axis at right angles to the line joining the knobs *A* and *B*, and produces a strong magnetic field between them. The pole pieces of the magnet are movable and properly formed so as to protrude between the brass knobs, in order to make the field as intense as possible; but to prevent the discharge from jumping to the magnet the pole pieces are protected by a layer of mica, *M M*, of sufficient thickness. s_1 s_1 and s_2 s_2 are screws for fastening the wires. On each side one of the screws is for large and the other for small wires. *L L* are screws for fixing in position the rods *R R*, which support the knobs.

In another arrangement with the magnet I take the discharge between the rounded pole pieces themselves, which in such case are insulated and preferably provided with polished brass caps.

The employment of an intense magnetic field is of advantage principally when the induction coil or transformer which charges the condenser is operated by currents of very low frequency. In such a case the number of the fundamental discharges between the knobs may be so small as to render the currents produced in the secondary unsuitable for many experiments. The intense magnetic field then serves to blow out the arc between the knobs as soon as it is formed, and the fundamental discharges occur in quicker succession.

Instead of the magnet, a draught or blast of air may be employed with some advantage. In this case the arc is preferably established between the knobs *A B*, in Fig. 2 (the knobs *a b* being generally joined, or entirely done away with), as in this disposition the arc is long and unsteady, and is easily affected by the draught.

When a magnet is employed to break the arc, it is better to

choose the connection indicated diagrammatically in Fig. 5, as in this case the currents forming the arc are much more powerful, and the magnetic field exercises a greater influence. The use of the magnet permits, however, of the arc being replaced by a vacuum tube, but I have encoun-

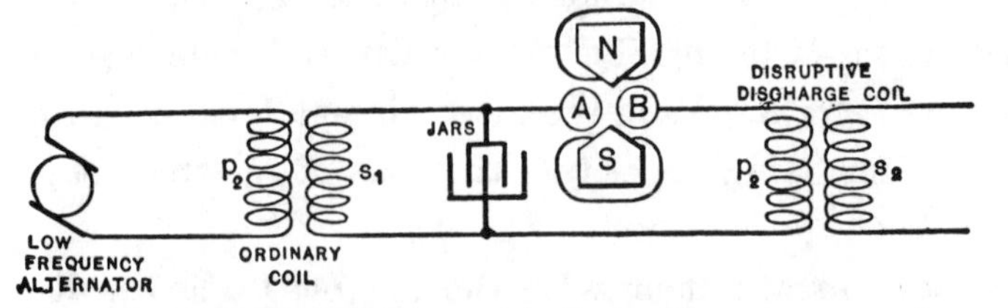

FIG. 5.—ARRANGEMENT WITH LOW-FREQUENCY ALTERNATOR AND IMPROVED DISCHARGER.

tered great difficulties in working with an exhausted tube.

The other form of discharger used in these and similar experiments is indicated in Figs. 6 and 7. It consists of a number of brass pieces *c c* (Fig. 6), each of which comprises

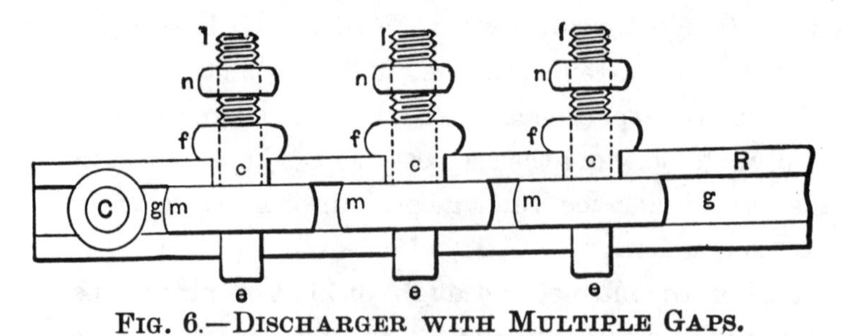

FIG. 6.—DISCHARGER WITH MULTIPLE GAPS.

a spherical middle portion *m* with an extension *e* below—which is merely used to fasten the piece in a lathe when polishing up the discharging surface—and a column above, which consists of a knurled flange *f* surmounted by a threaded stem *l* carrying a nut *n*, by means of which a wire is fastened to the column. The flange *f* conveniently serves for holding the brass piece when fastening the wire, and also for turning it in any position when it becomes necessary to present a fresh disharging surface. Two stout strips of hard rubber *R R*, with planed grooves *g g* (Fig. 7) to fit the middle portion of the pieces *c c*, serve to clamp the latter and hold them firmly in position by means of two bolts *C C* (of which only one is shown) passing through the ends of the strips.

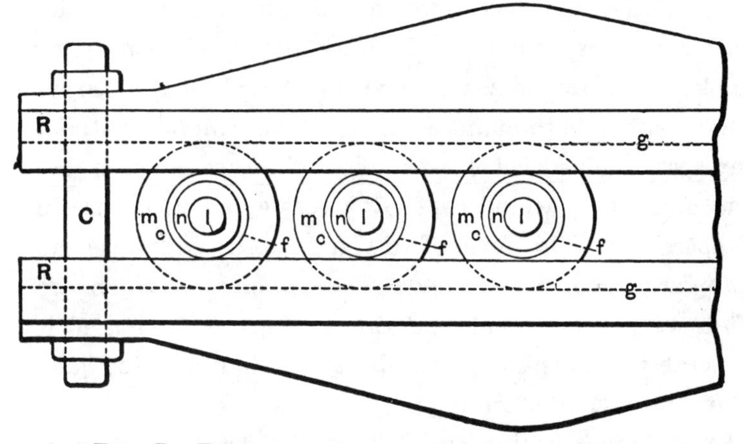

FIG. 7.—DISCHARGER WITH MULTIPLE GAPS.

In the use of this kind of discharger I have found three principal advantages over the ordinary form. First, the dielectric strength of a given total width of air space is greater when a great many small air gaps are used instead of one, which permits of working with a smaller length of air gap, and that means smaller loss and less deterioration of the metal; secondly by reason of splitting the arc up into smaller arcs, the polished surfaces are made to last much longer; and, thirdly, the apparatus affords some gauge in the experiments. I usually set the pieces by putting between them sheets of uniform thickness at a certain very small distance which is known from the experiments of Sir William Thomson to require a certain electromotive force to be bridged by the spark.

It should, of course, be remembered that the sparking distance is much diminished as the frequency is increased. By taking any number of spaces the experimenter has a rough idea of the electromotive force, and he finds it easier to repeat an experiment, as he has not the trouble of setting the knobs again and again. With this kind of discharger I have been able to maintain an oscillating motion without any spark being visible with the naked eye between the knobs, and they would not show a very appreciable rise in temperature. This form of discharge also lends itself to many arrangements of condensers and circuits which are often very convenient and time-saving. I have used it preferably in a disposition similar to that indicated in Fig. 2, when the currents forming the arc are small.

I may here mention that I have also used dischargers with single or multiple air gaps, in which the discharge surfaces were rotated with great speed. No particular advantage was, however, gained by this method, except in cases where the currents from the condenser were large and the keeping cool of the surfaces was necessary, and in cases when, the discharge not being oscillating of itself, the arc as soon as established was broken by the air current, thus starting the vibration at intervals in rapid succession. I have also used mechanical interrupters in many ways. To avoid the difficulties with frictional contacts, the preferred plan adopted was to establish the arc and rotate through it at great speed a rim of mica provided with many holes and fastened to a steel plate. It is understood, of course, that the employment of a magnet, air current, or other interrupter, produces an effect worth noticing, unless the self-induction, capacity and resistance are so related that there are oscillations set up upon each interruption.

I will now endeavor to show you some of the most noteworthy of these discharge phenomena.

I have stretched across the room two ordinary cotton covered wires, each about 7 metres in length. They are supported on insulating cords at a distance of about 30 centimetres. I attach now to each of the terminals of the coil one of the wires and set the coil in action. Upon turning the lights off in the room you see the wires strongly illuminated by the streams issuing abundantly from their whole surface in spite of the cotton covering, which may even be very thick. When the experiment is performed under good conditions, the light from the wires is sufficiently intense to allow distinguishing the objects in a room. To produce the best result it is, of course, necessary to adjust carefully the capacity of the jars, the arc between the knobs and the length of the wires. My experience is that calculation of the length of the wires leads, in such case, to no result whatever. The experimenter will do best to take the wires at the start very long, and then adjust by cutting off first long pieces, and then smaller and smaller ones as he approaches the right length.

A convenient way is to use an oil condenser of very small capacity, consisting of two small adjustable metal plates, in connection with this and similar experiments. In such case I take wires rather short and set at the beginning the condenser plates at maximum distance. If the streams for the wires increase by approach of the plates, the length of the wires is about right; if they diminish the wires are too long for that frequency and potential. When a condenser is used in connection with experiments with such a coil, it should be an oil condenser by all means, as in using an air condenser considerable energy might be wasted. The wires leading to the plates in the oil should be very thin, heavily coated with some insulating compound, and provided with a conducting covering—this preferably extending under the surface of the oil. The conducting cover should not be too near the terminals, or ends, of the wire, as a park would be apt to jump from the wire to it. The conducting coating is used to diminish the air losses, in virtue of its action as an electrostatic screen. As to the size of the vessel containing the oil, and the size of the plates, the experimenter gains at once an idea from a rough trial. The size of the plates *in oil* is, however, calculable, as the dielectric losses are very small.

In the preceding experiment it is of considerable interest to know what relation the quantity of the light emitted bears to the frequency and potential of the electric impulses. My opinion is that the heat as well as light effects produced should be proportionate, under otherwise equal conditions of test, to the product of frequency and square of potential, but the experimental verification of the law, whatever it may be, would be exceedingly difficult. One thing is certain, at any rate, and that is, that in augmenting the potential and frequency we rapidly intensify the streams; and, though it may be very sanguine, it is surely not altogether hopeless to expect that we may succeed in producing a practical illuminant on these lines. We would then be simply using burners or flames, in which there would be no chemical process, no consumption of material,

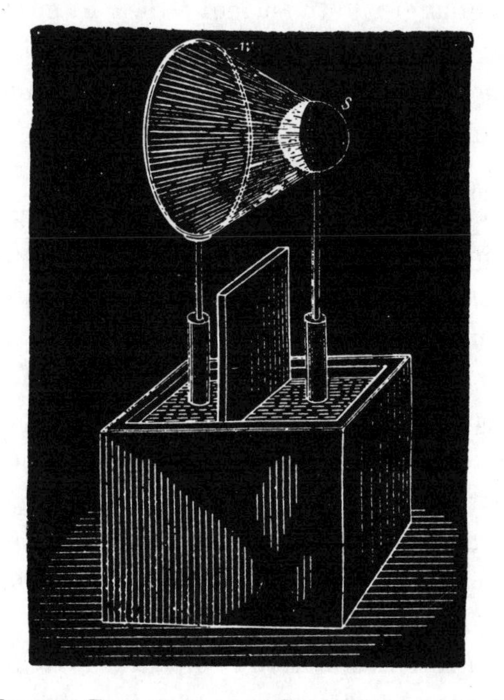

FIG. 8.—EFFECT PRODUCED BY CONCENTRATING STREAMS.

but merely a transfer of energy, and which would, in all probability emit more light and less heat than ordinary flames.

The luminous intensity of the streams is, of course, considerably increased when they are focused upon a small surface. This may be shown by the following experiment:

I attach to one of the terminals of the coil a wire *w* (Fig. 8), bent in a circle of about 30 centimetres in diameter, and to the other terminal I fasten a small brass sphere *s*, the surface of the wire being preferably equal to the surface of the sphere, and the centre of the latter being in a line at right angles to the plane of the wire circle and passing through its centre. When the discharge is established under proper conditions, a luminous hollow cone is formed, and in the dark one-half of the brass sphere is strongly illuminated, as shown in the cut.

By some artifice or other, it is easy to concentrate the streams upon small surfaces and to produce very strong light effects. Two thin wires may thus be rendered intensely luminous.

In order to intensify the streams the wires should be very thin and short; but as in this case their capacity would be generally too small for the coil—at least, for such a one as the present—it is necessary to augment the capacity to the required value, while, at the same time, the surface of the wires remains very small. This may be done in many ways.

Here, for instance, I have two plates, *R R*, of hard rubber (Fig. 9), upon which I have glued two very thin wires *w w*, so as to form a name. The wires may be bare or covered with the best insulation—it is immaterial for the success of the experiment. Well insulated wires, if anything, are preferable. On the back of each plate, indicated by the shaded portion, is a tinfoil coating *t t*. The plates are placed in line at a sufficient distance to prevent a spark passing from one to the other wire. The two tinfoil coatings I have joined by a conductor *C*, and the two wires I presently connect to the terminals of the coil. It is now easy, by varying the strength and frequency of the currents through the primary,

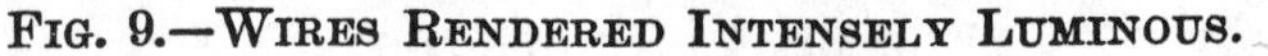
FIG. 9.—WIRES RENDERED INTENSELY LUMINOUS.

to find a point at which the capacity of the system is best suited to the conditions, and the wires become so strongly luminous that, when the light in the room is turned off the name formed by them appears in brilliant letters.

It is perhaps preferable to perform this experiment with a coil operated from an alternator of high frequency, as then, owing to the harmonic rise and fall, the streams are very uniform, though they are less abundant than when produced with such a coil as the present. This experiment, however, may be performed with low frequencies, but much less satisfactorily.

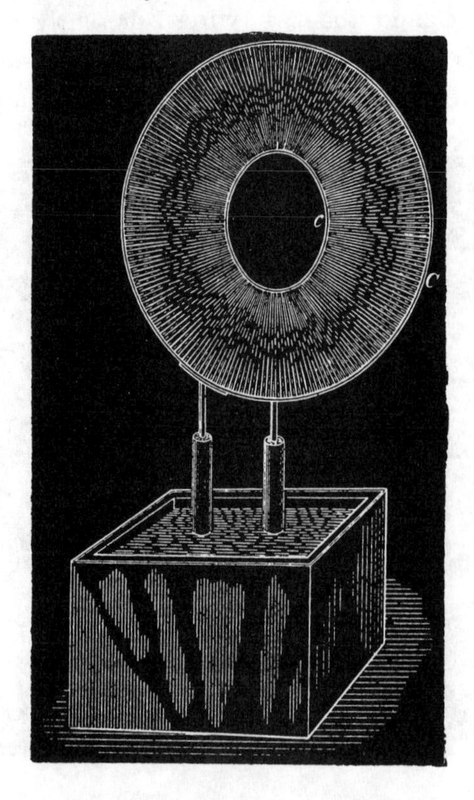

FIG. 10.—LUMINOUS DISCS.

When two wires, attached to the terminals of the coil, are set at the proper distance, the streams between them may be so intense as to produce a continuous luminous sheet. To show this phenomenon I have here two circles, *C* and *c* (Fig. 10), of rather stout wire, one being about

80 centimetres and the other 30 centimetres in diameter. To each of the terminals of the coil I attach one of the circles. The supporting wires are so bent that the circles may be placed in the same plane, coinciding as nearly as possible. When the light in the room is turned off and the coil set to work, you see the whole space between the wires uniformly filled with streams, forming a luminous disc, which could be seen from a considerable distance, such is the intensity of the streams. The outer circle could have been much larger than the present one; in fact, with this coil I have used much larger circles, and I have been able to produce a strongly luminous sheet, covering an area of more than one square metre, which is a remarkable effect with this very small coil. To avoid uncertainty, the circle has been taken smaller, and the area is now about 0.43 square metre.

The frequency of the vibration, and the quickness of succession of the sparks between the knobs, affect to a marked degree the appearance of the streams. When the frequency is very low, the air gives way in more or less the same manner, as by a steady difference of potential, and the streams consist of distinct threads, generally mingled with thin sparks, which probably correspond to the successive discharges occurring between the knobs. But when the frequency is extremely high, and the arc of the discharge produces a very *loud* but *smooth* sound—showing both that oscillation takes place and that the sparks succeed each other with great rapidity—then the luminous streams formed are perfectly uniform. To reach this result very small coils and jars of small capacity should be used. I take two tubes of thick Bohemian glass, about 5 centimetres in diameter and 20 centimetres long. In each of the tubes I slip a primary of very thick copper wire. On the top of each tube I wind a secondary of much thinner gutta-percha covered wire. The two secondaries I connect in series, the primaries preferably in multiple arc. The tubes are then placed in a large glass vessel, at a distance of 10 to 15 centimetres from each other, on insulating supports, and the vessel is filled with boiled-out oil, the oil reaching about an inch above the tubes. The free ends of the secondary are lifted out of the oil and placed parallel to each other at a distance of about 10 centimetres. The ends which are scraped should be dipped in the oil. Two four-pint jars joined in series may be used to discharge through the primary. When the necessary adjustments in the length and distance of the wires above the oil and in the arc of discharge are made, a luminous sheet is produced between the wires which is perfectly smooth and textureless, like the ordinary discharge through a moderately exhausted tube.

I have purposely dwelt upon this apparently insignificant experiment. In trials of this kind the experimenter arrives at the startling conclusion that, to pass ordinary luminous discharges through gases, no particular degree of exhaustion is needed, but that the gas may be at ordinary or even greater pressure. To accomplish this, a very high frequency is essential; a high potential is likewise required, but this is a merely incidental necessity. These experiments teach us that, in endeavoring to discover novel methods of producing light by the agitation of atoms, or molecules, of a gas, we need not limit our research to the vacuum tube, but may look forward quite seriously to the possibility of obtaining the light effects without the use of any vessel whatever, with air at ordinary pressure.

Such discharges of very high frequency, which render luminous the air at ordinary pressures, we have probably often occasion to witness in Nature. I have no doubt that if, as many believe, the aurora borealis is produced by sudden cosmic disturbances, such as eruptions at the sun's surface, which set the electrostatic charge of the earth in an extremely rapid vibration, the red glow observed is not confined to the upper rarefied strata of the air, but the discharge traverses, by reason of its very high frequency, also the dense atmosphere in the form of a *glow*, such as we ordinarily produce in a slightly exhausted tube. If the frequency were very low, or even more so, if the charge were not at all vibrating, the dense air would break down as in a lightning discharge. Indications of such breaking down of the lower dense strata of the air have been repeatedly observed at the occurrence of this marvelous phenomenon; but if it does occur, it can only be attributed to the fundamental disturbances, which are few in number, for the vibration produced by them would be far too rapid to allow a disruptive break. It is the original and irregular impulses which affect the instruments; the superimposed vibrations probably pass unnoticed.

When an ordinary low frequency discharge is passed through moderately rarefied air, the air assumes a purplish hue. If by some means or other we increase the intensity of the molecular, or atomic, vibration, the gas changes to a white color. A similar change occurs at ordinary pressures with electric impulses of very high frequency. If the molecules of the air around a wire are moderately agitated, the brush formed is reddish or violet; if the vibration is rendered sufficiently intense, the streams become white. We may accomplish this in various ways. In the experiment before shown with the two wires across the room, I have endeavored to secure the result by pushing to a high value both the frequency and potential; in the experiment with the thin wires glued on the rubber plate I have concentrated the action upon a very small surface—in other words, I have worked with a great electric density.

A most curious form of discharge is observed with such a coil when the frequency and potential are pushed to the extreme limit. To perform the experiment, every part of the coil should be heavily insulated, and only two small spheres—or, better still, two sharp-edged metal discs (*d d*, Fig. 11) of no more than a few centimetres in diameter—should be exposed to the air. The coil here used is immersed in oil, and the ends of the secondary reaching out of the oil are covered with an air-tight cover of hard rubber of great thickness. All cracks, if there are any, should be carefully stopped up, so that the brush discharge cannot form anywhere except on the small spheres or plates which are exposed to the air. In this case, since there are no large plates or other bodies of capacity attached to the terminals, the coil is capable of an extremely rapid vibration. The potential may be raised by increasing, as far as the experimenter judges proper, the rate of change of the primary current. With a coil not widely differing from the present, it is best to connect the two primaries in multiple arc; but if the secondary should have a much greater number of turns the primaries should preferably be used in series, as otherwise the vibration might be too fast for the secondary. It occurs under these conditions that misty white streams break forth from the

FIG. 11.—PHANTOM STREAMS.

edges of the discs and spread out phantom-like into space. With this coil, when fairly well produced, they are about 25 to 30 centimetres long. When the hand is held against them no sensation is produced, and a spark, causing a shock, jumps from the terminal only upon the hand being brought much nearer. If the oscillation of the primary current is rendered intermittent by some means or other, there is a corresponding throbbing of the streams, and now the hand or other conducting object may be brought in still greater proximity to the terminal without a spark being caused to jump.

Among the many beautiful phenomena which may be produced with such a coil I have here selected only those which appear to possess some features of novelty, and lead us to some conclusions of interest. One will not find it at all difficult to produce in the laboratory, by means of it, many other phenomena which appeal to the eye even more than these here shown, but present no particular feature of novelty.

Early experimenters describe the display of sparks produced by an ordinary large induction coil upon an insulating plate separating the terminals. Quite recently Siemens performed some experiments in which fine effects were obtained, which were seen by many with interest. No doubt large coils, even if operated with currents of low frequencies, are capable of producing beautiful effects. But the largest coil ever made could not, by far, equal the magnificent display of streams and sparks obtained from such a disruptive discharge coil when properly adjusted. To give an idea, a coil such as the present one will cover easily a plate of 1 metre in diameter completely with the streams. The best way to perform such experiments is to take a very thin rubber or a glass plate and glue on one side of it a narrow ring of tinfoil of very large diameter, and on the other a circular washer, the centre of the latter coinciding with that of the ring, and the surfaces of both being preferably equal, so as to keep the coil well balanced. The washer and ring should be connected to the terminals by heavily insulated thin wires. It is easy in observing the effect of the capacity to produce a sheet of uniform streams, or a fine network of thin silvery threads, or a mass of loud brilliant sparks, which completely cover the plate.

Since I have advanced the idea of the conversion by means of the disruptive discharge, in my paper before the American Institute of Electrical Engineers at the beginning of the past year, the interest excited in it has been considerable. It affords us a means for producing any potentials by the aid of inexpensive coils operated from ordinary systems of distribution, and—what is perhaps more appreciated—it enables us to convert currents of any frequency into currents of any other lower or higher frequency. But its chief value will perhaps be found in the help which it will afford us in the investigations of the phenomena of phosphorescence, which a disruptive discharge coil is capable of exciting in innumerable cases where ordinary coils, even the largest, would utterly fail.

Considering its probable uses for many practical purposes, and its possible introduction into laboratories for scientific research, a few additional remarks as to the construction of such a coil will perhaps not be found superfluous.

It is, of course, absolutely necessary to employ in such a coil wires provided with the best insulation.

Good coils may be produced by employing wires covered with several layers of cotton, boiling the coil a long time in pure wax, and cooling under moderate pressure. The advantage of such a coil is that it can be easily handled, but it cannot probably give as satisfactory results as a coil immersed in pure oil. Besides, it seems that the presence of a large body of wax affects the coil disadvantageously, whereas this does not seem to be the case with oil. Perhaps it is because the dielectric losses in the liquid are smaller.

I have tried at first silk and cotton covered wires with oil immersion, but I have been gradually led to use gutta-percha covered wires, which proved most satisfactory. Gutta-percha insulation adds, of course, to the capacity of the coil, and this, especially if the coil be large, is a great disadvantage when extreme frequencies are desired ; but, on the other hand, gutta-percha will withstand much more than an equal thickness of oil, and this advantage should be secured at any price. Once the coil has been immersed, it should never be taken out of the oil for more than a few hours, else the gutta-percha will crack up and the coil will not be worth half as much as before. Gutta-percha is probably slowly attacked by the oil, but after an immersion of eight to nine months I have found no ill effects.

I have obtained in commerce two kinds of gutta-percha wire: in one the insulation sticks tightly to the metal, in the other it does not. Unless a special method is followed to expel all air, it is much safer to use the first kind. I wind the coil within an oil tank so that all interstices are filled up with the oil. Between the layers I use cloth boiled out thoroughly in oil, calculating the thickness according to the difference of potential between the turns. There seems not to be a very great difference whatever kind of oil is used ; I use paraffine or linseed oil.

To exclude more perfectly the air, an excellent way to proceed, and easily practicable with small coils, is the following: Construct a box of hard wood of very thick boards which have been for a long time boiled in oil. The boards should be so joined as to safely withstand the external air pressure. The coil being placed and fastened in position within the box, the latter is closed with a strong lid, and covered with closely fitting metal sheets, the joints of which are soldered very carefully. On the top two small holes are drilled, passing through the metal sheet and the wood, and in these holes two small glass tubes are inserted and the joints made air-tight. One of the tubes is connected to a vacuum pump, and the other with a vessel containing a sufficient quantity of boiled-out oil. The latter tube has a very small hole at the bottom, and is provided with a stopcock. When a fairly good vacuum has been obtained, the stopcock is opened and the oil slowly fed in. Proceeding in this manner, it is impossible that any big bubbles, which are the principal danger, should remain between the turns. The air is most completely excluded, probably better than by boiling out, which, however, when gutta-percha coated wires are used, is not practicable.

For the primaries I use ordinary line wire with a thick cotton coating. Strands of very thin insulated wires properly interlaced would, of course, be the best to employ for the primaries, but they are not to be had.

In an experimental coil the size of the wires is not of great importance. In the coil here used the primary is No. 12 and the secondary No. 24 Brown & Sharpe gauge wire; but the sections may be varied considerably. It would only imply different adjustments; the results aimed at would not be materially affected.

I have dwelt at some length upon the various forms of brush discharge because, in studying them, we not only observe phenomena which please our eye, but also afford us food for thought, and lead us to conclusions of practical importance. In the use of alternating currents of very high tension, too much precaution cannot be taken to prevent the brush discharge. In a main conveying such currents, in an induction coil or transformer, or in a condenser, the brush discharge is a source of great danger to the insulation. In a condenser especially the gaseous matter must be most carefully expelled, for in it the charged surfaces are near each other, and if the potentials are high, just as sure as a weight will fall if let go, so the insulation will give way if a single gaseous bubble of some size be present, whereas, if all gaseous matter were carefully excluded, the condenser would safely withstand a much higher difference of potential. A main conveying alternating currents of very high tension may be injured merely by a blow hole or small crack in the insulation, the more so as a blowhole is apt to contain gas at low pressure; and as it appears almost impossible to completely obviate such little imperfections, I am led to believe that in our future distribution of electrical energy by currents of very high tension liquid insulation will be used. The cost is a great drawback, but if we employ an oil as an insulator the distribution of electrical energy with something like 100,000 volts, and even more, become, at least with higher frequencies, so easy that they could be hardly called engineering feats. With oil insulation and alternate current motors transmissions of power can be effected with safety and upon an industrial basis at distances of as much as a thousand miles.

A peculiar property of oils, and liquid insulation in general, when subjected to rapidly changing electric stresses, is to disperse any gaseous bubbles which may be present, and diffuse them through its mass, generally long before any injurious break can occur. This feature may be easily observed with an ordinary induction coil by taking the primary out, plugging up the end of the tube upon which the secondary is wound, and filling it with some fairly transparent insulator, such as paraffine oil. A primary of a diameter something like six millimetres smaller than the inside of the tube may be inserted in the oil. When the coil is set to work one may see, looking from the top through the oil, many luminous points—air bubbles which are caught by inserting the primary, and which are rendered luminous in consequence of the violent bombardment. The occluded air, by its impact against the oil, heats it; the oil begins to circulate, carrying some of the air along with it, until the bubbles are dispersed and the luminous points disappear. In this manner, unless large bubbles are occluded in such way that circulation is rendered impossible, a damaging break is averted, the only effect being a moderate warming up of the oil. If, instead of the liquid, a solid insulation, no matter how thick, were used, a breaking through and injury of the apparatus would be inevitable.

The exclusion of gaseous matter from any apparatus in

which the dielectric is subjected to more or less rapidly changing electric forces is, however, not only desirable in order to avoid a possible injury of the apparatus, but also on account of economy. In a condenser, for instance, as long as only a solid or only a liquid dielectric is used, the loss is small; but if a gas under ordinary or small pressure be present the loss may be very great. Whatever the nature of the force acting in the dielectric may be, it seems that in a solid or liquid the molecular displacement produced by the force is small: hence the product of force and displacement is insignificant, unless the force be very great; but in a gas the displacement, and therefore this product, is considerable; the molecules are free to move, they reach high speeds, and the energy of their impact is lost in heat or otherwise. If the gas be strongly compressed, the displacement due to the force is made smaller, and the losses are reduced.

In most of the succeeding experiments I prefer, chiefly on account of the regular and positive action, to employ the alternator before referred to. This is one of the several machines constructed by me for the purposes of these investigations. It has 384 pole projections, and is capable of giving currents of a frequency of about 10,000 per second. This machine has been illustrated and briefly described in my first paper before the American Institute of Electrical Engineers, May 20, 1891, to which I have already referred. A more detailed description, sufficient to enable any engineer to build a similar machine, will be found in several electrical journals of that period.

The induction coils operated from the machine are rather

small, containing from 5,000 to 15,000 turns in the secondary. They are immersed in boiled-out linseed oil, contained in wooden boxes covered with zinc sheet.

I have found it advantageous to reverse the usual position of the wires, and to wind, in these coils, the primaries on the top; this allowing the use of a much bigger primary, which, of course, reduces the danger of overheating and increases the output of the coil. I make the primary on each side at least one centimetre shorter than the secondary, to prevent the breaking through on the ends, which would surely occur unless the insulation on the top of the secondary be very thick, and this, of course, would be disadvantageous.

When the primary is made movable, which is necessary in some experiments, and many times convenient for the purposes of adjustment, I cover the secondary with wax, and turn it off in a lathe to a diameter slightly smaller than the inside of the primary coil. The latter I provide with a handle reaching out of the oil, which serves to shift it in any pssition along the secondary.

I will now venture to make, in regard to the general manipulation of induction coils, a few observations bearing upon points which have not been fully appreciated in earlier experiments with such coils, and are even now often overlooked.

The secondary of the coil possesses usually such a high self-induction that the current through the wire is inappreciable, and may be so even when the terminals are joined by a conductor of small resistance, If capacity is added to the terminals, the self-induction is counteracted, and a stronger current is made to flow through the secondary, though its terminals are insulated from each other. To one entirely unacquainted with the properties of alternating currents nothing will look more puzzling. This feature was illustrated in the experiment performed at the beginning with the top plates of wire gauze attached to the terminals and the rubber plate. When the plates of wire gauze were close together, and a small arc passed between them, the arc *prevented* a strong current from passing through the secondary, because it did away with the capacity on the terminals; when the rubber plate was inserted between, the capacity of the condenser formed counteracted the self-induction of the secondary, a stronger current passed now, the coil performed more work, and the discharge was by far more powerful.

The first thing, then, in operating the induction coil is to combine capacity with the secondary to overcome the self-induction. If the frequencies and potentials are very high gaseous matter should be carefully kept away from the charged surfaces. If Leyden jars are used, they should be immersed in oil, as otherwise considerable dissipation may occur if the jars are greatly strained. When high frequencies are used, it is of equal importance to combine a condenser with the primary. One may use a condenser connected to the ends of the primary or to the terminals of the alternator, but the latter is not to be recommended, as the machine might be injured. The best way is undoubtedly to use the condenser in series with the primary and with the alternator, and to adjust its capacity so as to annul the self-induction of both the latter. The condenser should be adjustable by very small steps, and for a finer adjustment a small oil condenser with movable plates may be used conveniently.

I think it best at this juncture to bring before you a phenomenon, observed by me some time ago, which to the purely scientific investigator may perhaps appear more interesting than any of the results which I have the privilege to present to you this evening.

It may be quite properly ranked among the brush phenomena—in fact, it is a brush, formed at, or near, a single terminal in high vacuum.

In bulbs provided with a conducting terminal, though it be of aluminium, the brush has but an ephemeral existence, and cannot, unfortunately, be indefinitely preserved in its most sensitive state, even in a bulb devoid of any conducting electrode. In studying the phenomenon, by all means a bulb having no leading-in wire should be used. I have found it best to use bulbs constructed as indicated in Figs. 12 and 13.

In Fig. 12 the bulb comprises an incandescent lamp globe *L*, in the neck of which is sealed a barometer tube *b*, the end of which is blown out to form a small sphere *s*. This sphere should be sealed as closely as possible in the centre of the large globe. Before sealing, a thin tube *t*, of aluminium sheet, may be slipped in the barometer tube, but it is not important to employ it.

The small hollow sphere *s* is filled with some conducting powder, and a wire *w* is cemented in the neck for the purpose of connecting the conducting powder with the generator.

The construction shown in Fig. 13 was chosen in order to remove from the brush any conducting body which might possibly affect it. The bulb consists in this case of a lamp globe *L*, which has a neck *n*, provided with a tube *b* and

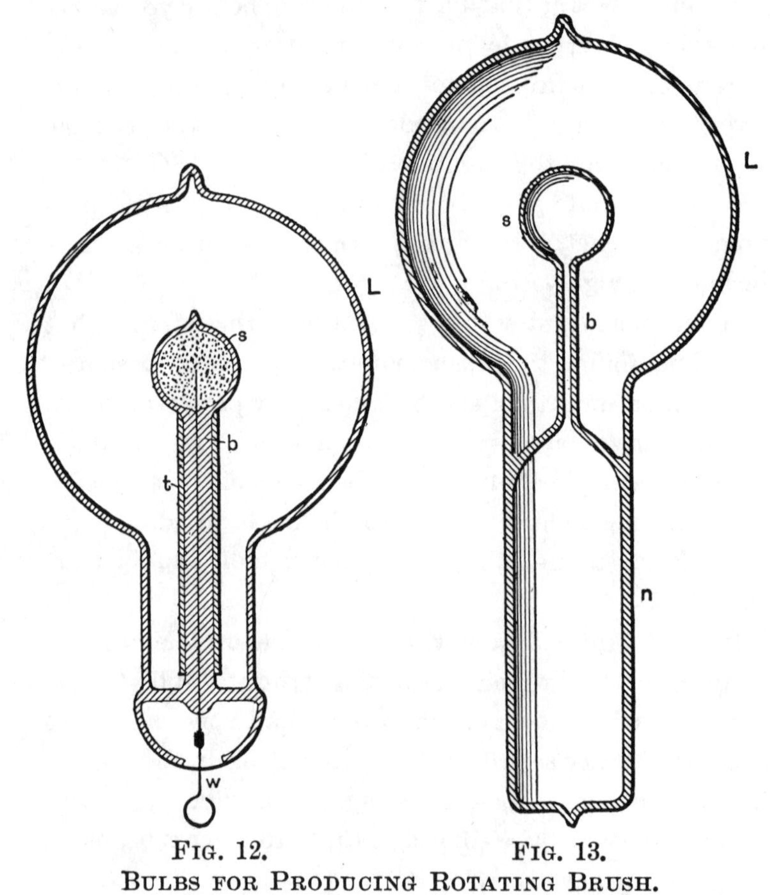

FIG. 12. FIG. 13.
BULBS FOR PRODUCING ROTATING BRUSH.

small sphere *s*, sealed to it, so that two entirely independent compartments are formed, as indicated in the drawing. When the bulb is in use, the neck *n* is provided with a tinfoil coating, which is connected to the generator and acts

Main entrance to Colorado Springs Laboratory in the early phase of development. Tesla is looking through the door (Tesla's own photograph now at the Nikola Tesla Museum, Belgrade)

inductively upon the moderately rarefied and highly conducting gas inclosed in the neck. From there the current passes through the tube *b* into the small sphere *s*, to act by induction upon the gas contained in the globe *L*.

It is of advantage to make the tube *t* very thick, the hole through it very small, and to blow the sphere *s* very thin. It is of the greatest importance that the sphere *s* be placed in the centre of the globe *L*.

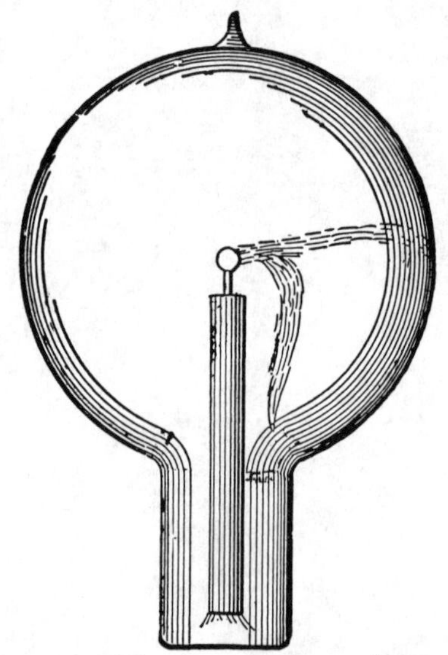

FIG. 14.—FORMS AND PHASES OF THE ROTATING BRUSH.

Figs. 14, 15 and 16 indicate different forms, or stages, of the brush. Fig. 14 shows the brush as it first appears in a bulb provided with a conducting terminal; but, as in such a bulb it very soon disappears—often after a few minutes—I will confine myself to the description of the phenomenon as seen in a bulb without conducting electrode. It is observed under the following conditions:

When the globe *L* (Figs. 12 and 13) is exhausted to a

very high degree, generally the bulb is not excited upon connecting the wire *w* (Fig. 12) or the tinfoil coating of the bulb (Fig. 13) to the terminal of the induction coil. To excite it, it is usually sufficient to grasp the globe *L* with the

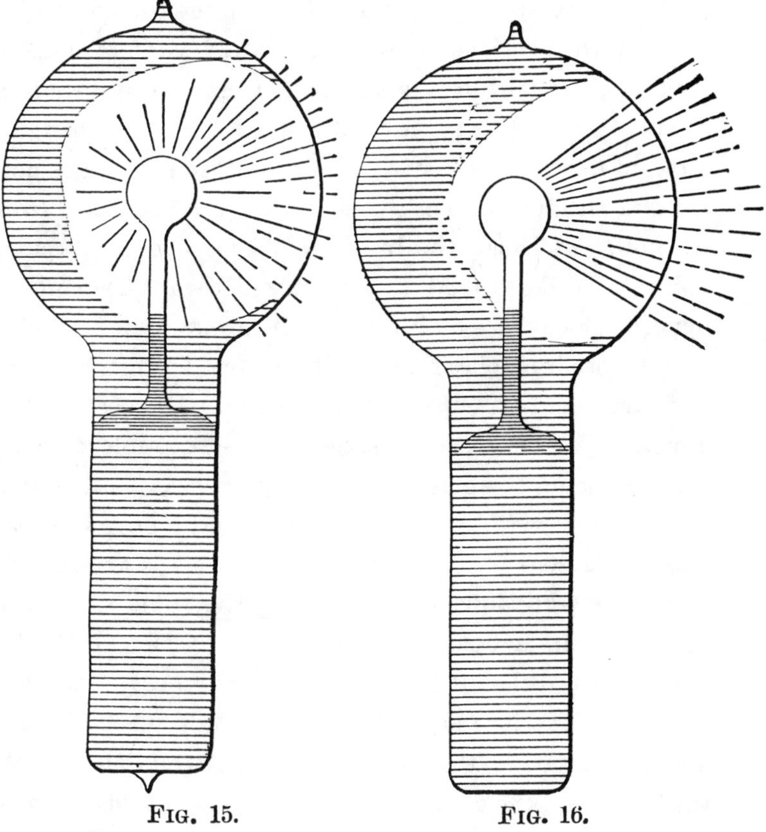

FIG. 15. FIG. 16.
FORMS AND PHASES OF THE ROTATING BRUSH.

hand. An intense phosphorescence then spreads at first over the globe, but soon gives place to a white, misty light. Shortly afterward one may notice that the luminosity is unevenly distributed in the globe, and after passing the current for some time the bulb appears as in Fig. 15. From this stage the phenomenon will gradually pass to that indicated in Fig. 16, after some minutes, hours, days or weeks, according as the bulb is worked. Warming the bulb or increasing the potential hastens the transit.

When the brush assumes the form indicated in Fig. 16, it may be brought to a state of extreme sensitiveness to electrostatic and magnetic influence. The bulb hanging straight down from a wire, and all objects being remote from it, the approach of the observer at a few paces from the bulb will cause the brush to fly to th opposite side, and if he walks around the bulb it will always keep on the opposite side. It may begin to spin around the terminal long before it reaches that sensitive stage. When it begins to turn around principally, but also before, it is affected by a magnet, and at a certain stage it is susceptible to magnetic influence to an astonishing degree. A small permanent magnet, with its poles at a distance of no more than two centimetres, will affect it visibly at a distance of two metres, slowing down or accelerating the rotation according to how it is held relatively to the brush. I think I have observed that at the stage when it is most sensitive to magnetic, it is not most sensitive to electrostatic, influence. My explanation is, that the electrostatic attraction between the brush and the glass of the bulb, which retards the rotation, grows much quicker than the magnetic influence when the intensity of the stream is increased.

When the bulb hangs with the globe *L* down, the rotation is always clockwise. In the southern hemisphere it would occur in the opposite direction and on the equator the brush should not turn at all. The rotation may be reversed by a magnet kept at some distance. The brush rotates best, seemingly, when it is at right angles to the lines of force of the earth. It very likely rotates, when at its maximum speed, in synchronism with the alternations, say 10,000 times a second. The rotation can be slowed down or accelerated by the approach or receding of the observer, or any conducting body, but it cannot be reversed by putting the bulb in any position. When it is in the state of the highest sensitiveness and the potential or frequency be varied the sensitiveness is rapidly diminished. Changing either of these but little will generally stop the rotation. The sensitiveness is likewise affected by the variations of temperature. To attain great sensitiveness it is necessary to have the small sphere *s* in the centre of the globe *L*, as otherwise the electrostatic action of the glass of the globe will tend to stop the rotation. The sphere *s* should be small and of uniform thickness; any dissymmetry of course has the effect to diminish the sensitiveness.

The fact that the brush rotates in a definite direction in a permanent magnetic field seems to show that in alternating currents of very high frequency the positive and negative impulses are not equal, but that one always preponderates over the other.

Of course, this rotation in one direction may be due to the action of two elements of the same current upon each other, or to the action of the field produced by one of the elements upon the other, as in a series motor, without necessarily one impulse being stronger than the other. The fact that the brush turns, as far as I could observe, in any position, would speak for this view. In such case it would turn at any point of the earth's surface. But, on the other hand, it is then hard to explain why a permanent magnet should reverse the rotation, and one must assume the preponderance of impulses of one kind.

As to the causes of the formation of the brush or stream, I think it is due to the electrostatic action of the globe and the dissymmetry of the parts. If the small bulb *s* and the globe *L* were perfect concentric spheres, and the glass throughout of the same thickness and quality, I think the brush would not form, as the tendency to pass would be equal on all sides. That the formation of the stream is due to an irregularity is apparent from the fact that it has the tendency to remain in one position, and rotation occurs most generally only when it is brought out of this position by electrostatic or magnetic influence. When in an extremely sensitive state it rests in one position, most curious experiments may be performed with it. For instance, the experimenter may, by selecting a proper position, approach the hand at a certain considerable distance to the bulb, and he may cause the brush to pass off by merely stiffening the muscles of the arm. When it begins to rotate slowly, and the hands are held at a proper distance, it is impossible to make even the slightest motion without producing a visible effect upon the brush. A metal plate connected to the other terminal of the coil affects it at a great distance, slowing down the rotation often to one turn a second.

I am firmly convinced that such a brush, when we learn how to produce it properly, will prove a valuable aid in the investigation of the nature of the forces acting in an electrostatic or magnetic field. If there is any motion which is measurable going on in the space, such a brush ought to reveal it. It is, so to speak, a beam of light, frictionless, devoid of inertia.

I think that it may find practical applications in telegraphy. With such a brush it would be possible to send dispatches across the Atlantic, for instance, with any speed, since its sensitiveness may be so great that the slightest changes will affect it. If it were possible to make the stream more intense and very narrow, its deflections could be easily photographed.

I have been interested to find whether there is a rotation of the stream itself, or whether there is simply a stress traveling around in the bulb. For this purpose I mounted a light mica fan so that its vanes were in the path of the brush. If the stream itself was rotating the fan would be spun around. I could produce no distinct rotation of the fan, although I tried the experiment repeatedly; but as the fan exerted a noticeable influence on the stream, and the apparent rotation of the latter was, in this case, never quite satisfactory, the experiment did not appear to be conclusive.

I have been unable to produce the phenomenon with the disruptive discharge coil, although every other of these phenomena can be well produced by it—many, in fact, much better than with coils operated from an alternator.

It may be possible to produce the brush by impulses of one direction, or even by a steady potential, in which case it would be still more sensitive to magnetic influence.

In operating an induction coil with rapidly alternating currents, we realize with astonishment, for the first time, the great importance of the relation of capacity, self-induction and frequency as regards the general result. The effects of capacity are the most striking, for in these experiments, since the self-induction and frequency both are high, the critical capacity is very small, and need be but slightly varied to produce a very considerable change. The experimenter may bring his body in contact with the terminals of the secondary of the coil, or attach to one or both terminals insulated bodies of very small bulk, such as bulbs, and he may produce a considerable rise or fall of potential, and greatly affect the flow of the current through the primary. In the experiment before shown, in which a brush appears at a wire attached to one terminal, and the wire is vibrated when the experimenter brings his insulated body in contact with the other terminal of the coil, the sudden rise of potential was made evident.

I may show you the behavior of the coil in another manner which possesses a feature of some interest. I have here a little light fan of aluminium sheet, fastened to a needle and arranged to rotate freely in a metal piece screwed to one of the terminals of the coil. When the coil is set to work, the molecules of the air are rhythmically attracted and repelled. As the force with which they are repelled is greater than that with which they are attracted, it results that there is a repulsion exerted on the surfaces of the fan. If the fan were made simply of a metal sheet, the repulsion would be equal on the opposite sides, and would produce no effect. But if one of the opposing surfaces is screened, or if, generally speaking, the bombardment on this side is weakened in some way or other, there remains the repulsion exerted upon the other, and the fan is set in rotation. The screening is best effected by fastening upon one of the opposing sides of the fan insulated conducting coatings, or, if the fan is made in the shape of an ordinary propeller screw, by fastening on one side, and close to it, an insulated metal plate. The static screen may, however, be omitted, and simply a thickness of insulating material fastened to one of the sides of the fan.

To show the behavior of the coil, the fan may be placed upon the terminal and it will readily rotate when the coil is operated by currents of very high frequency. With a steady potential, of course, and even with alternating currents of very low frequency, it would not turn, because of the very slow exchange of air and, consequently, smaller bombardment; but in the latter case it might turn if the potential were excessive. With a pin wheel, quite the opposite rule holds good ; it rotates best with a steady potential, and the effort is the smaller the higher the frequency. Now, it is very easy to adjust the conditions so that the potential is normally not sufficient to turn the fan, but that by connecting the other terminal of the coil with an insulated body it rises to a much greater value, so as to rotate the fan, and it is likewise possible to stop the rotation by connecting to the terminal a body of different size, thereby diminishing the potential.

Instead of using the fan in this experiment, we may use the "electric" radiometer with similar effect. But in this case it will be found that the vanes will rotate only at high exhaustion or at ordinary pressures; they will not rotate at moderate pressures, when the air is highly conducting.

This curious observation was made conjointly by Professor Crookes and myself. I attribute the result to the high conductivity of the air, the molecules of which then do not act as independent carriers of electric charges, but act all together as a single conducting body. In such case, of course, if there is any repulsion at all of the molecules from the vanes, it must be very small. It is possible, however, that the result is in part due to the fact that the greater part of the discharge passes from the leading-in wire through the highly conducting gas, instead of passing off from the conducting vanes.

In trying the preceding experiment with the electric radiometer the potential should not exceed a certain limit, as then the electrostatic attraction between the vanes and the glass of the bulb may be so great as to stop the rotation.

A most curious feature of alternate currents of high frequencies and potentials is that they enable us to perform many experiments by the use of one wire only. In many respects this feature is of great interest.

In a type of alternate current motor invented by me some years ago I produced rotation by inducing, by means of a single alternating current passed through a motor circuit, in the mass or other circuits of the motor, secondary currents, which, jointly with the primary or inducing current, created a moving field of force. A simple but crude form of such a motor is obtained by winding upon an iron core a primary, and close to it a secondary coil, joining the ends of the latter and placing a freely movable metal disc within the influence of the field produced by both. The iron core is employed for obvious reasons, but it is not essential to the operation. To improve the motor, the iron core is made to encircle the armature. Again to improve, the secondary coil is made to overlap partly the primary, so that it cannot free itself from a strong inductive action of the latter, repel its lines as it may. Once more to improve, the proper difference of phase is obtained between the primary and secondary currents by a condenser, self-induction, resistance or equivalent windings.

I had discovered, however, that rotation is produced by means of a single coil and core; my explanation of the phenomenon, and leading thought in trying the experiment, being that there must be a true time lag in the magnetization of the core. I remember the pleasure I had when, in the writings of Professor Ayrton, which came later to my hand, I found the idea of the time lag advocated. Whether there is a true time lag, or whether the retardation is due to eddy currents circulating in minute paths, must remain an open question, but the fact is that a coil wound upon an iron core and traversed by an alternating current creates a moving field of force, capable of setting an armature in rotation. It is of some interest, in conjunction with the historical Arago experiment, to mention that in lag or phase motors I have produced rotation in the opposite direction to the moving field, which means that in that experiment the magnet may not rotate, or may even rotate in the opposite direction to the moving disc. Here, then, is a motor (diagrammatically illustrated in Fig. 17), comprising a coil and iron core, and a freely movable copper disc in proximity to the latter.

motor and generator being insulated in space. To produce rotation it is generally (but not absolutely) necessary to connect the free end of the motor coil to an insulated body of some size. The experimenter's body is more than sufficient. If he touches the free terminal with an object held in the hand, a current passes through the coil and the copper disc is set in rotation. If an exhausted tube is put in series with the coil, the tube lights brilliantly, showing the passage of a strong current. Instead of the experimenter's body, a small metal sheet suspended on a cord may be used with the same result. In this case the plate acts as a condenser in series with the coil. It counteracts the self-induction of the latter and allows a strong current to pass. In such a combination, the greater the self-induction of the coil the smaller need be the plate, and this means that a lower frequency, or eventually a lower potential, is required to operate the motor. A single coil wound upon a core has a high self-induction; for this reason principally, this type of motor was chosen to perform the experiment. Were a secondary closed coil wound upon the core, it would tend to diminish the self-induction, and then it would be necessary to employ a much higher frequency and potential. Neither would be advisable, for a higher potential would endanger the insulation of the small primary coil, and a higher frequency would result in a materially diminished torque.

It should be remarked that when such a motor with a closed secondary is used, it is not at all easy to obtain rotation with excessive frequencies, as the secondary cuts off almost completely the lines of the primary—and this, of

To demonstrate a novel and interesting feature, I have, for a reason which I will explain, selected this type of motor. When the ends of the coil are connected to the terminals of an alternator the disc is set in rotation. But it is not this experiment, now well known, which I desire

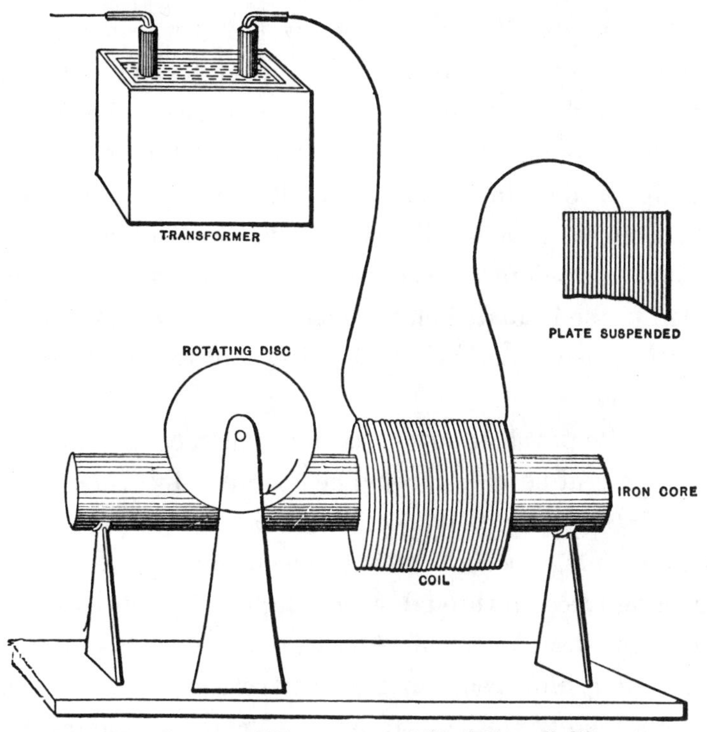

FIG. 17.—SINGLE WIRE AND "NO-WIRE" MOTOR.

to perform. What I wish to show you is that this motor rotates with *one single* connection between it and the generator; that is to say, one terminal of the motor is connected to one terminal of the generator—in this case the secondary of a high-tension induction coil—the other terminals of course, the more, the higher the frequency—and allows the passage of but a minute current. In such a case, unless the secondary is closed through a condenser, it is almost essential, in order to produce rotation, to make the primary and secondary coils overlap each other more or less.

But there is an additional feature of interest about this motor, namely, it is, not necessary to have even a single connection between the motor and generator, except, perhaps, through the ground; for not only is an insulated plate capable of giving off energy into space, but it is likewise capable of deriving it from an alternating electrostatic field, though in the latter case the available energy is much smaller. In this instance one of the motor terminals is connected to the insulated plate or body located within the alternating electrostatic field, and the other terminal preferably to the ground.

It is quite possible, however, that such "no-wire" motors, as they might be called, could be operated by conduction through the rarefied air at considerable distances. Alternate currents, especially of high frequencies, pass with astonishing freedom through even slightly rarefied gases. The upper strata of the air are rarefied. To reach a number of miles out into space requires the overcoming of difficulties of a merely mechanical nature. There is no doubt that with the enormous potentials obtainable by the use of high frequencies and oil insulation luminous discharges might be passed through many miles of rarefied air, and that, by thus directing the energy of many hundreds or thousands of horse-power, motors or lamps might be operated at considerable distances from stationary sources. But such schemes are mentioned merely as possibilities. We shall have no need to transmit power in this way. We shall have no need to *transmit* power at all. Ere many generations pass, our machinery will be driven by a power obtainable at any point of the universe. This idea is not novel. Men have been led to it long ago by instinct or reason. It has been expressed in many ways, and in many places, in the history of old and new. We find it in the delightful myth of Antheus, who derives power from the earth; we find it among the subtile speculations of one of your splendid mathematicians, and in many hints and statements of thinkers of the present time. Throughout space there is energy. Is this energy static or kinetic? If static our hopes are in vain; if kinetic—and this we know it is, for certain—then it is a mere question of time when men will succeed in attaching their machinery to the very wheelwork of nature. Of all, living or dead, Crookes came nearest to doing it. His radiometer will turn in the light of day and in the darkness of the night; it will turn everywhere where there is heat, and heat is everywhere. But, unfortunately, this beautiful little machine, while it goes down to posterity as the most interesting, must likewise be put on record as the most inefficient machine ever invented!

The preceding experiment is only one of many equally interesting experiments which may be performed by the use of only one wire with alternate currents of high potential and frequency. We may connect an insulated line to a source of such currents, we may pass an inappreciable current over the line, and on any point of the same we are able to obtain a heavy current, capable of fusing a thick copper wire. Or we may, by the help of some artifice, decompose a solution in any electrolytic cell by connecting only one pole of the cell to the line or source of energy. Or we may, by attaching to the line, or only bringing into its vicinity, light up an incandescent lamp, an exhausted tube, or a phosphorescent bulb.

However impracticable this plan of working may appear in many cases, it certainly seems practicable, and even recommendable, in the production of light. A perfected lamp would require but little energy, and if wires were used at all we ought to be able to supply that energy without a return wire.

It is now a fact that a body may be rendered incandescent or phosphorescent by bringing it either in single contact or merely in the vicinity of a source of electric impulses of the proper character, and that in this manner a quantity of light sufficient to afford a practical illuminant may be produced. It is, therefore, to say the least, worth while to attempt to determine the best conditions and to invent the best appliances for attaining this object.

Some experiences have already been gained in this direction, and I will dwell on them briefly, in the hope that they might prove useful.

The heating of a conducting body inclosed in a bulb, and connected to a source of rapidly alternating electric impulses, is dependent on so many things of a different nature, that it would be difficult to give a generally applicable rule under which the maximum heating occurs. As regards the size of the vessel, I have lately found that at ordinary or only slightly differing atmospheric pressures, when air is a good insulator, and hence practically the same amount of energy by a certain potential and frequency is given off from the body, whether the bulb be small or large, the body is brought to a higher temperature if inclosed in a small bulb, because of the better confinement of heat in this case.

At lower pressures, when air becomes more or less conducting, or if the air be sufficiently warmed as to become conducting, the body is rendered more intensely incandescent in a large bulb, obviously because, under otherwise equal conditions of test, more energy may be given off from the body when the bulb is large.

At very high degrees of exhaustion, when the matter in the bulb becomes "radiant," a large bulb has still an advantage, but a comparatively slight one, over the small bulb.

Finally, at excessively high degrees of exhaustion, which cannot be reached except by the employment of special means, there seems to be, beyond a certain and rather small size of vessel, no perceptible difference in the heating.

These observations were the result of a number of experiments, of which one, showing the effect of the size of the bulb at a high degree of exhaustion, may be described and shown here, as it presents a feature of interest. Three spherical bulbs of 2 inches, 3 inches and 4 inches diameter were taken, and in the centre of each was mounted an equal length of an ordinary incandescent lamp filament of uniform thickness. In each bulb the piece of filament was fastened to the leading-in wire of platinum, contained in a glass stem sealed in the bulb; care being taken, of course, to make everything as nearly alike as possible. On each glass stem in the inside of the bulb was slipped a highly polished tube made of aluminium sheet, which fitted the stem and was held on it by spring pressure. The function of this aluminium tube will be explained subsequently. In each bulb an equal length of filament protruded above the metal tube. It is sufficient to say now that under these conditions equal lengths of filament of the same thickness—in other words, bodies of equal bulk—were brought to incandescence. The three bulbs were sealed to a glass tube, which was connected to a Sprengel pump. When a high vacuum had been reached, the glass tube carrying the bulbs was sealed off. A current was then turned on successively on each bulb, and it was found that the filaments came to about the same brightness, and, if anything, the smallest bulb, which was placed midway between the two larger ones, may have been slightly brighter. This result was expected, for when either of the bulbs was connected to the coil the luminosity spread through the other two, hence the three bulbs constituted really one vessel. When all the three bulbs were connected in multiple arc to the coil, in the largest of them the filament glowed brightest, in the next smaller it was a little less bright, and in the smallest it only came to redness. The bulbs were then sealed off and separately tried. The brightness of the filaments was now such as would have been expected on the supposition that the energy given off was proportionate to the surface of the bulb, this surface in each case representing one of the coatings of a condenser. Accordingly, there was less difference between the largest and the middle sized than between the latter and the smallest bulb.

An interesting observation was made in this experiment. The three bulbs were suspended from a straight bare wire connected to a terminal of the coil, the largest bulb being placed at the end of the wire, at some distance from it the smallest bulb, and an equal distance from the latter the middle-sized one. The carbons glowed then in both the larger bulbs about as expected, but the smallest did not get its share by far. This observation led me to exchange the position of the bulbs, and I then observed that whichever of the bulbs was in the middle it was by far less bright than it was in any other position. This mystifying result was, of course, found to be due to the electrostatic action between the bulbs. When they were placed at a considerable distance, or when they were attached to the corners of an equilateral triangle of copper wire, they glowed about in the order determined by their surfaces.

As to the shape of the vessel, it is also of some importance, especially at high degrees of exhaustion. Of all the possible constructions, it seems that a spherical globe with the refractory body mounted in its centre is the best to employ. In experience it has been demonstrated that in such a globe a refractory body of a given bulk is more easily brought to incandescence than when otherwise shaped bulbs are used. There is also an advantage in giving to the incandescent body the shape of a sphere, for self-evident reasons. In any case the body should be mounted in the centre, where the atoms rebounding from the glass collide.

This object is best attained in the spherical bulb ; but it is also attained in a cylindrical vessel with one or two straight filaments coinciding with its axis, and possibly also in parabolical or spherical bulbs with the refractory body or bodies placed in the focus or foci of the same; though the latter is not probable, as the electrified atoms should in all cases rebound normally from the surface they strike, unless the speed were excessive, in which case they *would* probably follow the general law of reflection. No matter what shape the vessel may have, if the exhaustion be low, a filament mounted in the globe is brought to the same degree of incandescence in all parts; but if the exhaustion be high and the bulb be spherical or pear-shaped, as usual, focal points form and the filament is heated to a higher degree at or near such points.

To illustrate the effect, I have here two small bulbs which are alike, only one is exhausted to a low and the other to a very high degree. When connected to the coil, the filament in the former glows uniformly throughout all its length; whereas in the latter, that portion of the filament which is in the centre of the bulb glows far more intensely than the rest. A curious point is that the phenomenon occurs even if two filaments are mounted in a bulb, each being connected to one terminal of the coil, and, what is still more curious, if they be very near together, provided the vacuum be very high. I noted in experiments with such bulbs that the filaments would give way usually at a certain point, and in the first trials I attributed it to a defect in the carbon. But when the phenomenon occurred many times in succession I recognized its real cause.

In order to bring a refractory body inclosed in a bulb to incandescence, it is desirable, on account of economy, that all the energy supplied to the bulb from the source should reach without loss the body to be heated; from there, and from nowhere else, it should be radiated. It is, of course, out of the question to reach this theoretical result, but it is possible by a proper construction of the illuminating device to approximate it more or less.

For many reasons, the refractory body is placed in the centre of the bulb, and it is usually supported on a glass stem containing the leading-in wire. As the potential of this wire is alternated, the rarefied gas surrounding the stem is acted upon inductively, and the glass stem is violently bombarded and heated. In this manner by far the greater portion of the energy supplied to the bulb—especially when exceedingly high frequencies are used—may be lost for the purpose contemplated. To obviate this loss, or at least to reduce it to a minimum, I usually screen the rarefied gas surrounding the stem from the inductive action of the leading-in wire by providing the stem with a tube or coating of conducting material. It seems beyond doubt that the best among metals to employ for this purpose is aluminium, on account of its many remarkable properties. Its only fault is that it is easily fusible, and, therefore, its distance from the incandescing body should be properly estimated. Usually, a thin tube, of a diameter somewhat smaller than that of the glass stem, is made of the finest aluminium sheet, and slipped on the stem. The tube is conveniently prepared by wrapping around a rod fastened in a lathe a piece of aluminium sheet of the proper size, grasping the sheet firmly with clean chamois leather or blotting paper, and spinning the rod very fast. The sheet is wound tightly around the rod, and a highly polished tube of one or three layers of the sheet is obtained. When slipped on the stem, the pressure is generally sufficient to prevent it from slipping off, but, for safety, the lower edge of the sheet may be turned inside. The upper inside corner of the sheet—that is, the one which is nearest to the refractory incandescent body—should be cut out diagonally, as it often happens that, in consequence of the intense heat, this corner turns toward the inside and comes very near to, or in contact with, the wire, or filament, supporting the refractory body. The greater part of the energy supplied to the bulb is then used up in heating the metal tube, and the bulb is rendered useless for the purpose. The aluminium sheet should project above the glass stem more or less—one inch or so—or else, if the glass be too close to the incandescing body, it may be strongly heated and become more or less conducting, whereupon it may be ruptured, or may, by its conductivity, establish a good electrical connection between the metal tube and the leading-in wire, in which case, again, most of the energy will be lost in heating the former. Perhaps the best way is to make the top of the glass tube, for about an inch, of a much smaller diameter. To still further reduce the danger arising from the heating of the glass stem, and also with the view of preventing an electrical connection between the metal tube and the electrode, I preferably wrap the stem with several layers of thin mica, which extends at least as far as the metal tube. In some bulbs I have also used an outside insulating cover.

The preceding remarks are only made to aid the experimenter in the first trials, for the difficulties which he encounters he may soon find means to overcome in his own way.

To illustrate the effect of the screen, and the advantage of using it, I have here two bulbs of the same size, with their stems, leading-in wires and incandescent lamp filaments tied to the latter, as nearly alike as possible. The stem of one bulb is provided with an aluminium tube, the stem of the other has none. Originally the two bulbs were joined by a tube which was connected to a Sprengel pump. When a high vacuum had been reached, first the connecting tube, and then the bulbs, were sealed off; they are therefore of the same degree of exhaustion. When they are separately connected to the coil giving a certain potential, the carbon filament in the bulb provided with the aluminium screen is rendered highly incandescent, while the filament in the other bulb may, with the same potential, not even come to redness, although in reality the latter bulb takes generally more energy than the former. When they are both connected together to the terminal, the difference is even more apparent, showing the importance of the screening. The metal tube placed on the stem containing the leading-in wire performs really two distinct functions: First; it acts more or less as an electrostatic screen, thus economizing the energy supplied to the bulb; and, second, to whatever extent it may fail to act electrostatically, it acts mechanically, preventing the bombardment, and consequently intense heating and possible deterioration of the slender support of the refractory incandescent body, or of the glass stem containing the leading-in wire. I say *slender* support, for it is evident that in order to confine the heat more completely to the incandescing body its support should be very thin, so as to carry away the smallest possible amount of heat by conduction. Of all the supports used I have found an ordinary incandescent lamp filament to be the best, principally because among conductors it can withstand the highest degrees of heat.

The effectiveness of the metal tube as an electrostatic screen depends largely on the degree of exhaustion.

At excessively high degrees of exhaustion—which are reached by using great care and special means in connection with the Sprengel pump—when the matter in the globe is in the ultra-radiant state, it acts most perfectly. The shadow of the upper edge of the tube is then sharply defined upon the bulb.

At a somewhat lower degree of exhaustion, which is about the ordinary "non-striking" vacuum, and generally as long as the matter moves predominantly in straight lines, the screen still does well. In elucidation of the preceding remark it is necessary to state that what is a "non-striking" vacuum for a coil operated, as ordinarily, by impulses, or currents, of low frequency, is not, by far, so when the coil is operated by currents of very high frequency. In such case the discharge may pass with great freedom through the rarefied gas through which a low-frequency discharge may not pass, even though the potential be much higher. At ordinary atmospheric pressures just the reverse rule holds good: the higher the frequency, the less the spark discharge is able to jump between the terminals, especially if they are knobs or spheres of some size.

Finally, at very low degrees of exhaustion, when the gas is well conducting, the metal tube not only does not act as an electrostatic screen, but even is a drawback, aiding to a considerable extent the dissipation of the energy laterally from the leading-in wire. This, of course, is to be expected. In this case, namely, the metal tube is in good electrical connection with the leading-in wire, and most of the bombardment is directed upon the tube. As long as the electrical connection is not good, the conducting tube is always of some advantage, for although it may not greatly economize energy, still it protects the support of the refractory button, and is a means for concentrating more energy upon the same.

To whatever extent the aluminium tube performs the function of a screen, its usefulness is therefore limited to very high degrees of exhaustion when it is insulated from the electrode—that is, when the gas as a whole is non-conducting, and the molecules, or atoms, act as independent carriers of electric charges.

In addition to acting as a more or less effective screen, in the true meaning of the word, the conducting tube or coating may also act, by reason of its conductivity, as a sort of equalizer or dampener of the bombardment against the stem. To be explicit, I assume the action as follows: Suppose a rhythmical bombardment to occur against the conducting tube by reason of its imperfect action as a screen, it certainly must happen that some molecules, or atoms, strike the tube sooner than others. Those which come first in contact with it give up their superfluous charge, and the tube is electrified, the electrification instantly spreading over its surface. But this must diminish the energy lost in the bombardment for two reasons: first, the charge given up by the atoms spreads over a great area, and hence the electric density at any point is small, and the atoms are repelled with less energy than they would be if they would strike against a good insulator; secondly, as the tube is electrified by the atoms which first come in contact with it, the progress of the following atoms against the tube is more or less checked by the repulsion which the electrified tube must exert upon the similarly electrified atoms. This repulsion may perhaps be sufficient to prevent a large portion of the atoms from striking the tube, but at any rate it must diminish the energy of their impact. It is clear that when the exhaustion is very low, and the rarefied gas well conducting, neither of the above effects can occur, and, on the other hand, the fewer the atoms, with the greater freedom they move; in other words, the higher the degree of exhaustion, up to a limit, the more telling will be both the effects.

What I have just said may afford an explanation of the phenomenon observed by Prof. Crookes, namely, that a discharge through a bulb is established with much greater facility when an insulator than when a conductor is present in the same. In my opinion, the conductor acts as a dampener of the motion of the atoms in the two ways pointed out; hence, to cause a visible discharge to pass through the bulb, a much higher potential is needed if a conductor, especially of much surface, be present.

For the sake of clearness of some of the remarks before made, I must now refer to Figs. 18, 19 and 20, which illustrate various arrangements with a type of bulb most generally used.

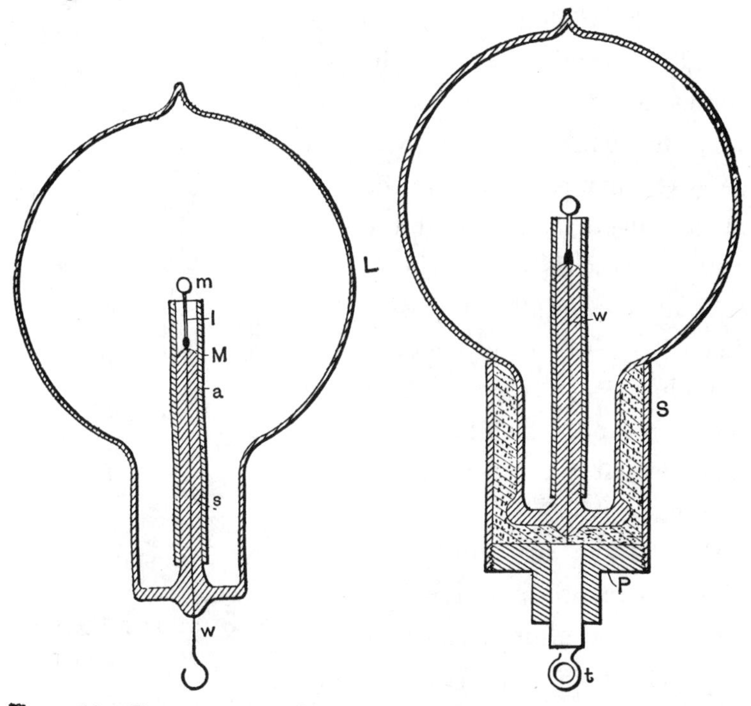

FIG. 18.—BULB WITH MICA TUBE AND ALUMINIUM SCREEN.

FIG. 19.—IMPROVED BULB WITH SOCKET AND SCREEN.

Fig. 18 is a section through a spherical bulb *L*, with the glass stem *s*, containing the leading-in wire *w*, which has a lamp filament *l* fastened to it, serving to support the refractory button *m* in the centre. *M* is a sheet of thin mica wound in several layers around the stem *s*, and *a* is the aluminium tube.

Fig. 19 illustrates such a bulb in a somewhat more advanced stage of perfection. A metallic tube *S* is fastened by means of some cement to the neck of the tube. In the tube is screwed a plug *P*, of insulating material, in the centre of which is fastened a metallic terminal *t*, for the connection to the leading-in wire *w*. This terminal must be well insulated from the metal tube *S*, therefore, if the cement used is conducting—and most generally it is sufficiently so—the space between the plug *P* and the neck of the bulb should be filled with some good insulating material, as mica powder.

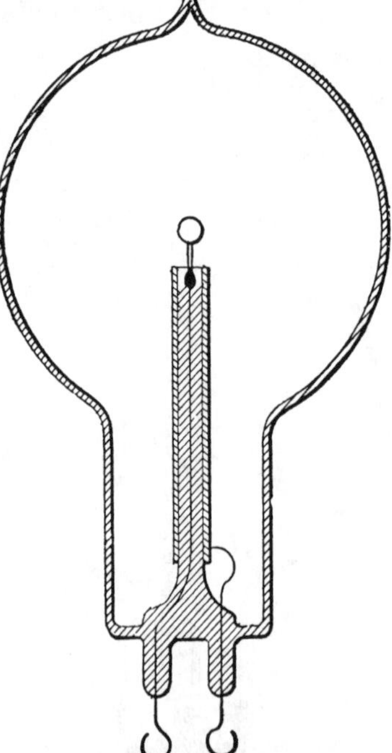

FIG. 20.—BULB FOR EXPERIMENTS WITH CONDUCTING TUBE.

Fig. 20 shows a bulb made for experimental purposes. In this bulb the aluminium tube is provided with an external connection, which serves to investigate the effect of the tube under various conditions. It is referred to chiefly to suggest a line of experiment followed.

Since the bombardment against the stem containing the leading-in wire is due to the inductive action of the latter upon the rarefied gas, it is of advantage to reduce this action as far as practicable by employing a very thin wire, surrounded by a very thick insulation of glass or other material, and by making the wire passing through the rarefied gas as short as practicable. To combine these features I employ a large tube *T* (Fig. 21), which protrudes into the bulb to some distance, and carries on the top a very short glass stem *s*, into which is sealed the leading-in wire *w*, and I protect the top of the glass stem against the heat by a small, aluminium tube *a* and a layer of mica underneath the same, as usual. The wire *w*, passing through the large tube to the outside of the bulb, should be well insulated—with a glass tube, for instance—and the space between ought to be filled out with some excellent insulator. Among many insulating powders I have tried, I have found that mica powder is the best to employ. If this precaution is not taken, the tube *T*, protruding into the bulb, will surely be cracked in consequence of the heating by the brushes which are apt to form in the upper part of the tube, near the exhausted globe, especially if the vacuum be excellent, and therefore the potential necessary to operate the lamp very high.

Fig. 22 illustrates a similar arrangement, with a large tube *T* protruding into the part of the bulb containing the refractory button *m*. In this case the wire leading from the outside into the bulb is omitted, the energy required being supplied through condenser coatings *C C*. The insulating packing *P* should in this construction be tightly fitting to the glass, and rather wide, or otherwise the discharge might avoid passing through the wire *w*, which connects the inside condenser coating to the incandescent button *m*.

The molecular bombardment against the glass stem in the bulb is a source of great trouble. As illustration I will cite a phenomenon only too frequently and unwillingly observed. A bulb, preferably a large one, may be taken,

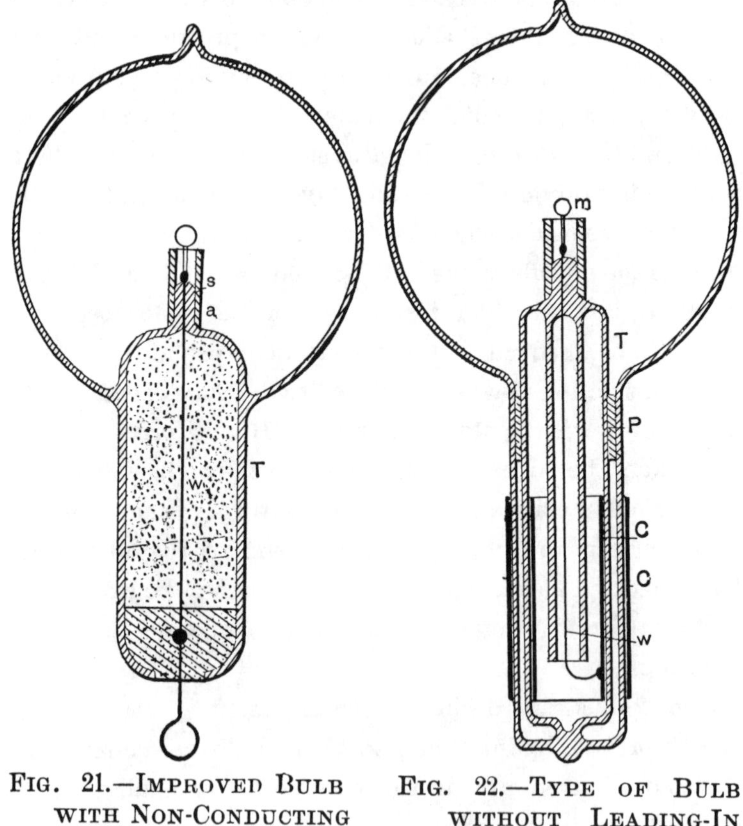

FIG. 21.—IMPROVED BULB WITH NON-CONDUCTING BUTTON.

FIG. 22.—TYPE OF BULB WITHOUT LEADING-IN WIRE.

and a good conducting body, such as a piece of carbon, may be mounted in it upon a platinum wire sealed in the glass stem. The bulb may be exhausted to a fairly high degree, nearly to the point when phosphorescence begins to appear.

When the bulb is connected with the coil, the piece of carbon, if small, may become highly incandescent at first, but its brightness immediately diminishes, and then the discharge may break through the glass somewhere in the middle of the stem, in the form of bright sparks, in spite of the fact that the platinum wire is in good electrical connection with the rarefied gas through the piece of carbon or metal at the top. The first sparks are singularly bright, recalling those drawn from a clear surface of mercury. But, as they heat the glass rapidly, they, of course, lose their brightness, and cease when the glass at the ruptured place becomes incandescent, or generallv sufficiently hot to conduct. When observed for the first time the phenomenon must appear very curious, and shows in a striking manner how radically different alternate currents, or impulses, of high frequency behave, as compared with steady currents, or currents of low frequency. With such currents—namely, the latter—the phenomenon would of course not occur. When frequencies such as are obtained by mechanical means are used, I think that the rupture of the glass is more or less the consequence of the bombardment, which warms it up and impairs its insulating power; but with frequencies obtainable with condensers I have no doubt that the glass may give way without previous heating. Although this appears most singular at first, it is in reality what we might expect to occur. The energy supplied to the wire leading into the bulb is given off partly by direct action through the carbon button, and partly by inductive action through the glass surrounding the wire. The case is thus analogous to that in which a condenser shunted by a conductor of low resistance is connected to a source of alternating currents. As long as the frequencies are low, the conductor gets the most, and the condenser is perfectly safe; but when the frequency becomes excessive, the *rôle* of the conductor may become quite insignificant. In the

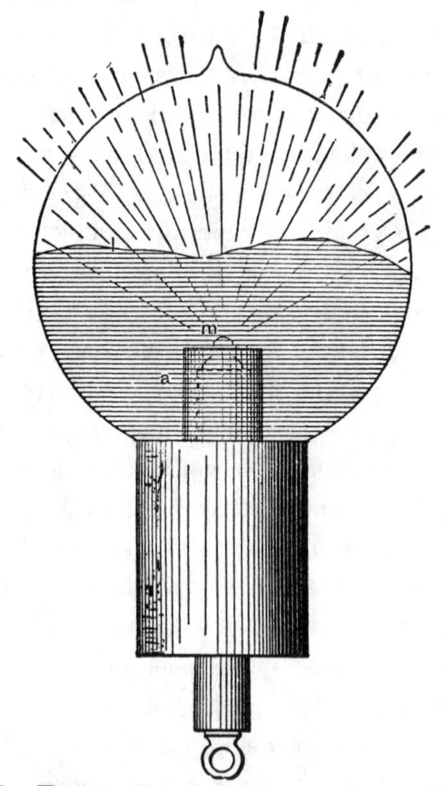

FIG. 23.—EFFECT PRODUCED BY A RUBY DROP.

latter case the difference of potential at the terminals of the condenser may become so great as to rupture the dielectric, notwithstanding the fact that the terminals are joined by a conductor of low resistance.

It is, of course, not necessary, when it is desired to produce the incandescence of a body inclosed in a bulb by means of these currents, that the body should be a conductor, for even a perfect non-conductor may be quite as readily heated. For this purpose it is sufficient to surround a conducting electrode with a non-conducting material, as, for instance, in the bulb described before in Fig. 21, in which a thin incandescent lamp filament is coated with a non-conductor, and supports a button of the same material on the top. At the start the bombardment goes on by inductive action through the non-conductor, until the same is sufficiently heated to become conducting, when the bombardment continues in the ordinary way.

A different arrangement used in some of the bulbs constructed is illustrated in Fig. 23. In this instance a non-conductor *m* is mounted in a piece of common arc light carbon so as to project some small distance above the latter. The carbon piece is connected to the leading-in wire passing through a glass stem, which is wrapped with several layers of mica. An aluminium tube *a* is employed as usual for screening. It is so arranged that it reaches very nearly as high as the carbon and only the non-conductor *m* projects a little above it. The bombardment goes at first against the upper surface of carbon, the lower parts being protected by the aluminium tube. As soon, however, as the non-conductor *m* is heated it is rendered good conducting, and then it becomes the centre of the bombardment, being most exposed to the same.

I have also constructed during these experiments many such single-wire bulbs with or without internal electrode, in which the radiant matter was projected against, or focused upon, the body to be rendered incandescent. Fig. 24 illustrates one of the bulbs used. It consists of a spherical globe *L*, provided with a long neck *n*, on the top, for increasing the action in some cases by the application of an external conducting coating. The globe *L* is blown out on the bottom into a very small bulb *b*, which serves to hold it firmly in a socket *S* of insulating material into which it is cemented. A fine lamp filament *f*, supported on a wire *w*, passes through the centre of the globe *L*. The filament is rendered incandescent in the middle portion, where the bombardment proceeding from the lower inside surface of the globe is most intense. The lower portion of the globe, as far as the socket *S* reaches, is rendered conducting, either by a tinfoil coating or otherwise, and the external electrode is connected to a terminal of the coil.

The arrangement diagrammatically indicated in Fig. 24 was found to be an inferior one when it was desired to render incandescent a filament or button supported in the centre of the globe, but it was convenient when the object was to excite phosphorescence.

In many experiments in which bodies of a different kind were mounted in the bulb as, for instance, indicated in Fig. 23, some observations of interest were made.

It was found, among other things, that in such cases, no matter where the bombardment began, just as soon as a high temperature was reached there was generally one of the bodies which seemed to take most of the bombardment upon itself, the other, or others, being thereby relieved. This quality appeared to depend principally on the point of fusion, and on the facility with which the body was "evaporated," or, generally speaking, disintegrated—meaning by the latter term not only the throwing off of atoms, but likewise of larger lumps. The observation made was in accordance with generally accepted notions. In a highly exhausted bulb electricity is carried off from the electrode by independent carriers, which are partly the atoms, or molecules, of the residual atmosphere, and partly the atoms, molecules, or lumps thrown off from the electrode. If the electrode is composed of bodies of different character, and if one of these is more easily disintegrated than the others, most of the electricity supplied is carried off from that body, which is then brought to a higher temperature than the others, and this the more, as upon an increase of the temperature the body is still more easily disintegrated.

It seems to me quite probable that a similar process takes place in the bulb even with a homogeneous electrode, and I think it to be the principal cause of the disintegration. There is bound to be some irregularity, even if the surface is highly polished, which, of course, is impossible with most of the refractory bodies employed as electrodes. Assume that a point of the electrode gets hotter, instantly most of the discharge passes through that point, and a minute patch is probably fused and evaporated. It is now possible that in consequence of the violent disintegration the spot attacked sinks in temperature, or that a counter force is created, as in an arc; at any rate, the local tearing off meets with the limitations incident to the experiment, whereupon the same process occurs on another place. To the eye the electrode appears uniformly brilliant, but there are upon it points constantly shifting and wandering around, of a temperature far above the mean, and this materially hastens the process of deterioration. That some such thing occurs, at least when the electrode is at a lower temperature, sufficient experimental evidence can be obtained in the following manner: Exhaust a bulb to a very high degree, so that with a fairly high potential the discharge cannot pass—that is, not a *luminous* one, for a weak invisible discharge occurs always, in all probability. Now raise slowly and carefully the potential, leaving the primary current on no more than for an instant. At a certain point, two, three, or half a dozen phosphorescent spots will appear on the globe. These places of the glass are evidently more violently bombarded than others, this being due to the unevenly distributed electric density, necessitated, of course, by sharp projections, or, generally speaking, irregularities of the electrode. But the luminous patches are constantly changing in position, which is especially well observable if one manages to produce very few, and this indicates that the configuration of the electrode is rapidly changing.

From experiences of this kind I am led to infer that, in order to be most durable, the refractory button in the bulb should be in the form of a sphere with a highly polished surface. Such a small sphere could be manufactured from a diamond or some other crystal, but a better way would be to fuse, by the employment of extreme degrees of temperature, some oxide—as, for instance, zirconia—into a small drop, and then keep it in the bulb at a temperature somewhat below its point of fusion.

Interesting and useful results can no doubt be reached in the direction of extreme degrees of heat. How can such high temperatures be arrived at? How are the highest degrees of heat reached in nature? By the impact of stars, by high speeds and collisions. In a collision any rate of heat generation may be attained. In a chemical process we are limited. When oxygen and hydrogen combine, they fall, metaphorically speaking, from a definite height. We cannot go very far with a blast, nor by confining heat in a furnace, but in an exhausted bulb we can concentrate any amount of energy upon a minute button. Leaving practicability out of consideration, this, then, would be the means which, in my opinion, would enable us to reach the highest temperature. But a great difficulty when proceeding in this way is encountered, namely, in most cases the body is carried off before it can fuse and form a drop. This difficulty exists principally with an oxide such as zirconia, because it cannot be compressed in so hard a cake that it would not be carried off quickly. I endeavored repeatedly to fuse zirconia, placing it in a cup or arc light carbon as indicated in Fig. 23. It glowed with a most intense light, and the stream of the particles projected out of the carbon cup was of a vivid white; but whether it was compressed in a cake or made into a paste with carbon, it was carried off before it could be fused. The carbon cup containing the zirconia had to be mounted very low in the neck of a large bulb, as the heating of the glass by the projected particles of the oxide was so rapid that in the first trial the bulb was cracked almost in an instant when the current was turned on. The heating of the glass by the projected particles was found to be always greater when the carbon cup contained a body which was rapidly carried off—I presume because in such cases, with the same potential, higher speeds were reached, and also because, per unit of time, more matter was projected—that is, more particles would strike the glass.

The before-mentioned difficulty did not exist, however, when the body mounted in the carbon cup offered great resistance to deterioration. For instance, when an oxide was first fused in an oxygen blast and then mounted in the bulb, it melted very readily into a drop.

Generally during the process of fusion magnificent light effects were noted, of which it would be difficult to give an adequate idea. Fig. 23 is intended to illustrate the effect observed with a ruby drop. At first one may see a narrow funnel of white light projected against the top of the globe, where it produces an irregularly outlined phosphorescent patch. When the point of the ruby fuses the phosphorescence becomes very powerful; but as the atoms are projected with much greater speed from the surface of the drop, soon the glass gets hot and "tired," and now only the outer edge of the patch glows. In this manner an intensely phosphorescent, sharply defined line, *l*, corresponding to the outline of the drop, is produced, which spreads slowly over the globe as the drop gets larger. When the mass begins to boil, small bubbles and cavities are formed, which cause dark colored spots to sweep across the globe. The bulb may be turned downward without fear of the drop falling off, as the mass possesses considerable viscosity.

I may mention here another feature of some interest,

which I believe to have noted in the course of these experiments, though the observations do not amount to a certitude. It *appeared* that under the molecular impact caused by the rapidly alternating potential the body was fused and maintained in that state at a lower temperature in a highly exhausted bulb than was the case at normal pressure and application of heat in the ordinary way—that is, at least, judging from the quantity of the light emitted. One of the experiments performed may be mentioned here by way of illustration. A small piece of pumice stone was stuck on a platinum wire, and first melted to it in a gas burner. The wire was next placed between two pieces of charcoal and a burner applied so as to produce an intense heat, sufficient to melt down the pumice stone into a small glass-like button. The platinum wire had to be taken of sufficient thickness to prevent its melting in the fire. While in the charcoal fire, or when held in a burner to get a better idea of the degree of heat, the button glowed with great brilliancy. The wire with the button was then mounted in a bulb, and upon exhausting the same to a high degree, the current was turned on slowly so as to prevent the cracking of the button. The button was heated to the point of fusion, and when it melted it did not, apparently, glow with the same brilliancy as before, and this would indicate a lower temperature. Leaving out of consideration the observer's possible, and even probable, error, the question is, can a body under these conditions be brought from a solid to a liquid state with evolution of *less* light?

When the potential of a body is rapidly alternated it is

certain that the structure is jarred. When the potential is very high, although the vibrations may be few—say 20,000 per second—the effect upon the structure may be considerable. Suppose, for example, that a ruby is melted into a drop by a steady application of energy. When it forms a drop it will emit visible and invisible waves, which will be in a definite ratio, and to the eye the drop will appear to be of a certain brilliancy. Next, suppose we diminish to any degree we choose the energy steadily supplied, and, instead, supply energy which rises and falls according to a certain law. Now, when the drop is formed, there will be emitted from it three different kinds of vibrations—the ordinary visible, and two kinds of invisible waves: that is, the ordinary dark waves of all lengths, and, in addition, waves of a well defined character. The latter would not exist by a steady supply of the energy; still they help to jar and loosen the structure. If this really be the case, then the ruby drop will emit relatively less visible and more invisible waves than before. Thus it would seem that when a platinum wire, for instance, is fused by currents alternating with extreme rapidity, it emits at the point of fusion less light and more invisible radiation than it does when melted by a steady current, though the total energy used up in the process of fusion is the same in both cases. Or, to cite another example, a lamp filament is not capable of withstanding as long with currents of extreme frequency as it does with steady currents, assuming that it be worked at the same luminous intensity. This means that for rapidly alternating currents the filament should be shorter and thicker. The higher the frequency—that is, the greater the departure from the steady flow—the worse it would be for the filament. But if the truth of this remark were demonstrated, it would be erroneous to conclude that such a refractory button as used in these bulbs would be deteriorated quicker by currents of extremely high frequency than by steady or low frequency currents. From experience I may say that just the opposite holds good: the button withstands the bombardment better with currents of very high frequency. But this is due to the fact that a high frequency discharge passes through a rarefied gas with much greater freedom than a steady or low frequency discharge, and this will say that with the former we can work with a lower potential or with a less violent impact. As long, then, as the gas is of no consequence, a steady or low frequency current is better; but as soon as the action of the gas is desired and important, high frequencies are preferable.

In the course of these experiments a great many trials were made with all kinds of carbon buttons. Electrodes made of ordinary carbon buttons were decidedly more durable when the buttons were obtained by the application of enormous pressure. Electrodes prepared by depositing carbon in well known ways did not show up well; they blackened the globe very quickly. From many experiences I conclude that lamp filaments obtained in this manner can be advantageously used only with low potentials and low frequency currents. Some kinds of carbon withstand so well that, in order to bring them to the point of fusion, it is necessary to employ very small buttons. In this case the observation is rendered very difficult on account of the intense heat produced. Nevertheless there can be no doubt that all kinds of carbon are fused under the molecular bombardment, but the liquid state must be one of great instability. Of all the bodies tried there were two which withstood best—diamond and carborundum. These two showed up about equally, but the latter was preferable, for many reasons. As it is more than likely that this body is not yet generally known, I will venture to call your attention to it.

It has been recently produced by Mr. E. G. Acheson, of Monongahela City, Pa., U. S. A. It is intended to replace ordinary diamond powder for polishing precious stones, etc., and I have been informed that it accomplishes this object quite successfully. I do not know why the name "carborundum" has been given to it, unless there is something in the process of its manufacture which justifies this selection. Through the kindness of the inventor, I obtained a short while ago some samples which I desired to test in regard to their qualities of phosphorescence and capability of withstanding high degrees of heat.

Carborundum can be obtained in two forms—in the form of "crystals" and of powder. The former appear to the naked eye dark colored, but are very brilliant; the latter is of nearly the same color as ordinary diamond powder, but very much finer. When viewed under a microscope the samples of crystals given to me did not appear to have any definite form, but rather resembled pieces of broken up egg coal of fine quality. The majority were opaque, but there were some which were transparent and colored. The crystals are a kind of carbon containing some impurities; they are extremely hard, and withstand for a long time even an oxygen blast. When the blast is directed against them they at first form a cake of some compactness, probably in consequence of the fusion of impurities they contain. The mass withstands for a very long time the blast without further fusion ; but a slow carrying off, or burning, occurs, and, finally, a small quantity of a glass-like residue is left, which, I suppose, is melted alumina. When compressed strongly they conduct very well, but not as well as ordinary carbon. The powder, which is obtained from the crystals in some way, is practically non-conducting. It affords a magnificent polishing material for stones.

The time has been too short to make a satisfactory study of the properties of this product, but enough experience has been gained in a few weeks I have experimented upon it to say that it does possess some remarkable properties in many respects. It withstands excessively high degrees of heat, it is little deteriorated by molecular bombardment, and it does not blacken the globe as ordinary carbon does. The only difficulty which I have found in its use in connection with these experiments was to find some binding material which would resist the heat and the effect of the bombardment as successfully as carborundum itself does.

I have here a number of bulbs which I have provided with buttons of carborundum. To make such a button of carborundum crystals I proceed in the following manner: I take an ordinary lamp filament and dip its point in tar, or some other thick substance or paint which may be readily carbonized. I next pass the point of the filament through the crystals, and then hold it vertically over a hot plate. The tar softens and forms a drop on the point of the filament, the crystals adhering to the surface of the drop. By regulating the distance from the plate the tar is slowly dried out and the button becomes solid. I then once more dip the button in tar and hold it again over a plate until the tar is evaporated, leaving only a hard mass which firmly binds the crystals. When a larger button is required I repeat the process several times, and I generally also cover the filament a certain distance below the button with crystals. The button being mounted in a bulb, when a good vacuum has been reached, first a weak and then a strong discharge is passed through the bulb to carbonize the tar and expel all gases, and later it is brought to a very intense incandescence.

When the powder is used I have found it best to proceed as follows: I make a thick paint of carborundum and tar, and pass a lamp filament through the paint. Taking then most of the paint off by rubbing the filament against a piece of chamois leather, I hold it over a hot plate until the tar evaporates and the coating becomes firm. I repeat this process as many times as it is necessary to obtain a certain thickness of coating. On the point of the coated filament I form a button in the same manner.

There is no doubt that such a button—properly prepared under great pressure—of carborundum, especially of powder of the best quality, will withstand the effect of the bombardment fully as well as anything we know. The difficulty is that the binding material gives way, and the carborundum is slowly thrown off after some time. As it does not seem to blacken the globe in the least, it might be found useful for coating the filaments of ordinary incandescent lamps, and I think that it is even possible to produce thin threads or sticks of carborundum which will replace the ordinary filaments in an incandescent lamp. A carborundum coating seems to be more durable than other coatings, not only because the carborundum can withstand high degrees of heat, but also because it seems to unite with the carbon better than any other material I have tried. A coating of zirconia or any other oxide, for instance, is far more quickly destroyed. I prepared buttons of diamond dust in the same manner as of carborundum, and these came in durability nearest to those prepared of carborundum, but the binding paste gave way much more quickly in the diamond buttons : this, however, I attributed to the size and irregularity of the grains of the diamond.

It was of interest to find whether carborundum possesses the quality of phosphorescence. One is, of course, prepared to encounter two difficulties: first, as regards the rough product, the "crystals," they are good conducting, and it is a fact that conductors do not phosphoresce ; second, the powder, being exceedingly fine, would not be apt to exhibit very prominently this quality, since we know that when crystals, even such as diamond or ruby, are finely powdered, they lose the property of phosphorescence to a considerable degree.

The question presents itself here, can a conductor phosphoresce ? What is there in such a body as a metal, for instance, that would deprive it of the quality of phosphorescence, unless it is that property which characterizes it as a conductor? for it is a fact that most of the phosphorescent bodies lose that quality when they are sufficiently heated to become more or less conducting. Then, if a metal be in a large measure, or perhaps entirely, deprived of that property, it should be capable of phosphorescence. Therefore it is quite possible that at some extremely high frequency, when behaving practically as a non-conductor, a metal or any other conductor might exhibit the quality of phosphorescence, even though it be entirely incapable of phosphorescing under the impact of a low-frequency discharge. There is, however, another possible way how a conductor might at least *appear* to phosphoresce.

Considerable doubt still exists as to what really is phosphorescence, and as to whether the various phenomena comprised under this head are due to the same causes. Suppose that in an exhausted bulb, under the molecular impact, the surface of a piece of metal or other conductor is rendered strongly luminous, but at the same time it is found that it remains comparatively cool, would not this luminosity be called phosphorescence? Now such a result, theoretically at least, is possible, for it is a mere question of potential or speed. Assume the potential of the electrode, and consequently the speed of the projected atoms, to be sufficiently high, the surface of the metal piece against which the atoms are projected would be rendered highly incandescent, since the process of heat generation would be incomparably faster than that of radiating or conducting away from the surface of the collision. In the eye of the observer a single impact of the atoms would cause an instantaneous flash, but if the impacts were repeated with sufficient rapidity they would produce a continuous impression upon his retina. To him then the surface of the metal would appear continuously incandescent and of constant luminous intensity, while in reality the light would be either intermittent or at least changing periodically in intensity. The metal piece would rise in temperature until equilibrium was attained—that is, until the energy continuously radiated would equal that intermittently supplied. But the supplied energy might under such conditions not be sufficient to bring the body to any more than a very moderate mean temperature, especially if the frequency of the atomic impacts be very low—just enough that the fluctuation of the intensity of the light emitted could not be detected by the eye. The body would now, owing to the manner in which the energy is supplied, emit a strong light, and yet be at a comparatively very low mean temperature. How could the observer call the luminosity thus produced? Even if the analysis of the light would teach him something definite, still he would probably rank it under the phenomena of phosphorescence. It is conceivable that in such a way both conducting and nonconducting bodies may be maintained at a certain luminous intensity, but the energy required would very greatly vary with the nature and properties of the bodies.

These and some foregoing remarks of a speculative nature were made merely to bring out curious features of alternate currents or electric impulses. By their help we may cause a body to emit *more* light, while at a certain mean temperature, than it would emit if brought to that temperature by a steady supply; and, again, we may bring a body to the point of fusion, and cause it to emit *less* light than when fused by the application of energy in ordinary ways. It all depends on how we supply the energy, and what kind of vibrations we set up: in one case the vibrations are more, in the other less, adapted to affect our sense of vision.

Some effects, which I had not observed before, obtained with carborundum in the first trials, I attributed to phosphorescence, but in subsequent experiments it appeared that it was devoid of that quality. The crystals possess a noteworthy feature. In a bulb provided with a single electrode in the shape of a small circular metal disc, for instance, at a certain degree of exhaustion the electrode is covered with a milky film, which is separated by a dark space from the glow filling the bulb. When the metal disc is covered with carborundum crystals, the film is far more intense, and snow-white. This I found later to be merely an effect of the bright surface of the crystals, for when an aluminium electrode was highly polished it exhibited more or less the same phenomenon. I made a number of experiments with the samples of crystals obtained, principally because it would have been of special interest to find that they are capable of phosphorescence, on account of their being conducting. I could not produce phosphorescence distinctly, but I must remark that a decisive opinion cannot be formed until other experimenters have gone over the same ground.

The powder behaved in some experiments as though it contained alumina, but it did not exhibit with sufficient distinctness the red of the latter. Its dead color brightens considerably under the molecular impact, but I am now convinced it does not phosphoresce. Still, the tests with the powder are not conclusive, because powdered carborundum probably does not behave like a phosphorescent sulphide, for example, which could be finely powdered without impairing the phosphorescence, but rather like powdered ruby or diamond, and therefore it would be necessary, in order to make a decisive test, to obtain it in a large lump and polish up the surface.

If the carborundum proves useful in connection with these and similar experiments, its chief value will be found in the production of coatings, thin conductors, buttons, or other electrodes capable of withstanding extremely high degrees of heat.

The production of a small electrode capable of withstanding enormous temperatures I regard as of the greatest importance in the manufacture of light. It would enable us to obtain, by means of currents of very high frequencies, certainly 20 times, if not more, the quantity of light which is obtained in the present incandescent lamp by the same expenditure of energy. This estimate may appear to many exaggerated, but in reality I think it is far from being so. As this statement might be misunderstood I think it necessary to expose clearly the problem with which in this line of work we are confronted, and the manner in which, in my opinion, a solution will be arrived at.

Any one who begins a study of the problem will be apt to think that what is wanted in a lamp with an electrode is a very high degree of incandescence of the electrode. There he will be mistaken. The high incandescence of the button is a necessary evil, but what is really wanted is the high incandescence of the gas surrounding the button. In other words, the problem in such a lamp is to bring a mass of gas to the highest possible incandescence. The higher the incandescence, the quicker the mean vibration, the greater is the economy of the light production. But to maintain a mass of gas at a high degree of incandescence in a glass vessel, it will always be necessary to keep the incandescent mass away from the glass; that is, to confine it as much as possible to the central portion of the globe.

In one of the experiments this evening a brush was produced at the end of a wire. This brush was a flame, a source of heat and light. It did not emit much perceptible heat, nor did it glow with an intense light; but is it the less a flame because it does not scorch my hand? Is it the less a flame because it does not hurt my eye by its brilliancy? The problem is precisely to produce in the bulb such a flame, much smaller in size, but incomparably more powerful. Were there means at hand for producing electric impulses of a sufficiently high frequency, and for transmitting them, the bulb could be done away with, unless it were used to protect the electrode, or to economize the energy by confining the heat. But as such means are not at disposal, it becomes necessary to place the terminal in a bulb and rarefy the air in the same. This is done merely to enable the apparatus to perform the work which it is not capable of performing at ordinary air pressure. In the bulb we are able to intensify the action to any degree—so far that the brush emits a powerful light.

The intensity of the light emitted depends principally on the frequency and potential of the impulses, and on the electric density on the surface of the electrode. It is of the greatest importance to employ the smallest possible button, in order to push the density very far. Under the violent impact of the molecules of the gas surrounding it, the small electrode is of course brought to an extremely high temperature, but around it is a mass of highly incandescent gas, a flame photosphere, many hundred times the volume of the electrode. With a diamond, carborundum or zirconia button the photosphere can be as much as one thousand times the volume of the button. Without much reflecting one would think that in pushing so far the incandescence of the electrode it would be instantly volatilized. But after a careful consideration he would find that, theoretically, it should not occur, and in this fact—which, however, is experimentally demonstrated—lies principally the future value of such a lamp.

At first, when the bombardment begins, most of the work is performed on the surface of the button, but when a highly conducting photosphere is formed the button is comparatively relieved. The higher the incandescence of the photosphere the more it approaches in conductivity to that of the electrode, and the more, therefore, the solid and the gas form one conducting body. The consequence is that the further is forced the incandescence the more work, comparatively, is performed on the gas, and the less on the electrode. The formation of a powerful photosphere is consequently the very means for protecting the electrode. This protection, of course, is a relative one, and it should not be thought that by pushing the incandescence higher the electrode is actually less deteriorated, Still, theoretically, with extreme frequencies, this result must be reached, but probably at a temperature too high for most of the refractory bodies known. Given, then, an electrode which can withstand to a very high limit the effect of the bombardment and outward strain, it would be safe no matter how much it is forced beyond that limit. In an incandescent lamp quite different considerations apply. There the gas is not at all concerned: the whole of the work is performed on the filament; and the life of the lamp diminishes so rapidly with the increase of the degree of incandescence that economical reasons compel us to work it at a low incandescence. But if an incandescent lamp is operated with currents of very high frequency, the action of the gas cannot be neglected, and the rules for the most economical working must be considerably modified.

In order to bring such a lamp with one or two electrodes to a great perfection, it is necessary to employ impulses of very high frequency. The high frequency secures, among others, two chief advantages, which have a most important bearing upon the economy of the light production. First, the deterioration of the electrode is reduced by reason of the fact that we employ a great many small impacts, instead of a few violent ones, which shatter quickly the structure; secondly, the formation of a large photosphere is facilitated.

In order to reduce the deterioration of the electrode to the minimum, it is desirable that the vibration be harmonic, for any suddenness hastens the process of destruction. An electrode lasts much longer when kept at incandescence by currents, or impulses, obtained from a high-frequency alternator, which rise and fall more or less harmonically, than by impulses obtained from a disruptive discharge coil. In the latter case there is no doubt that most of the damage is done by the fundamental sudden discharges.

One of the elements of loss in such a lamp is the bombardment of the globe. As the potential is very high, the molecules are projected with great speed; they strike the glass, and usually excite a strong phosphorescence. The effect produced is very pretty, but for economical reasons it would be perhaps preferable to prevent, or at least reduce to the minimum, the bombardment against the globe, as in such case it is, as a rule, not the object to excite phosphorescence, and as some loss of energy results from the bombardment. This loss in the bulb is principally dependent on the potential of the impulses and on the electric density on the surface of the electrode. In employing very high frequencies the loss of energy by the bombardment is greatly reduced, for, first, the potential needed to perform a given amount of work is much smaller; and, secondly, by producing a highly conducting photosphere around the electrode, the same result is obtained as though the electrode were much larger, which is equivalent to a smaller electric density. But be it by the diminution of the maximum potential or of the density, the gain is effected in the same manner, namely, by avoiding violent shocks, which strain the glass much beyond its limit of elasticity. If the frequency could be brought high enough, the loss due to the imperfect elasticity of the glass would be entirely negligible. The loss due to bombardment of the globe may, however, be reduced by using two electrodes instead of one. In such case each of the electrodes may be connected to one of the terminals; or else, if it is preferable to use only one wire, one electrode may be connected to one terminal and the other to the ground or to an insulated body of some surface, as, for instance, a shade on the lamp. In the latter case, unless some judgment is used, one of the electrodes might glow more intensely than the other.

But on the whole I find it preferable when using such high frequencies to employ only one electrode and one connecting wire. I am convinced that the illuminating device of the near future will not require for its operation more than one lead, and, at any rate, it will have no leading-in wire, since the energy required can be as well transmitted through the glass. In experimental bulbs the leading-in wire is most generally used on account of convenience, as in employing condenser coatings in the manner indicated in Fig. 22, for example, there is some difficulty in fitting the parts, but these difficulties would not exist if a great many bulbs were manufactured; otherwise the energy can be conveyed through the glass as well as through a wire, and with these high frequencies the losses are very small. Such illuminating devices will necessarily involve the use of very high potentials, and this, in the eyes of practical men, might be an objectionable feature. Yet, in reality, high potentials are not objectionable — certainly not in the least as far as the safety of the devices is concerned.

There are two ways of rendering an electric appliance safe. One is to use low potentials, the other is to determine the dimensions of the apparatus so that it is safe no matter how high a potential is used. Of the two the latter seems to me the better way, for then the safety is absolute, unaffected by any possible combination of circumstances which might render even a low-potential appliance dangerous to life and property. But the practical conditions require not only the judicious determination of the dimensions of the apparatus; they likewise necessitate the employment of energy of the proper kind. It is easy, for instance, to construct a transformer capable of giving, when operated from an ordinary alternate current machine of low tension, say 50,000 volts, which might be required to light a highly exhausted phosphorescent tube, so that, in spite of the high potential, it is perfectly safe, the shock from it producing no inconvenience. Still, such a transformer would be expensive, and in itself inefficient; and, besides, what energy was obtained from it would not be economically used for the production of light. The economy demands the employment of energy in the form of extremely rapid vibrations. The problem of producing light has been likened to that of maintaining a certain high-pitch note by means of a bell. It should be said a *barely audible* note; and even these words would not express it, so wonderful is the sensitiveness of the eye. We may deliver powerful blows at long intervals, waste a good deal of energy, and still not get what we want; or we may keep up the note by delivering frequent gentle taps, and get nearer to the object sought by the expenditure of much less energy. In the production of light, as far as the illuminating device is concerned, there can be only one rule—that is, to use as high frequencies as can be obtained; but the means for the production and conveyance of impulses of such character impose, at present at least, great limitations. Once it is decided to use very high frequencies, the return wire becomes unnecessary, and all the appliances are simplified. By the use of obvious means the same result is obtained as though the return wire were used. It is sufficient for this purpose to bring in contact with the bulb, or merely in the vicinity of the same, an insulated body of some surface. The surface need, of course, be the smaller, the higher the frequency and potential used, and necessarily, also, the higher the economy of the lamp or other device.

This plan of working has been resorted to on several occasions this evening. So, for instance, when the incandescence of a button was produced by grasping the bulb with the hand, the body of the experimenter merely served to intensify the action. The bulb used was similar to that illustrated in Fig. 19, and the coil was excited to a small potential, not sufficient to bring the button to incandescence when the bulb was hanging from the wire; and incidentally, in order to perform the experiment in a more suitable manner, the button was taken so large that a perceptible time had to elapse before, upon grasping the bulb, it could be rendered incandescent. The contact with the bulb was, of course, quite unnecessary. It is easy, by using a rather large bulb with an exceedingly small electrode, to adjust the conditions so that the latter is brought to bright incandescence by the mere approach of the experimenter within

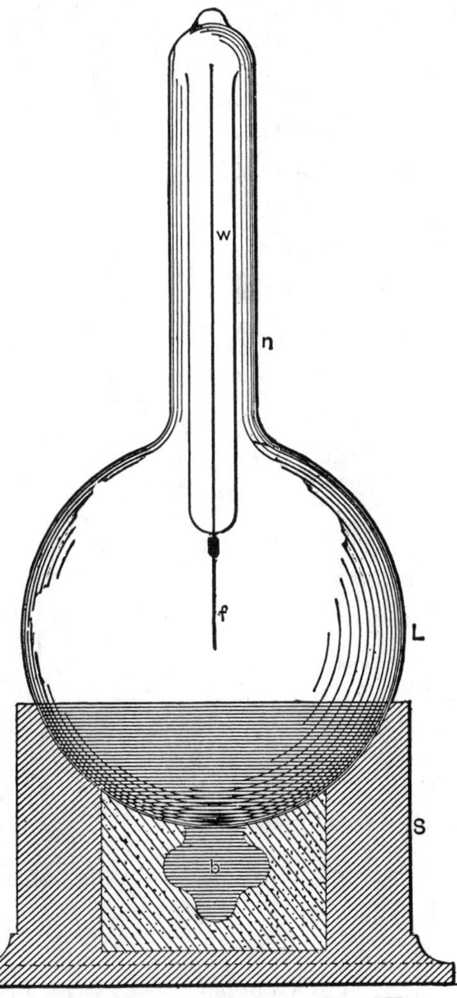

FIG. 24.—BULB WITHOUT LEADING-IN WIRE, SHOWING EFFECT OF PROJECTED MATTER.

a few feet of the bulb, and that the incandescence subsides upon his receding.

In another experiment, when phosphorescence was excited, a similar bulb was used. Here again, originally, the potential was not sufficient to excite phosphorescence until the action was intensified—in this case, however, to present a different feature, by touching the socket with a metallic object held in the hand. The electrode in the bulb was a carbon button so large that it could not be brought to incandescence, and thereby spoil the effect produced by phosphorescence.

Again, in another of the early experiments, a bulb was used as illustrated in Fig. 12. In this instance, by touching the bulb with one or two fingers, one or two shadows of the stem inside were projected against the glass, the touch of the finger producing the same result as the application of an external negative electrode under ordinary circumstances.

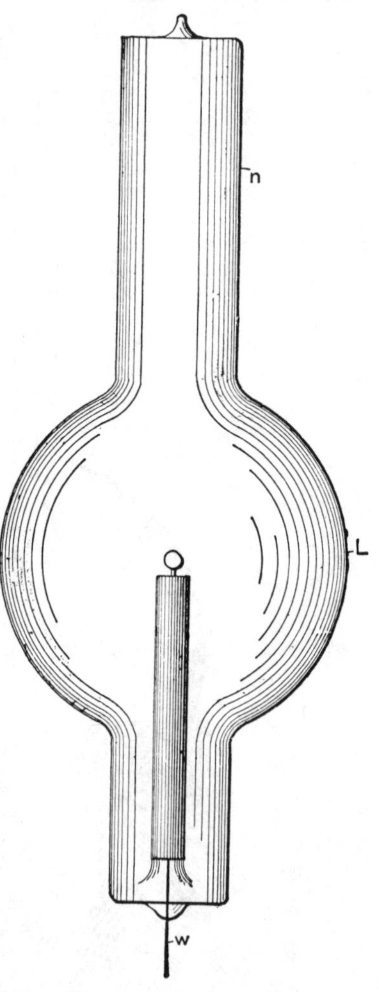

FIG. 25.—IMPROVED EXPERIMENTAL BULB.

In all these experiments the action was intensified by

augmenting the capacity at the end of the lead connected to the terminal. As a rule, it is not necessary to resort to such means, and would be quite unnecessary with still higher frequencies; but when it *is* desired, the bulb, or tube, can be easily adapted to the purpose.

In Fig. 24, for example, an experimental bulb L is shown, which is provided with a neck n on the top for the application of an external tinfoil coating, which may be connected to a body of larger surface. Such a lamp as illus-

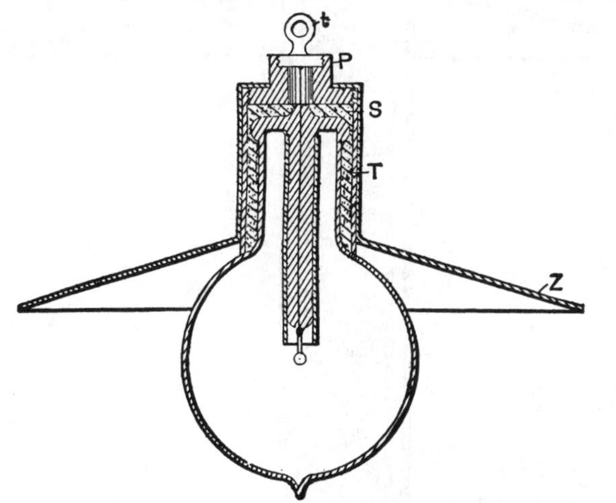

FIG. 26.—IMPROVED BULB WITH INTENSIFYING REFLECTOR.

trated in Fig. 25 may also be lighted by connecting the tinfoil coating on the neck n to the terminal, and the leading-in wire w to an insulated plate. If the bulb stands in a socket upright, as shown in the cut, a shade of conducting material may be slipped in the neck n, and the action thus magnified.

A more perfected arrangement used in some of these bulbs is illustrated in Fig. 26. In this case the construction of the bulb is as shown and described before, when reference was made to Fig. 19. A zinc sheet *Z*, with a tubular extension *T*, is slipped over the metallic socket *S*. The bulb hangs downward from the terminal *t*, the zinc sheet

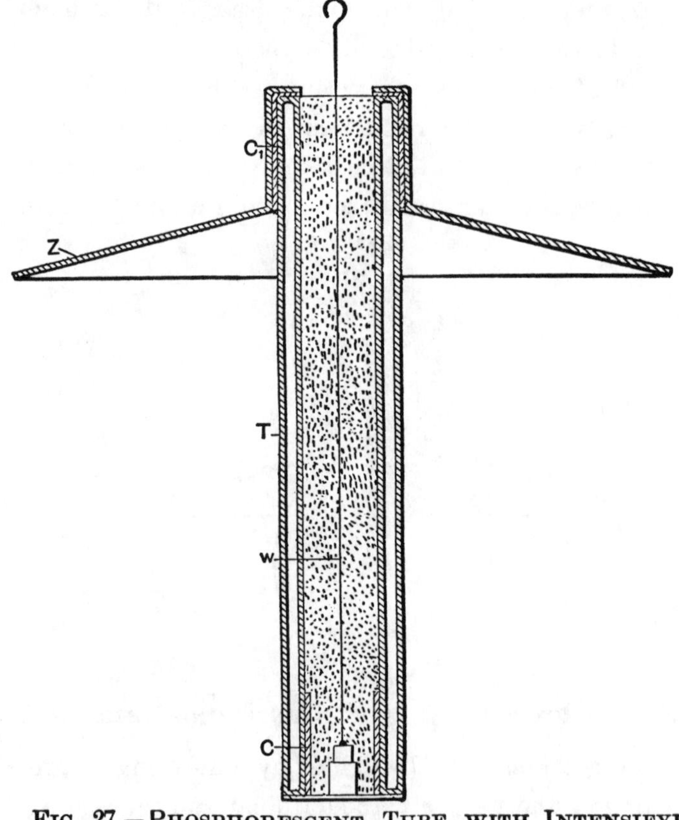

FIG. 27.—PHOSPHORESCENT TUBE WITH INTENSIFYING REFLECTOR.

Z, performing the double office of intensifier and reflector. The reflector is separated from the terminal *t* by an extension of the insulating plug *P*.

A similar disposition with a phosphorescent tube is illusstrated in Fig. 27. The tube T is prepared from two short tubes of a different diameter, which are sealed on the ends. On the lower end is placed an outside conducting coating C, which connects to the wire w. The wire has a hook on the upper end for suspension, and passes through the centre of the inside tube, which is filled with some good and tightly packed insulator. On the outside of the upper end of the tube T is another conducting coating C_1, upon which is slipped a metallic reflector Z, which should be separated by a thick insulation from the end of wire w.

The economical use of such a reflector or intensifier would require that all energy supplied to an air condenser should be recoverable, or, in other words, that there should not be any losses, neither in the gaseous medium nor through its action elsewhere. This is far from being so, but, fortunately, the losses may be reduced to anything desired. A few remarks are necessary on this subject, in order to make the experiences gathered in the course of these investigations perfectly clear.

Suppose a small helix with many well insulated turns, as in experiment Fig. 17, has one of its ends connected to one of the terminals of the induction coil, and the other to a metal plate, or, for the sake of simplicity, a sphere, insulated in space. When the coil is set to work, the potential of the sphere is alternated, and the small helix now behaves as though its free end were connected to the other terminal of the induction coil. If an iron rod be held within the small helix it is quickly brought to a high temperature, indicating the passage of a strong current through the helix. How does the insulated sphere act in this case?

It can be a condenser, storing and returning the energy supplied to it, or it can be a mere sink of energy, and the conditions of the experiment determine whether it is more one or the other. The sphere being charged to a high potential, it acts inductively upon the surrounding air, or whatever gaseous medium there might be. The molecules, or atoms, which are near the sphere are of course more attracted, and move through a greater distance than the farther ones. When the nearest molecules strike the sphere they are repelled, and collisions occur at all distances within the inductive action of the sphere. It is now clear that, if the potential be steady, but little loss of energy can be caused in this way, for the molecules which are nearest to the sphere, having had an additional charge imparted to them by contact, are not attracted until they have parted, if not with all, at least with most of the additional charge, which can be accomplished only after a great many collisions. From the fact that with a steady potential there is but little loss in dry air, one must come to such a conclusion. When the potential of the sphere, instead of being steady, is alternating, the conditions are entirely different. In this case a rhythmical bombardment occurs, no matter whether the molecules after coming in contact with the sphere lose the imparted charge or not; what is more, if the charge is not lost, the impacts are only the more violent. Still if the frequency of the impulses be very small, the loss caused by the impacts and collisions would not be serious unless the potential were excessive. But when extremely high frequencies and more or less high potentials are used, the loss may be very great. The total energy lost per unit of time is proportionate to the product of the number of impacts per second, or the frequency and the energy lost in each impact. But the energy of an impact must be proportionate to the square of the electric density of the sphere, since the charge imparted to the molecule is proportionate to that density. I conclude from this that the total energy lost must be proportionate to the product of the frequency and the square of the electric density; but this law needs experimental confirmation. Assuming the preceding considerations to be true, then, by rapidly alternating the potential of a body immersed in an insulating gaseous medium, any amount of energy may be dissipated into space. Most of that energy then, I believe, is not dissipated in the form of long ether waves, propagated to considerable distance, as is thought most generally, but is consumed—in the case of an insulated sphere, for example—in impact and collisional losses—that is, heat vibrations—on the surface and in the vicinity of the sphere. To reduce the dissipation it is necessary to work with a small electric density—the smaller the higher the frequency.

But since, on the assumption before made, the loss is diminished with the square of the density, and since currents of very high frequencies involve considerable waste when transmitted through conductors, it follows that, on the whole, it is better to employ one wire than two. Therefore, if motors, lamps, or devices of any kind are perfected, capable of being advantageously operated by currents of extremely high frequency, economical reasons will make it advisable to use only one wire, especially if the distances are great.

When energy is absorbed in a condenser the same behaves as though its capacity were increased. Absorption always exists more or less, but generally it is small and of no consequence as long as the frequencies are not very great. In using extremely high frequencies, and, necessarily in such case. also high potentials, the absorption—or, what is here meant more particularly by this term, the loss of energy due to the presence of a gaseous medium—is an important factor to be considered, as the energy absorbed in the air condenser may be any fraction of the supplied energy. This would seem to make it very difficult to tell from the measured or computed capacity of an air condenser its actual capacity or vibration period, especially if the condenser is of very small surface and is charged to a very high potential. As many important results are dependent upon the correctness of the estimation of the vibration period, this subject demands the most careful scrutiny of other investigators. To reduce the probable error as much as possible in experiments of the kind alluded to, it is advisable to use spheres or plates of large surface, so as to make the density exceedingly small. Otherwise, when it is practicable, an oil condenser should be used in preference. In oil or other liquid dielectrics there are seemingly no such losses as in gaseous media. It being impossible to exclude entirely the gas in condensers with solid dielectrics, such condensers should be immersed in oil, for economical reasons if nothing else; they can then be strained to the utmost and will remain cool. In Leyden jars the loss due to air is comparatively small, as the tinfoil coatings are large, close together, and the charged surfaces not directly exposed; but when the potentials are very high, the loss may be more or less considerable at, or near, the upper edge of the foil, where the air is principally acted upon. If the jar be immersed in boiled-out oil, it will be capable of performing four times the amount of work which it can for any length of time when used in the ordinary way, and the loss will be inappreciable.

It should not be thought that the loss in heat in an air condenser is necessarily associated with the formation of *visible* streams or brushes. If a small electrode, inclosed in an unexhausted bulb, is connected to one of the terminals of the coil, streams can be seen to issue from the electrode and the air in the bulb is heated; if, instead of a small electrode, a large sphere is inclosed in the bulb, no streams are observed, still the air is heated.

Nor should it be thought that the temperature of an air condenser would give even an approximate idea of the loss in heat incurred, as in such case heat must be given off much more quickly, since there is, in addition to the ordinary radiation, a very active carrying away of heat by independent carriers going on, and since not only the apparatus, but the air at some distance from it is heated in consequence of the collisions which must occur.

Owing to this, in experiments with such a coil, a rise of temperature can be distinctly observed only when the body connected to the coil is very small. But with apparatus on a larger scale, even a body of considerable bulk would be heated, as, for instance, the body of a person ; and I think that skilled physicians might make observations of utility in such experiments, which, if the apparatus were judiciously designed, would not present the slightest danger.

A question of some interest, principally to meteórologists, presents itself here. How does the earth behave? The earth is an air condenser, but is it a perfect or a very imperfect one—a mere sink of energy? There can be little doubt that to such small disturbance as might be caused in an experiment the earth behaves as an almost perfect condenser. But it might be different when its charge is set in vibration by some sudden disturbance occurring in the heavens. In such case, as before stated, probably only little of the energy of the vibrations set up would be lost into space in the form of long ether radiations, but most of the energy, I think, would spend itself in molecular impacts and collisions, and pass off into space in the form of short heat, and possibly light, waves. As both the frequency of the vibrations of the charge and the potential are in all probability excessive, the energy converted into heat may be considerable. Since the density must be unevenly distributed, either in consequence of the irregularity of the earth's surface, or on account of the condition of the atmosphere in various places, the effect produced would accordingly vary from place to place. Considerable variations in the temperature and pressure of the atmosphere may in this manner be caused at any point of the surface of the earth. The variations may be gradual or very sudden, according to the nature of the general disturbance, and may produce rain and storms, or locally modify the weather in any way.

From the remarks before made one may see what an important factor of loss the air in the neighborhood of a charged surface becomes when the electric density is great and the frequency of the impulses excessive. But the action as explained implies that the air is insulating—that is, that it is composed of independent carriers immersed in an insulating medium. This is the case only when the air is at something like ordinary or greater, or at extremely small, pressure. When the air is slightly rarefied and con ducting, then true conduction losses occur also. In such case, of course, considerable energy may be dissipated into space even with a steady potential, or with impulses of low frequency, if the density is very great.

When the gas is at very low pressure, an electrode is heated more because higher speeds can be reached. If the gas around the electrode is strongly compressed, the displacements, and consequently the speeds, are very small, and the heating is insignificant. But if in such case the frequency could be sufficiently increased, the electrode would be brought to a high temperature as well as if the gas were at very low pressure; in fact, exhausting the bulb is only necessary because we cannot produce (and possibly not convey) currents of the required frequency.

Returning to the subject of electrode lamps, it is obviously of advantage in such a lamp to confine as much as possible the heat to the electrode by preventing the circulation of the gas in the bulb. If a very small bulb be taken, it would confine the heat better than a large one, but it might not be of sufficient capacity to be operated from the coil, or, if so, the glass might get too hot. A simple way to improve in this direction is to employ a globe of the required size, but to place a small bulb, the diameter of which is properly estimated, over the refractory button contained in the globe. This arrangement is illustrated in Fig. 28.

The globe L has in this case a large neck n, allowing

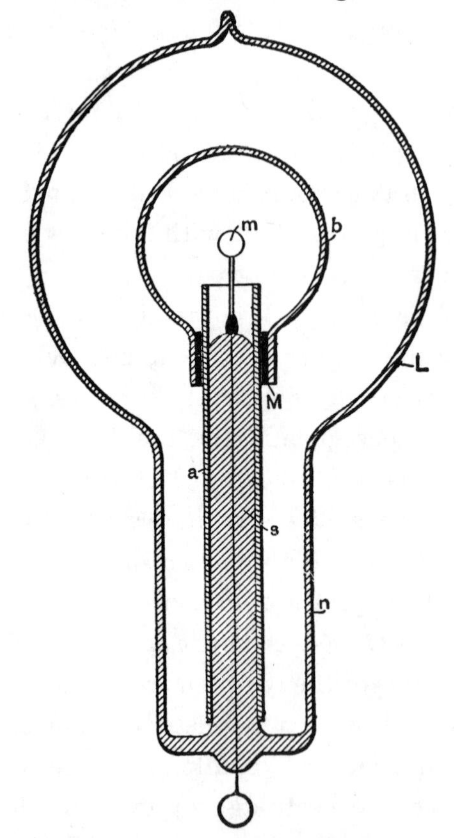

FIG. 28.—LAMP WITH AUXILIARY BULB FOR CONFINING THE ACTION TO THE CENTRE.

the small bulb b to slip through. Otherwise the construction is the same as shown in Fig. 18, for example. The small bulb is conveniently supported upon the stem s, carrying the refractory button m. It is separated from the aluminium tube a by several layers of mica M, in order to prevent the cracking of the neck by the rapid heating of the aluminium tube upon a sudden turning on of the current. The inside bulb should be as small as possible when it is desired to obtain light only by incandescence of the electrode. If it is desired to produce phosphorescence, the bulb should be larger, else it would be apt to get too hot, and the phosphorescence would cease. In this arrangement usually only the small bulb shows phosphorescence, as there is practically no bombardment against the outer globe. In some of these bulbs constructed as illustrated in Fig. 28 the small tube was coated with phosphorescent paint, and beautiful effects were obtained. Instead of making the inside bulb large, in order to avoid undue heating, it answers the purpose to make the electrode m larger. In this case the bombardment is weakened by reason of the smaller electric density.

Many bulbs were constructed on the plan illustrated in Fig. 29. Here a small bulb b, containing the refractory button m, upon being exhausted to a very high degree was sealed in a large globe L, which was then moderately exhausted and sealed off. The principal advantage of this construction was that it allowed of reaching extremely high vacua, and, at the same time use a large bulb. It was found, in the course of experiences with bulbs such as illustrated in Fig. 29, that it was well to make the stem s near the seal at e very thick, and the leading-in wire w thin, as it occurred sometimes that the stem at e was heated and the bulb was cracked. Often the outer globe L was exhausted only just enough to allow the discharge to pass through, and the space between the bulbs appeared crimson, pro ducing a curious effect. In some cases, when the exhaustion in globe *L* was very low, and the air good conducting, it was found necessary, in order to bring the button *m* to

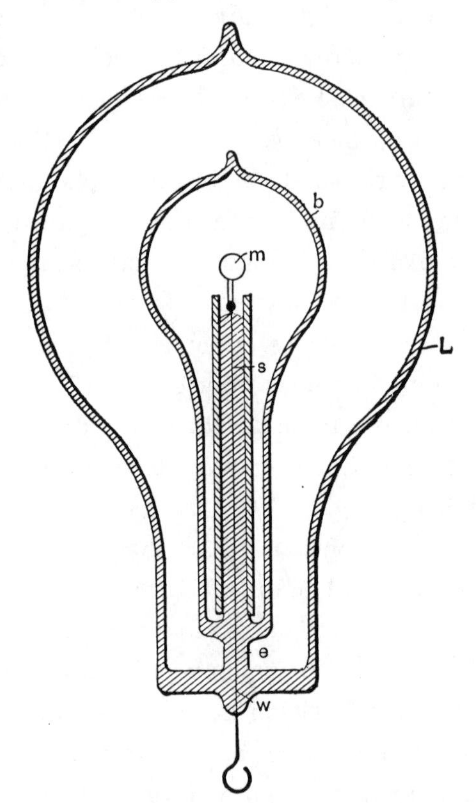

FIG. 29.—LAMP WITH INDEPENDENT AUXILIARY BULB.

high incandescence, to place, preferably on the upper part of the neck of the globe, a tinfoil coating which was connected to an insulated body, to the ground, or to the other terminal of the coil, as the highly conducting air weakened the effect somewhat, probably by being acted upon inductively from the wire *w*, where it entered the bulb at *e*. Another difficulty—which, however, is always present when the refractory button is mounted in a very small bulb —existed in the construction illustrated in Fig. 29, namely, the vacuum in the bulb *b* would be impaired in a comparatively short time.

The chief idea in the two last described constructions was to confine the heat to the central portion of the globe by preventing the exchange of air. An advantage is secured, but owing to the heating of the inside bulb and slow evaporation of the glass the vacuum is hard to maintain, even if the construction illustrated in Fig. 28 be chosen, in which both bulbs communicate.

But by far the better way—the ideal way—would be to reach sufficiently high frequencies. The higher the frequency the slower would be the exchange of the air, and I think that a frequency may be reached at which there would be no exchange whatever of the air molecules around the terminal. We would then produce a flame in which there would be no carrying away of material, and a queer flame it would be, for it would be rigid! With such high frequencies the inertia of the particles would come into play. As the brush, or flame, would gain rigidity in virtue of the inertia of the particles, the exchange of the latter would be prevented. This would necessarily occur, for, the number of the impulses being augmented, the potential energy of each would diminish, so that finally only atomic vibrations could be set up, and the motion of translation through measurable space would cease. Thus an ordinary gas burner connected to a source of rapidly alternating potential might have its efficiency augmented to a certain limit, and this for two reasons—because of the additional vibration imparted, and because of a slowing down of the process of carrying off. But the renewal being rendered difficult, and renewal being necessary to maintain the *burner*, a continued increase of the frequency of the impulses, assuming they could be transmitted to and impressed upon the flame, would result in the " extinction " of the latter, meaning by this term only the cessation of the chemical process.

I think, however, that in the case of an electrode immersed in a fluid insulating medium, and surrounded by independent carriers of electric charges, which can be acted upon inductively, a sufficiently high frequency of the impulses would probably result in a gravitation of the gas all around toward the electrode. For this it would be only necessary to assume that the independent bodies are irregularly shaped; they would then turn toward the electrode their side of the greatest electric density, and this would be a position in which the fluid resistance to approach would be smaller than that offered to the receding.

The general opinion, I do not doubt, is that it is out of the question to reach any such frequencies as might—assuming some of the views before expressed to be true—produce any of the results which I have pointed out as mere possibilities. This may be so, but in the course of these investigations, from the observation of many phenomena I have gained the conviction that these frequencies would be much lower than one is apt to estimate at first. In a flame we set up light vibrations by causing molecules, or atoms, to collide.

But what is the ratio of the frequency of the collisions and that of the vibrations set up? Certainly it must be incomparably smaller than that of the knocks of the bell and the sound vibrations, or that of the discharges and the oscillations of the condenser. We may cause the molecules of the gas to collide by the use of alternate electric impulses of high frequency, and so we may imitate the process in a flame; and from experiments with frequencies which we are now able to obtain, I think that the result is producible with impulses which are transmissible through a conductor.

In connection with thoughts of a similar nature, it appeared to me of great interest to demonstrate the rigidity of a vibrating gaseous column. Although with such low frequencies as, say 10,000 per second, which I was able to obtain without difficulty from a specially constructed alternator, the task looked discouraging at first, I made a series of experiments. The trials with air at ordinary pressure led to no result, but with air moderately rarefied I obtain what I think to be an unmistakable experimental evidence of the property sought for. As a result of this kind might lead able investigators to conclusions of importance I will describe one of the experiments performed.

It is well known that when a tube is slightly exhausted the discharge may be passed through it in the form of a thin luminous thread. When produced with currents of low frequency, obtained from a coil operated as usual, this thread is inert. If a magnet be approached to it, the part near the same is attracted or repelled, according to the direction of the lines of force of the magnet. It occurred to me that if such a thread would be produced with currents of very high frequency, it should be more or less rigid, and as it was visible it could be easily studied. Accordingly I prepared a tube about 1 inch in diameter and 1 metre long, with outside coating at each end. The tube was exhausted to a point at which by a little working the thread discharge could be obtained. It must be remarked here that the general aspect of the tube, and the degree of exhaustion, are quite different than when ordinary low frequency currents are used. As it was found preferable to work with one terminal, the tube prepared was suspended from the end of a wire connected to the terminal, the tinfoil coating being connected to the wire, and to the lower coating sometimes a small insulated plate was attached. When the thread was formed it extended through the upper part of the tube and lost itself in the lower end. If it possessed rigidity it resembled, not exactly an elastic cord stretched tight between two supports, but a cord suspended from a height with a small weight attached at the end. When the finger or a magnet was approached to the upper end of the luminous thread, it could be brought locally out of position by electrostatic or magnetic action ; and when the disturbing object was very quickly removed, an analogous result was produced, as though a suspended cord would be displaced and quickly released near the point of suspension. In doing this the luminous thread was set in vibration, and two very sharply marked nodes, and a third indistinct one, were formed. The vibration, once set up, continued for fully eight minutes, dying gradually out. The speed of the vibration often varied perceptibly, and it could be observed that the electrostatic attraction of the glass affected the vibrating thread; but it was clear that the electrostatic action was not the cause of the vibration, for the thread was most generally stationary, and could always be set in vibration by passing the finger quickly near the upper part of the tube. With a magnet the thread could be split in two and both parts vibrated. By approaching the hand to the lower coating of the tube, or insulated plate if attached, the vibration was quickened; also, as far as I could see, by raising the potential or frequency. Thus, either increasing the frequency or passing a stronger discharge of the same frequency corresponded to a tightening of the cord. I did not obtain any experimental evidence with condenser discharges. A luminous band excited in a bulb by repeated discharges of a Leyden jar must possess rigidity, and if deformed and suddenly released should vibrate. But probably the amount of vibrating matter is so small that in spite of the extreme speed the inertia cannot prominently assert itself. Besides, the observation in such a case is rendered extremely difficult on account of the fundamental vibration.

The demonstration of the fact—which still needs better experimental confirmation—that a vibrating gaseous column possesses rigidity, might greatly modify the views of thinkers. When with low frequencies and insignificant potentials indications of that property may be noted, how must a gaseous medium behave under the influence of enormous electrostatic stresses which may be active in the interstellar space, and which may alternate with inconceivable rapidity? The existence of such an electrostatic, rhythmically throbbing force—of a vibrating electrostatic field—would show a possible way how solids might have formed from the ultra-gaseous uterus, and how transverse and all kinds of vibrations may be transmitted through a gaseous medium filling all space. Then, ether might be a true fluid, devoid of rigidity, and at rest, it being merely necessary as a connecting link to enable interaction. What determines the rigidity of a body? It must be the speed and the amount of moving matter. In a gas the speed may be considerable, but the density is exceedingly small; in a liquid the speed would be likely to be small, though the density may be considerable; and in both cases the inertia resistance offered to displacement is practically *nil.* But place a gaseous (or liquid) column in an intense, rapidly alternating electrostatic field, set the particles vibrating with enormous speeds, then the inertia resistance asserts itself. A body might move with more or less freedom through the vibrating mass, but as a whole it would be rigid.

There is a subject which I must mention in connection with these experiments: it is that of high vacua. This is a subject the study of which is not only interesting, but useful, for it may lead to results of great practical importance. In commercial apparatus, such as incandescent lamps, operated from ordinary systems of distribution, a much higher vacuum than obtained at present would not secure a very great advantage. In such a case the work is performed on the filament and the gas is little concerned; the improvement, therefore, would be but trifling. But when we begin to use very high frequencies and potentials, the action of the gas becomes all important, and the degree of exhaustion materially modifies the results. As long as ordinary coils, even very large ones, were used, the study of the subject was limited, because just at a point when it became most interesting it had to be interrupted on account of the "non-striking" vacuum being reached. But presently we are able to obtain from a small disruptive discharge coil potentials much higher than even the largest coil was capable of giving, and, what is more, we can make the potential alternate with great rapidity. Both of these results enable us now to pass a luminous discharge through almost any vacua obtainable, and the field of our investigations is greatly extended. Think we as we may, of all the possible directions to develop a practical illuminant, the line of high vacua seems to be the most promising at present. But to reach extreme vacua the appliances must be much more improved, and ultimate perfection will not be attained until we shall have discarded the mechanical and perfected an *electrical* vacuum pump. Molecules and atoms can be thrown out of a bulb under the action of an enormous potential: *this* will be the principle of the vacuum pump of the future. For the present, we must secure the best results we can with mechanical appliances. In this respect, it might not be out of the way to say a few words about the method of, and apparatus for, producing excessively high degrees of exhaustion of which I have availed myself in the course of these investigations. It is very probable that other experimenters have used similar arrangements; but as it is possible that there may be an item of interest in their description, a few remarks, which will render this investigation more complete, might be permitted.

The apparatus is illustrated in a drawing shown in Fig. 30. *S* represents a Sprengel pump, which has been

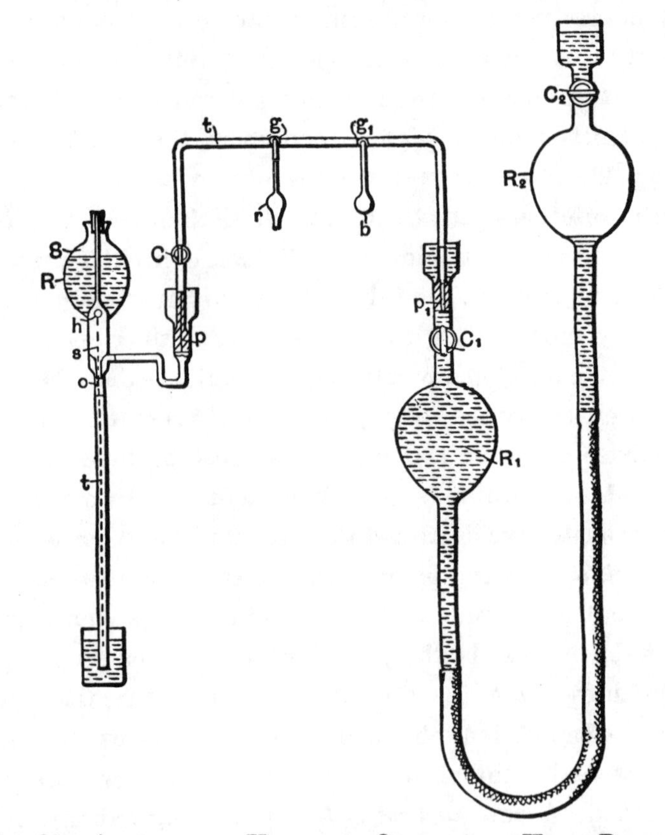

FIG. 30.—APPARATUS USED FOR OBTAINING HIGH DEGREES OF EXHAUSTION.

specially constructed to better suit the work required. The stop-cock which is usually employed has been omitted, and instead of it a hollow stopper *s* has been fitted in the neck of the reservoir R. This stopper has a small hole h, through which the mercury descends; the size of the outlet o being properly determined with respect to the section of the fall tube t, which is sealed to the reservoir instead of being connected to it in the usual manner. This arrangement overcomes the imperfections and troubles which often arise from the use of the stopcock on the reservoir and the connection of the latter with the fall tube.

The pump is connected through a U-shaped tube t to a very large reservoir R_1. Especial care was taken in fitting the grinding surfaces of the stoppers p and p_1, and both of these and the mercury caps above them were made exceptionally long. After the U-shaped tube was fitted and put in place, it was heated, so as to soften and take off the strain resulting from imperfect fitting. The U-shaped tube was provided with a stopcock C, and two ground connections g and g_1—one for a small bulb b, usually containing caustic potash, and the other for the receiver r, to be exhausted.

The reservoir R_1 was connected by means of a rubber tube to a slightly larger reservoir R_2, each of the two reservoirs being provided with a stopcock C_1 and C_2, respectively. The reservoir R_2 could be raised and lowered by a wheel and rack, and the range of its motion was so determined that when it was filled with mercury and the stopcock C_2 closed, so as to form a Torricellian vacuum in it when raised, it could be lifted so high that the mercury in reservoir R_1 would stand a little above stopcock C_1; and when this stopcock was closed and the reservoir R_2 descended, so as to form a Torricellian vacuum in reservoir R_1, it could be lowered so far as to completely empty the latter, the mercury filling the reservoir R_2 up to a little above stopcock C_2.

The capacity of the pump and of the connections was taken as small as possible relatively to the volume of reservoir, R_1, since, of course, the degree of exhaustion depended upon the ratio of these quantities.

With this apparatus I combined the usual means indicated by former experiments for the production of very high vacua. In most of the experiments it was convenient to use caustic potash. I may venture to say, in regard to its use, that much time is saved and a more perfect action of the pump insured by fusing and boiling the potash as soon as, or even before, the pump settles down. If this course is not followed the sticks, as ordinarily employed, may give moisture off at a certain very slow rate, and the pump may work for many hours without reaching a very high vacuum. The potash was heated either by a spirit lamp or by passing a discharge through it, or by passing a current through a wire contained in it. The advantage in the latter case was that the heating could be more rapidly repeated.

Generally the process of exhaustion was the following:— At the start, the stop-cocks C and C_1 being open, and all other connections closed, the reservoir R_2 was raised so far that the mercury filled the reservoir R_1 and a part of the narrow connecting U-shaped tube. When the pump was set to work, the mercury would, of course, quickly rise in the tube, and reservoir R_2 was lowered, the experimenter keeping the mercury at about the same level. The reservoir R_2 was balanced by a long spring which facilitated the operation, and the friction of the parts was generally sufficient to keep it almost in any position. When the Sprengel pump had done its work, the reservoir R_2 was further lowered and the mercury descended in R_1 and filled R_2, whereupon stopcock C_2 was closed. The air adhering to the walls of R_1 and that absorbed by the mercury was carried off, and to free the mercury of all air the reservoir R_2 was for a long time worked up and down. During this process some air, which would gather below stopcock C_2, was expelled from R_2 by lowering it far enough and opening the stopcock, closing the latter again before raising the reservoir. When all the air had been expelled from the mercury, and no air would gather in R_2 when it was lowered, the caustic potash was resorted to. The reservoir R_2 was now again raised until the mercury in R_1 stood above stopcock C_1. The caustic potash was fused and boiled, and the moisture partly carried off by the pump and partly re-absorbed; and this process of heating and cooling was repeated many times, and each time, upon the moisture being absorbed or carried off, the reservoir R_2 was for a long time raised and lowered. In this manner all the moisture was carried off from the mercury, and both the reservoirs were in proper condition to be used. The reservoir R_2 was then again raised to the top, and the pump was kept working for a long time. When the highest vacuum obtainable with the pump had been reached the potash bulb was usually wrapped with cotton which was sprinkled with ether so as to keep the potash at a very low temperature, then the reservoir R_2 was lowered, and upon reservoir R_1 being emptied the receiver r was quickly sealed up.

When a new bulb was put on, the mercury was always raised above stopcock C_1, which was closed, so as to always keep the mercury and both the reservoirs in fine condition, and the mercury was never withdrawn from R_1 except when the pump had reached the highest degree of exhaustion. It is necessary to observe this rule if it is desired to use the apparatus to advantage.

By means of this arrangement I was able to proceed very quickly, and when the apparatus was in perfect order it was possible to reach the phosphorescent stage in a small bulb in less than 15 minutes, which is certainly very quick work for a small laboratory arrangement requiring all in all about 100 pounds of mercury. With ordinary small bulbs the ratio of the capacity of the pump, receiver, and connections, and that of reservoir R was about 1-20, and the degrees of exhaustion reached were necessarily very high, though I am unable to make a precise and reliable statement how far the exhaustion was carried.

What impresses the investigator most in the course of these experiences is the behavior of gases when subjected to great rapidly alternating electrostatic stresses. But he must remain in doubt as to whether the effects observed are due wholly to the molecules, or atoms, of the gas which chemical analysis discloses to us, or whether there enters into play another medium of a gaseous nature, comprising atoms, or molecules, immersed in a fluid pervading the space. Such a medium surely must exist, and I am convinced that, for instance, even if air were absent, the surface and neighborhood of a body in space would be heated by rapidly alternating the potential of the body; but no such heating of the surface or neighborhood could occur if all free atoms were removed and only a homogeneous, incompressible, and elastic fluid—such as ether is supposed to be—would remain, for then there would be no impacts, no collisions. In such a case, as far as the body itself is concerned, only frictional losses in the inside could occur.

It is a striking fact that the discharge through a gas is established with ever increasing freedom as the frequency of the impulses is augmented. It behaves in this respect quite contrarily to a metallic conductor. In the latter the impedance enters prominently into play as the frequency is increased, but the gas acts much as a series of condensers would: the facility with which the discharge passes through seems to depend on the rate of change of potential. If it act so, then in a vacuum tube even of great length, and no matter how strong the current, self-induction could not assert itself to any appreciable degree. We have, then, as far as we can now see, in the gas a conductor which is capable of transmitting electric impulses of any frequency which we may be able to produce. Could the frequency be brought high enough, then a queer system of electric distribution, which would be likely to interest gas companies, might be realized: metal pipes filled with gas—the metal being the insulator, the gas the conductor—supplying phosphorescent bulbs, or perhaps devices as yet uninvented. It is certainly possible to take a hollow core of copper, rarefy the gas in the same, and by passing impulses of sufficiently high frequency through a circuit around it, bring the gas inside to a high degree of incandescence; but as to the nature of the forces there would be considerable uncertainty, for it would be doubtful whether with such impulses the copper core would act as a static screen. Such paradoxes and apparent impossibilities we encounter at every step in this line of work, and therein lies, to a great extent, the charm of the study.

I have here a short and wide tube which is exhausted to a high degree and covered with a substantial coating of bronze, the coating allowing barely the light to shine through. A metallic clasp, with a hook for suspending the tube, is fastened around the middle portion of the latter, the clasp being in contact with the bronze coating. I now want to light the gas inside by suspending the tube on a wire connected to the coil. Any one who would try the experiment for the first time, not having any previous experience, would probably take care to be quite alone when making the trial, for fear that he might become the joke of his assistants. Still, the bulb lights in spite of the metal coating, and the light can be distinctly perceived through the latter. A long tube covered with aluminium bronze lights when held in one hand—the other touching the terminal of the coil—quite powerfully. It might be objected that the coatings are not sufficiently conducting; still, even if they were highly resistant, they ought to screen the gas. They certainly screen it perfectly in a condition of rest, but not by far perfectly when the charge is surging in the coating. But the loss of energy which occurs within the tube, notwithstanding the screen, is occasioned principally by the presence of the gas. Were we to take a large hollow metallic sphere and fill it with a perfect incompressible fluid dielectric, there would be no loss inside of the sphere, and consequently the inside might be considered as perfectly screened, though the potential be very rapidly alternating. Even were the sphere filled with oil, the loss would be incomparably smaller than when the fluid is replaced by a gas, for in the latter case the force produces displacements; that means impact and collisions in the inside.

No matter what the pressure of the gas may be, it becomes an important factor in the heating of a conductor when the electric density is great and the frequency very high. That in the heating of conductors by lightning discharges air is an element of great importance, is almost as certain as an experimental fact. I may illustrate the action of the air by the following experiment: I take a short tube which is exhausted to a moderate degree and has a platinum wire running through the middle from one end to the other. I pass a steady or low frequency current through the wire, and it is heated uniformly in all parts. The heating here is due to conduction, or frictional losses, and the gas around the wire has—as far as we can see—no function to perform. But now let me pass sudden discharges, or a high frequency current, through the wire. Again the wire is heated, this time principally on the ends and least in the middle portion; and if the frequency of the impulses, or the rate of change, is high enough, the wire might as well be cut in the middle as not, for practically all the heating is due to the rarefied gas. Here the gas might only act as a conductor of no impedance diverting the current from the wire as the impedance of the latter is enormously increased, and merely heating the ends of the wire by reason of their resistance to the passage of the discharge. But it is not at all necessary that the gas in the tube should be conducting; it might be at an extremely low pressure, still the ends of the wire would be heated—as, however, is ascertained by experience—only the two ends would in such case not be electrically connected through the gaseous medium. Now what with these frequencies and potentials occurs in an exhausted tube occurs in the lightning discharges at ordinary pressure. We only need remember one of the facts arrived at in the course of these investigations, namely, that to impulses of very high frequency the gas at ordinary pressure behaves much in the same manner as though it were at moderately low pressure. I think that in lightning discharges frequently wires or conducting objects are volatilized merely because air is present, and that, were the conductor immersed in an insulating liquid, it would be safe, for then the energy would have to spend itself somewhere else. From the behavior of gases to sudden impulses of high potential I am led to conclude that there can be no surer way of diverting a lightning discharge than by affording it a passage through a volume of gas, if such a thing can be done in a practical manner.

There are two more features upon which I think it necessary to dwell in connection with these experiments—the "radiant state" and the "non-striking vacuum."

Any one who has studied Crookes' work must have received the impression that the "radiant state" is a property of the gas inseparably connected with an extremely high degree of exhaustion. But it should be remembered that the phenomena observed in an exhausted vessel are limited to the character and capacity of the apparatus which is made use of. I think that in a bulb a molecule, or atom,

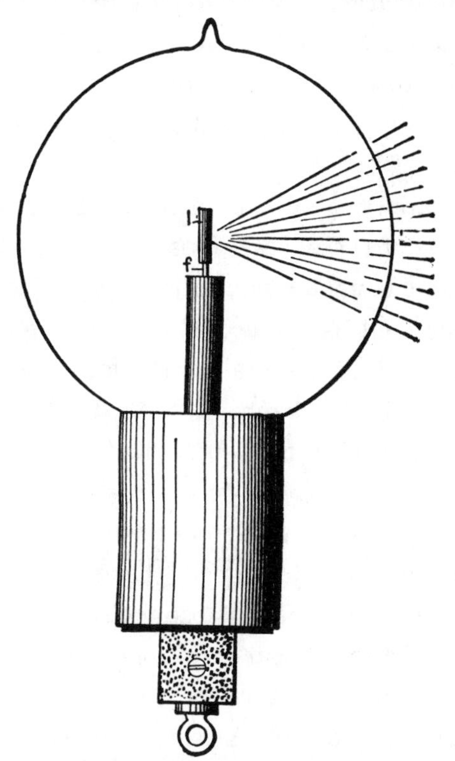

FIG. 31.—BULB SHOWING RADIANT LIME STREAM AT LOW EXHAUSTION.

does not precisely move in a straight line because it meets no obstacle, but because the velocity imparted to it is sufficient to propel it in a sensibly straight line. The mean free path is one thing, but the velocity—the energy associated with the moving body—is another, and under ordinary circumstances I believe that it is a mere question of potential or speed. A disruptive discharge coil, when the potential is pushed very far, excites phosphorescence and projects shadows, at comparatively low degrees of exhaustion. In a lightning discharge, matter moves in straight lines at ordinary pressure when the mean free path is exceedingly small, and frequently images of wires or other metallic objects have been produced by the particles thrown off in straight lines.

I have prepared a bulb to illustrate by an experiment the correctness of these assertions. In a globe *L* (Fig. 31, I have mounted upon a lamp filament *f* a piece of lime *l*. The lamp filament is connected with a wire which leads into the bulb, and the general construction of the latter is as indicated in Fig. 19, befcre described. The bulb being suspended from a wire connected to the terminal of the coil, and the latter being set to work, the lime piece *l* and the projecting parts of the filament *f* are bombarded. The degree of exhaustion is just such that with the potential the coil is capable of giving phosphorescence of the glass is produced, but disappears as soon as the vacuum is impaired. The lime containing moisture, and moisture being given off as soon as heating occurs, the phosphorescence lasts only for a few moments. When the lime has been sufficiently heated, enough moisture has been given off to impair materially the vacuum of the bulb. As the bombardment goes on, one point of the lime piece is more heated than other points, and the result is that finally practically all the discharge passes through that point which is intensely heated, and a white stream of lime particles (Fig. 31) then breaks forth from that point. This stream is composed of "radiant" matter, yet the degree of exhaustion is low. But the particles move in straight lines because the velocity imparted to them is great, and this is due to three causes—to the great electric density, the high temperature of the small point, and the fact that the particles of the lime are easily torn and thrown off—far more easily than those of carbon. With frequencies such as we are able to obtain, the particles are bodily thrown off and projected to a considerable distance; but with sufficiently high frequencies no such thing would occur: in such case only a stress would spread or a vibration would be propagated through the bulb. It would be out of the question to reach any such frequency on the assumption that the atoms move with the speed of light; but I believe that such a thing is impossible; for this an enormous potential would be required. With potentials which we are able to obtain, even with a disruptive discharge coil, the speed must be quite insignificant.

As to the "non-striking vacuum," the point to be noted is that it can occur only with low frequency impulses, and it is necessitated by the impossibility of carrying off enough energy with such impulses in high vacuum since the few atoms which are around the terminal upon coming in contact with the same are repelled and kept at a distance for a comparatively long period of time, and not enough work can be performed to render the effect perceptible to the eye. If the difference of potential between the terminals is raised, the dielectric breaks down. But with very high frequency impulses there is no necessity for such breaking down, since any amount of work can be performed by continually agitating the atoms in the exhausted vessel, provided the frequency is high enough. It is easy to reach—even with frequencies obtained from an alternator as here used—a stage at which the discharge does not pass between two electrodes in a narrow tube, each of these being connected to one of the terminals of the coil, but it is difficult to reach a point at which a luminous discharge would not occur around each electrode.

A thought which naturally presents itself in connection with high frequency currents, is to make use of their powerful electro-dynamic inductive action to produce light effects in a sealed glass globe. The leading-in wire is one of the defects of the present incandescent lamp, and if no other improvement were made, that imperfection at least should be done away with. Following this thought, I have carried on experiments in various directions, of which some were indicated in my former paper. I may here mention one or two more lines of experiment which have been followed up.

Many bulbs were constructed as shown in Fig. 32 and Fig. 33.

In Fig. 32 a wide tube *T* was sealed to a smaller W-shaped tube *U*, of phosphorescent glass. In the tube *T* was placed a coil *C* of aluminium wire, the ends of which were provided with small spheres t and t_1 of aluminium, and reached into the *U* tube. The tube *T* was slipped into a socket containing a primary coil through which usually the discharges of Leyden jars were directed, and the rarefied gas in the small *U* tube was excited to strong luminosity by the high-tension currents induced in the coil *C*. When Leyden jar discharges were used to induce cur-

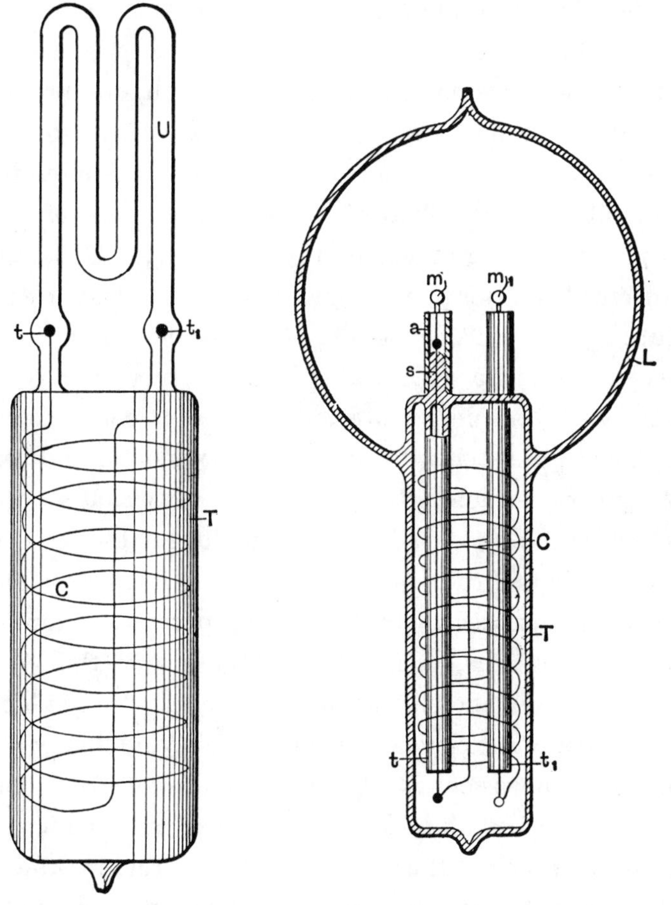

FIG 32.—ELECTRO-DYNAMIC INDUCTION TUBE.

FIG. 33.—ELECTRO-DYNAMIC INDUCTION LAMP.

rents in the coil *C*, it was found necessary to pack the tube *T* tightly with insulating powder, as a discharge would occur frequently between the turns of the coil, especially when the primary was thick and the air gap, through which the jars discharged, large, and no little trouble was experienced in this way.

In Fig. 33 is illustrated another form of the bulb constructed. In this case a tube T is sealed to a globe L. The tube contains a coil C, the ends of which pass through two small glass tubes t and t_1, which are sealed to the tube T. Two refractory buttons m and m_1 are mounted on lamp filaments which are fastened to the ends of the wires passing through the glass tubes t and t_1. Generally in bulbs made on this plan the globe L communicated with the tube T. For this purpose the ends of the small tubes t and t_1 were just a trifle heated in the burner, merely to hold the wires, but not to interfere with the communication. The tube T, with the small tubes, wires through the same, and the refractory buttons m and m_1, was first prepared, and then sealed to globe L, whereupon the coil C was slipped in and the connections made to its ends. The tube was then packed with insulating powder, jamming the latter as tight as possible up to very nearly the end, then it was closed and only a small hole left through which the remainder of the powder was introduced, and finally the end of the tube was closed. Usually in bulbs constructed as shown in Fig. 33 an aluminium tube a was fastened to the upper end s of each of the tubes t and t_1, in order to protect that end against the heat. The buttons m and m_1 could be brought to any degree of incandescence by passing the discharges of Leyden jars around the coil C. In such bulbs with two buttons a very curious effect is produced by the formation of the shadows of each of the two buttons.

Another line of experiment, which has been assiduously followed, was to induce by electro-dynamic induction a current or luminous discharge in an exhausted tube or bulb. This matter has received such able treatment at the hands of Prof. J. J. Thomson that I could add but little to what he has made known, even had I made it the special subject of this lecture. Still, since experiences in this line have gradually led me to the present views and results, a few words must be devoted here to this subject.

It has occurred, no doubt, to many that as a vacuum tube is made longer the electromotive force per unit length of the tube, necessary to pass a luminous discharge through the latter, gets continually smaller; therefore, if the exhausted tube be made long enough, even with low frequencies a luminous discharge could be induced in such a tube closed upon itself. Such a tube might be placed around a hall or on a ceiling, and at once a simple appliance capable of giving considerable light would be obtained. But this would be an appliance hard to manufacture and extremely unmanageable. It would not do to make the tube up of small lengths, because there would be with ordinary frequencies considerable loss in the coatings, and besides, if coatings were used, it would be better to supply the current directly to the tube by connecting the coatings to a transformer. But even if all objections of such nature were removed, still, with low frequencies the light conversion itself would be inefficient, as I have before stated. In using extremely high frequencies the length of the secondary—in other words, the size of the vessel—can be reduced as far as desired, and the efficiency of the light conversion is increased, provided that means are invented for efficiently obtaining such high frequencies. Thus one is led, from theoretical and practical considerations, to the use of high frequencies, and this means high electromotive forces and small currents in the primary. When he works with condenser charges—and they are the only means up to the present known for reaching these extreme frequencies—he gets to electromotive forces of several thousands of volts per turn of the primary. He cannot multiply the electro-dynamic inductive effect by taking more turns in the primary, for he arrives at the conclusion that the best way is to work with one single turn—though he must sometimes depart from this rule—and he must get along with whatever inductive effect he can obtain with one turn. But before he has long experimented with the extreme frequencies required to set up in a small bulb an electromotive force of several thousands of volts he realizes the great importance of electrostatic effects, and these effects grow relatively to the electro-dynamic in significance as the frequency is increased.

Now, if anything is desirable in this case, it is to increase the frequency, and this would make it still worse for the electro-dynamic effects. On the other hand, it is easy to exalt the electrostatic action as far as one likes by taking more turns on the secondary, or combining self-induction and capacity to raise the potential. It should also be remembered that, in reducing the current to the smallest value and increasing the potential, the electric impulses of high frequency can be more easily transmitted through a conductor.

These and similar thoughts determined me to devote more attention to the electrostatic phenomena, and to endeavor to produce potentials as high as possible, and alternating as fast as they could be made to alternate. I then found that I could excite vacuum tubes at considerable distance from a conductor connected to a properly constructed coil, and that I could, by converting the oscillatory current of a condenser to a higher potential, establish electrostatic alternating fields which acted through the whole extent of a room, lighting up a tube no matter where it was held in space. I thought I recognized that I had made a step in advance, and I have persevered in this line; but I wish to say that I share with all lovers of science and progress the one and only desire—to reach a result of utility to men in any direction to which thought or experiment may lead me. I think that this departure is the right one, for I cannot see, from the observation of the phenomena which manifest themselves as the frequency is increased, what there would remain to act between two circuits conveying, for instance, impulses of several hundred millions per second, except electrostatic forces. Even with such trifling frequencies the energy would be practically all potential, and my conviction has grown strong that, to whatever kind of motion light may be due, it is produced by tremendous electrostatic stresses vibrating with extreme rapidity.

Of all these phenomena observed with currents, or electric impulses, of high frequency, the most fascinating for an audience are certainly those which are noted in an electrostatic field acting through considerable distance, and the best an unskilled lecturer can do is to begin and finish with the exhibition of these singular effects. I take a tube in the hand and move it about, and it is lighted wherever I may hold it; throughout space the invisible forces act. But I may take another tube and it might not light, the vacuum being very high. I excite it by means of a disruptive discharge coil, and now it will light in the electrostatic field. I may put it away for a few weeks or months, still it retains the faculty of being excited. What change have I produced in the tube in the act of exciting it? If a motion imparted to the atoms, it is difficult to perceive how it can persist so long without being arrested by frictional losses; and if a strain exerted in the dielectric, such as a simple electrification would produce, it is easy to see how it may persist indefinitely, but very difficult to understand why such a condition should aid the excitation when we have to deal with potentials which are rapidly alternating.

Since I have exhibited these phenomena for the first time, I have obtained some other interesting effects. For instance, I have produced the incandescence of a button, filament, or wire enclosed in a tube. To get to this result it was necessary to economize the energy which is obtained from the field and direct most of it on the small body to be rendered incandescent. At the beginning the task appeared difficult, but the experiences gathered permitted me to reach the result easily. In Fig. 34 and Fig. 35 two such tubes are illustrated which are prepared for the occasion. In Fig. 34 a short tube T_1, sealed to another long tube T, is provided with a stem s, with a platinum wire sealed in the latter. A very thin lamp filament l is fastened to this wire, and connection to the outside is made through a thin copper wire w. The tube is provided with outside and inside coatings, C and C_1 respectively, and is filled as far as

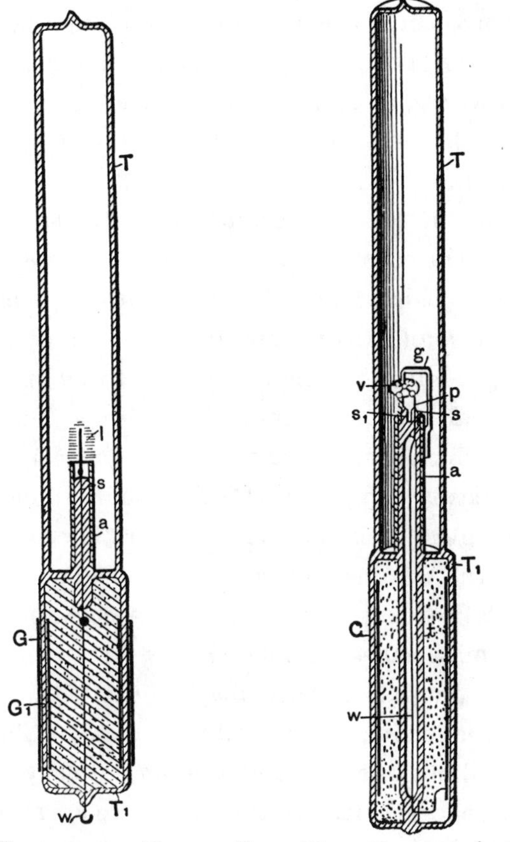

FIG. 34.—TUBE WITH FILAMENT RENDERED INCANDESCENT IN AN ELECTRO STATIC FIELD.

FIG. 35. — CROOKES' EXPERIMENT IN ELECTROSTATIC FIELD.

the coatings reach with conducting, and the space above with insulating powder. These coatings are merely used to enable me to perform two experiments with the tube—

namely, to produce the effect desired either by direct connection of the body of the experimenter or of another body to the wire *w*, or by acting inductively through the glass. The stem *s* is provided with an aluminium tube *a*, for purposes before explained, and only a small part of the filament reaches out of this tube. By holding the tube T_1 anywhere in the electrostatic field the filament is rendered incandescent.

A more interesting piece of apparatus is illustrated in Fig. 35. The construction is the same as before, only instead of the lamp filament a small platinum wire *p*, sealed in a stem *s*, and bent above it in a circle, is connected to the copper wire *w*, which is joined to an inside coating *C*. A small stem s_1 is provided with a needle, on the point of which is arranged to rotate very freely a very light fan of mica *v*. To prevent the fan from falling out, a thin stem of glass *g* is bent properly and fastened to the aluminium tube. When the glass tube is held anywhere in the electrostatic field the platinum wire becomes incandescent, and the mica vanes are rotated very fast.

Intense phosphorescence may be excited in a bulb by merely connecting it to a plate within the field, and the plate need not be any larger than an ordinary lamp shade. The phosphorescence excited with these currents is incomparably more powerful than with ordinary apparatus. A small phosphorescent bulb, when attached to a wire connected to a coil, emits sufficient light to allow reading ordinary print at a distance of five to six paces. It was of interest to see how some of the phosphorescent bulbs of Professor Crookes would behave with these currents, and he has had the kindness to lend me a few for the occasion. The effects produced are magnificent, especially by the sulphide of calcium and sulphide of zinc. From the disruptive discharge coil they glow intensely merely by holding them in the hand and connecting the body to the terminal of the coil.

To whatever results investigations of this kind may lead, their chief interest lies for the present in the possibilities they offer for the production of an efficient illuminating device. In no branch of electric industry is an advance more desired than in the manufacture of light. Every thinker, when considering the barbarous methods employed, the deplorable losses incurred in our best systems of light production, must have asked himself, What is likely to be the light of the future? Is it to be an incandescent solid, as in the present lamp, or an incandescent gas, or a phosphorescent body, or something like a burner, but incomparably more efficient?

There is little chance to perfect a gas burner; not, perhaps, because human ingenuity has been bent upon that problem for centuries without a radical departure having been made—though this argument is not devoid of force—but because in a burner the higher vibrations can never be reached except by passing through all the low ones. For how is a flame produced unless by a fall of lifted weights? Such process cannot be maintained without renewal, and renewal is repeated passing from low to high vibrations. One way only seems to be open to improve a burner, and that is by trying to reach higher degrees of incandescence. Higher incandescence is equivalent to a quicker vibration;

that means more light from the same material, and that, again, means more economy. In this direction some improvements have been made, but the progress is hampered by many limitations. Discarding, then, the burner, there remain the three ways first mentioned, which are essentially electrical.

Suppose the light of the immediate future to be a solid rendered incandescent by electricity. Would it not seem that it is better to employ a small button than a frail filament? From many considerations it certainly must be concluded that a button is capable of a higher economy, assuming, of course, the difficulties connected with the operation of such a lamp to be effectively overcome. But to light such a lamp we require a high potential; and to get this economically we must use high frequencies.

Such considerations apply even more to the production of light by the incandescence of a gas, or by phosphorescence. In all cases we require high frequencies and high potentials. These thoughts occurred to me a long time ago.

Incidentally we gain, by the use of very high frequencies, many advantages, such as a higher economy in the light production, the possibility of working with one lead, the possibility of doing away with the leading-in wire, etc.

The question is, how far can we go with frequencies? Ordinary conductors rapidly lose the facility of transmitting electric impulses when the frequency is greatly increased. Assume the means for the production of impulses of very great frequency brought to the utmost perfection, every one will naturally ask how to transmit them when the necessity arises. In transmitting such impulses through conductors we must remember that we have to deal with *pressure* and *flow*, in the ordinary interpretation of these terms. Let the pressure increase to an enormous value, and let the flow correspondingly diminish, then such impulses—variations merely of pressure, as it were—can no doubt be transmitted through a wire even if their frequency be many hundreds of millions per second. It would, of course, be out of question to transmit such impulses through a wire immersed in a gaseous medium, even if the wire were provided with a thick and excellent insulation for most of the energy would be lost in molecular bombardment and consequent heating. The end of the wire connected to the source would be heated, and the remote end would receive but a trifling part of the energy supplied. The prime necessity, then, if such electric impulses are to be used, is to find means to reduce as much as possible the dissipation.

The first thought is, employ the thinnest possible wire surrounded by the thickest practicable insulation. The next thought is to employ electrostatic screens. The insulation of the wire may be covered with a thin conducting coating and the latter connected to the ground. But this would not do, as then all the energy would pass through the conducting coating to the ground and nothing would get to the end of the wire. If a ground connection is made it can only be made through a conductor offering an enormous impedance, or though a condenser of extremely small capacity. This, however, does not do away with other difficulties.

If the wave length of the impulses is much smaller than

the length of the wire, then corresponding short waves will be sent up in the conducting coating, and it will be more or less the same as though the coating were directly connected to earth. It is therefore necessary to cut up the coating in sections much shorter than the wave length. Such an arrangement does not still afford a perfect screen, but it is ten thousand times better than none. I think it preferable to cut up the conducting coating in small sections, even if the current waves be much longer than the coating.

If a wire were provided with a perfect electrostatie screen, it would be the same as though all objects were removed from it at infinite distance. The capacity would then be reduced to the capacity of the wire itself, which would be very small. It would then be possible to send over the wire current vibrations of very high frequencies at enormous distance without affecting greatly the character of the vibrations. A perfect screen is of course out of the question, but I believe that with a screen such as I have just described telephony could be rendered practicable across the Atlantic. According to my ideas, the gutta-percha covered wire should be provided with a third conducting coating subdivided in sections. On the top of this should be again placed a layer of gutta-percha and other insulation, and on the top of the whole the armor. But such cables will not be constructed, for ere long intelligence—transmitted without wires—will throb through the earth like a pulse through a living organism. The wonder is that, with the present state of knowledge and the experiences gained, no attempt is being made to disturb the electrostatic or magnetic condition of the earth, and transmit, if nothing else, intelligence.

It has been my chief aim in presenting these results to point out phenomena or features of novelty, and to advance ideas which I am hopeful will serve as starting points of new departures. It has been my chief desire this evening to entertain you with some novel experiments. Your applause, so frequently and generously accorded, has told me that I have succeeded.

In conclusion, let me thank you most heartily for your kindness and attention, and assure you that the honor I have had in addressing such a distinguished audience, the pleasure I have had in presenting these results to a gathering of so many able men—and among them also some of those in whose work for many years past I have found enlightenment and constant pleasure—I shall never forget.

Chapter 4

MORE PATENTS (1889–1900)

(No Model.)

2 Sheets—Sheet 1.

N. TESLA.

THERMO MAGNETIC MOTOR.

No. 396,121.

Patented Jan. 15, 1889.

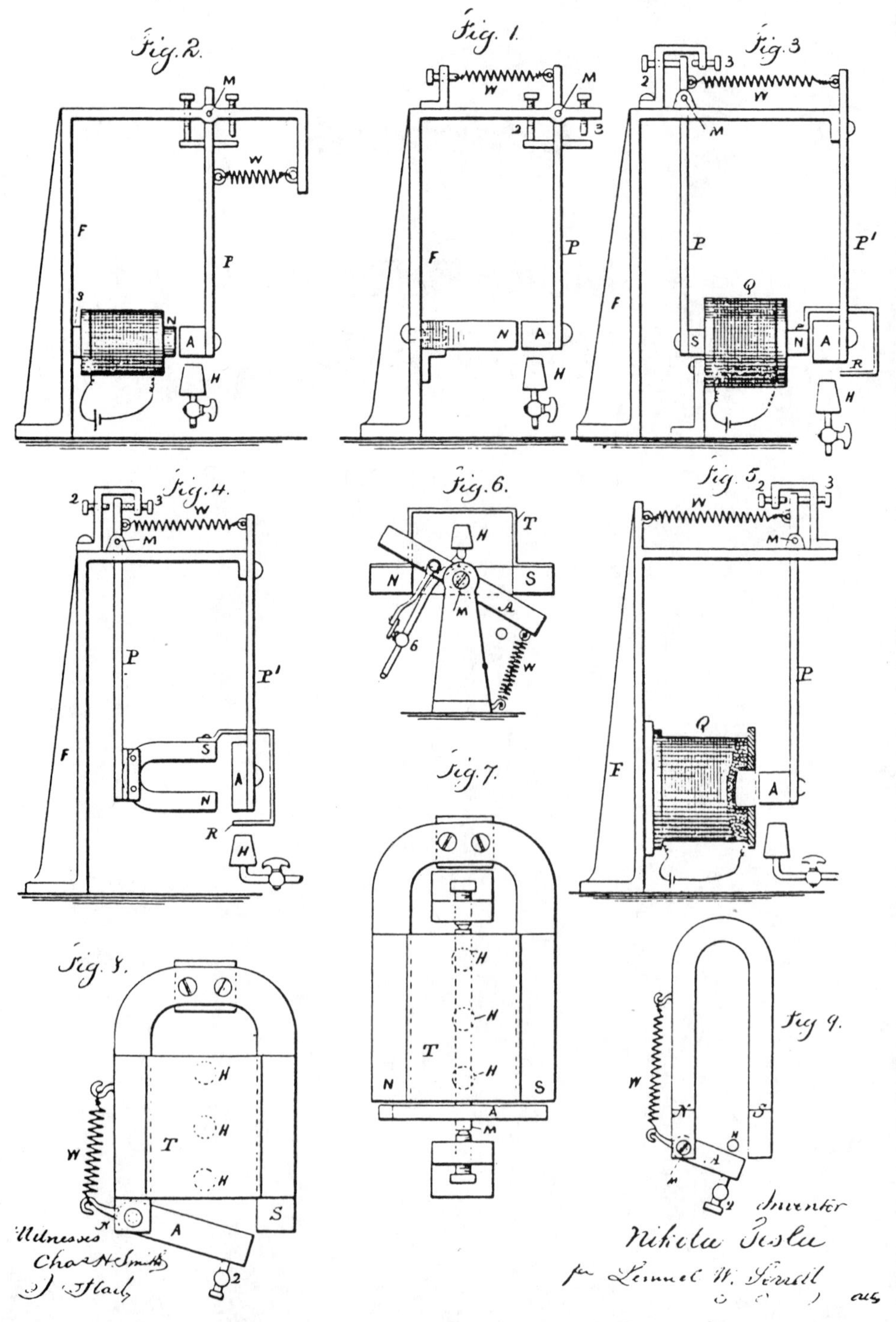

(No Model.)

N. TESLA.
ELECTRO MAGNETIC MOTOR.

No. 416,193. Patented Dec. 3, 1889.

Fig. 1

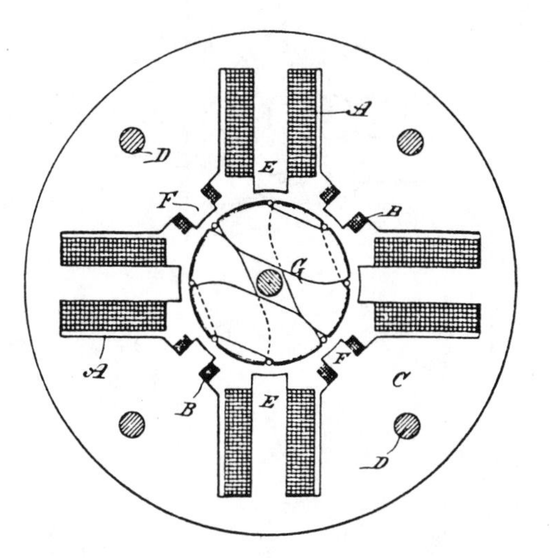

Fig. 2

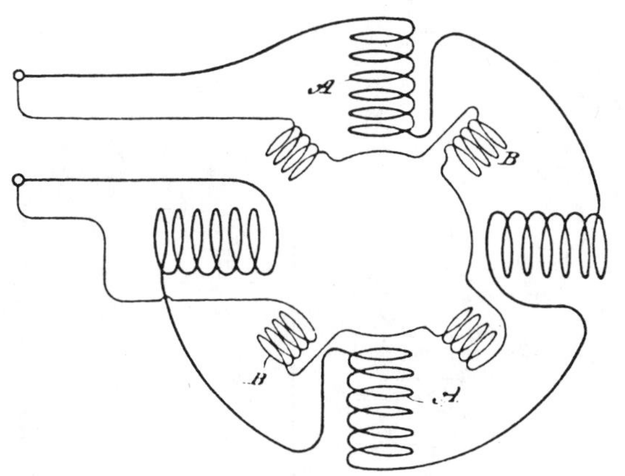

Witnesses:
Raphaël Netter
Robt. F. Gaylord

Inventor
Nikola Tesla
By
Duncan, Curtis & Page
Attorneys.

(No Model.)

N. TESLA.

METHOD OF ELECTRICAL POWER TRANSMISSION.

No. 405,859 Patented June 25, 1889.

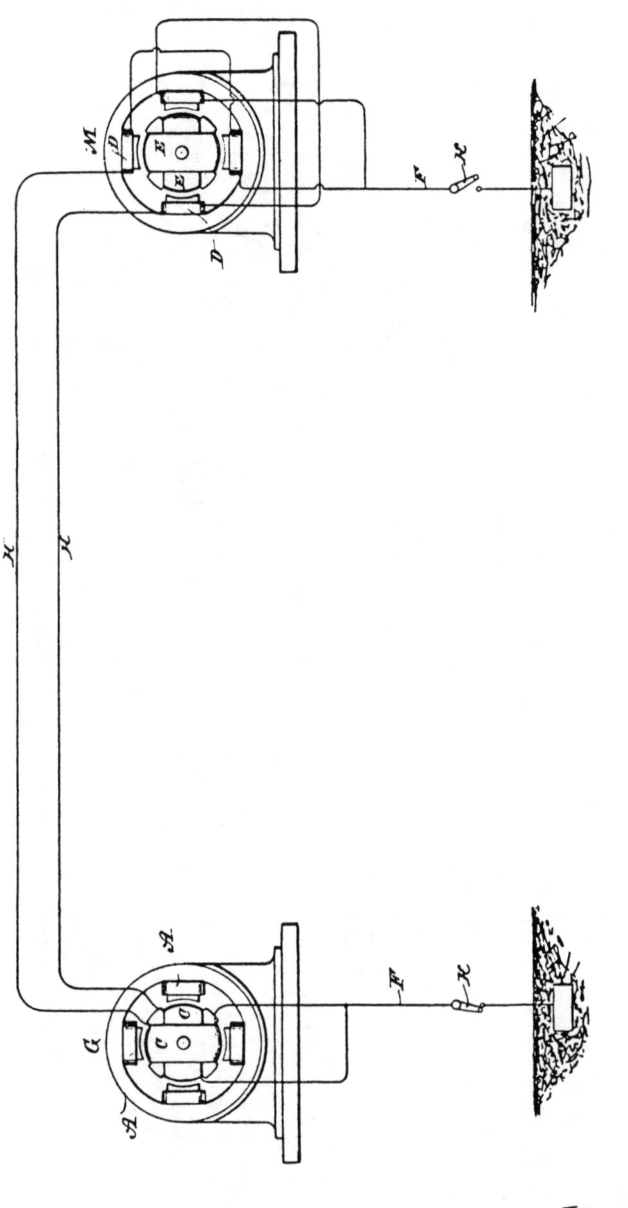

Witnesses:
Raphaël Netter
Robt. F. Gaylord

Inventor
Nikola Tesla
By
Duncan, Curtis & Page.
Attorneys.

(No Model.)

N. TESLA.

ELECTRO MAGNETIC MOTOR.

No. 405,858. Patented June 25, 1889.

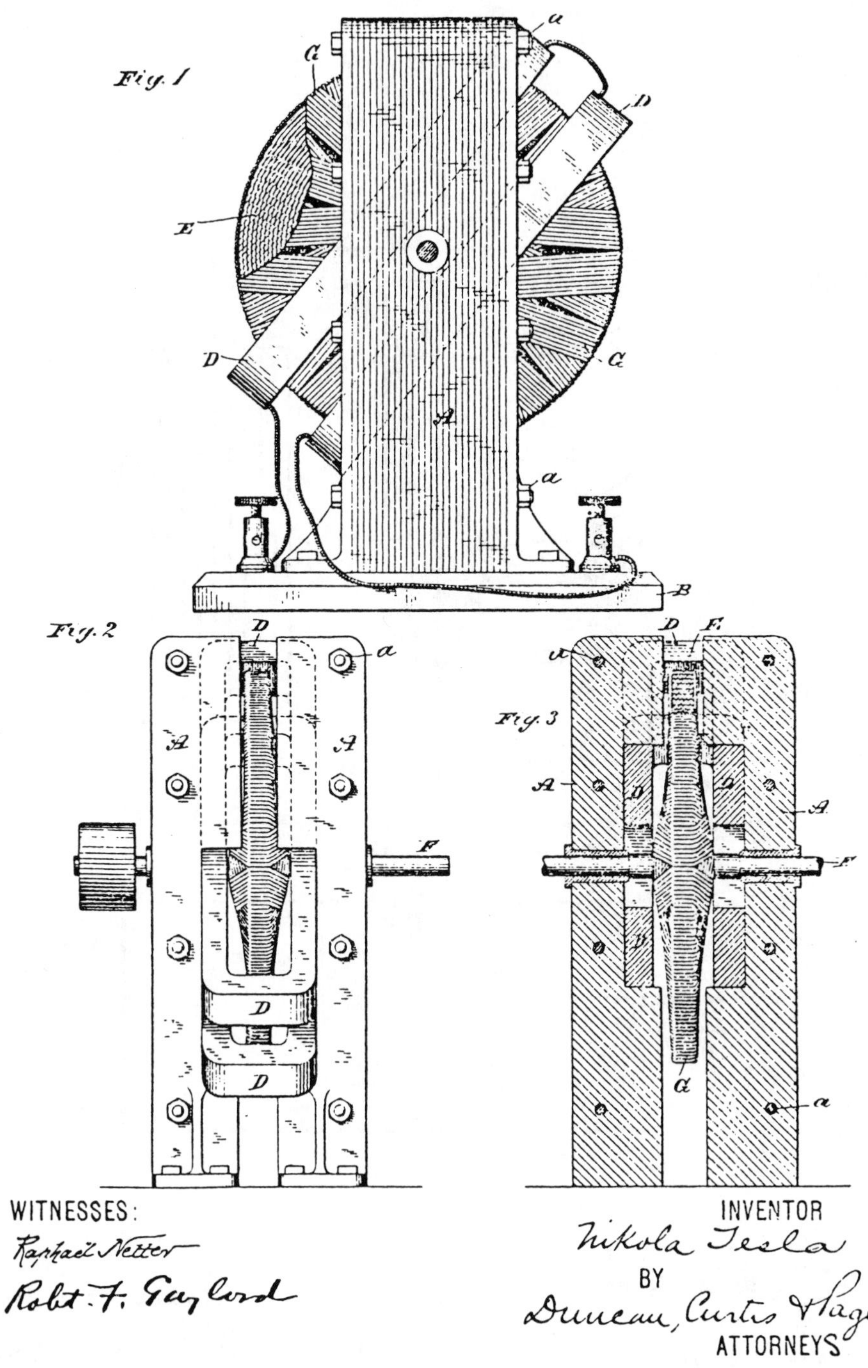

(No Model.) 2 Sheets—Sheet 1.

N. TESLA.

APPARATUS FOR PRODUCING OZONE.

No. 568,177. Patented Sept. 22, 1896.

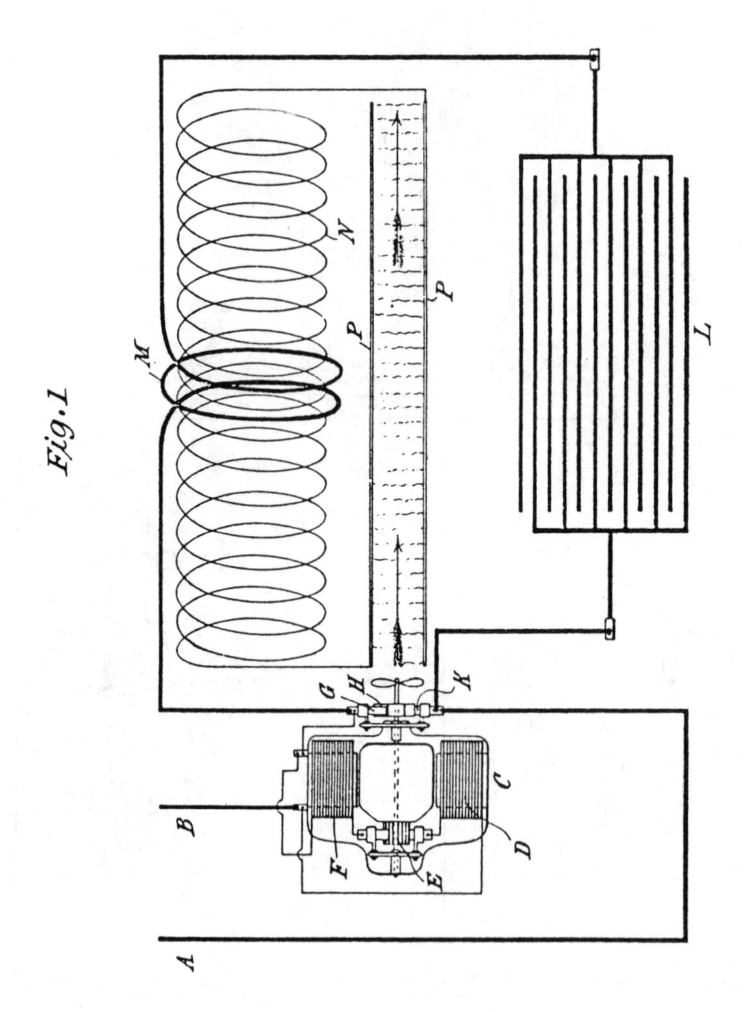

Witnesses:

Raphaël Netter

Drury W. Cooper

Nikola Tesla, Inventor

by Kerr, Curtis & Page. Att'ys.

(No Model.)

3 Sheets—Sheet 2.

N. TESLA.

METHOD OF OBTAINING DIRECT FROM ALTERNATING CURRENTS.

No. 413,353. Patented Oct. 22, 1889.

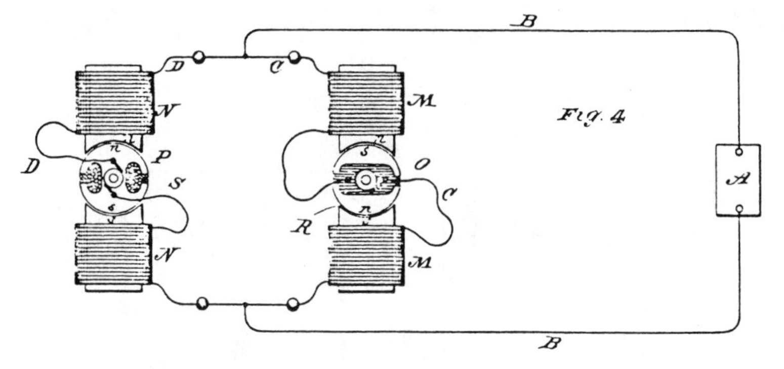

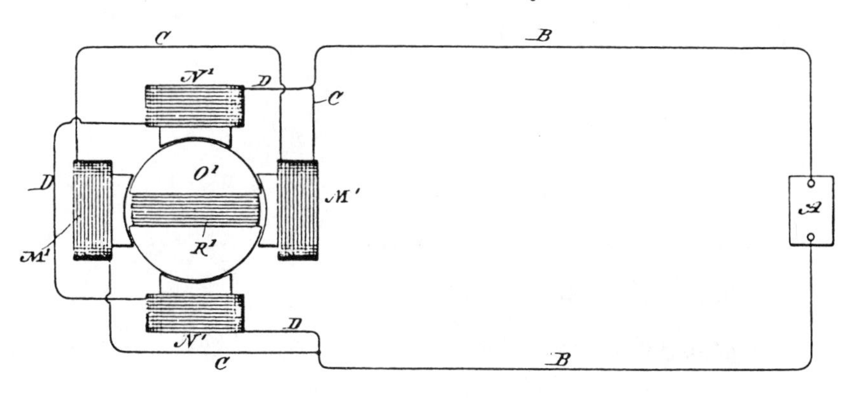

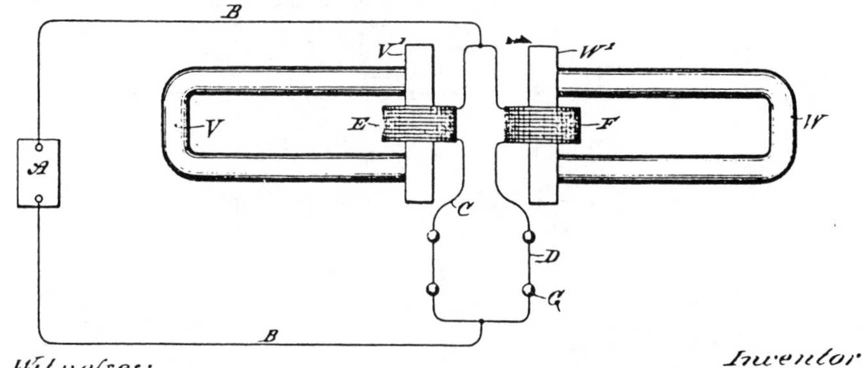

Witnesses:
Raphaël Netter
Frank E. Hartley

Inventor
Nikola Tesla
By
Duncan, Curtis & Page
Attorneys.

(No Model.) 2 Sheets—Sheet 2.

N. TESLA.

METHOD OF OPERATING ELECTRO MAGNETIC MOTORS.

No. 416,192. Patented Dec. 3, 1889.

Fig. 4

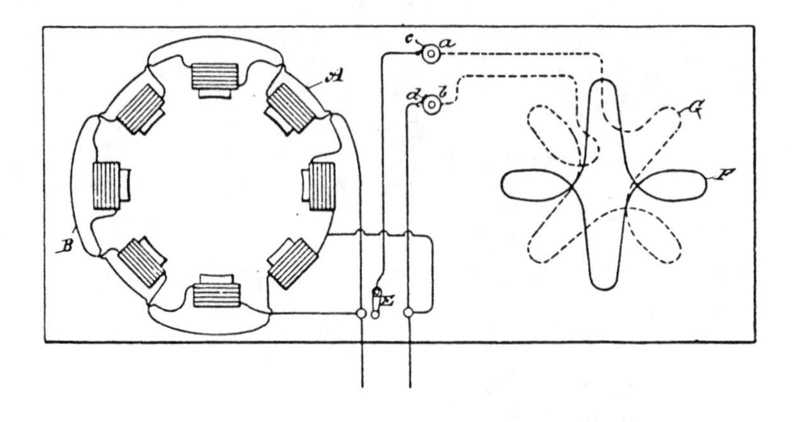

Fig. 5

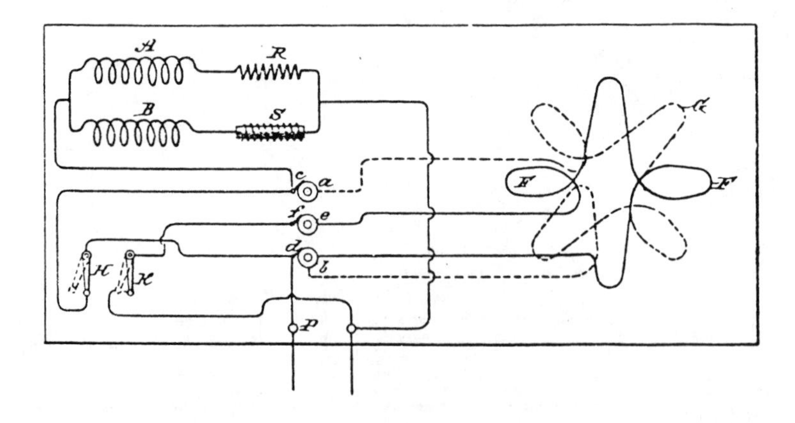

Witnesses:

Raphaël Netter

Frank E. Hartley

Inventor

Nikola Tesla

By

Duncan, Curtis & Page

Attorneys.

(No Model.) 3 Sheets—Sheet 1.

N. TESLA.

ELECTRO MAGNETIC MOTOR.

No. 416,195. Patented Dec. 3, 1889.

Fig. 1

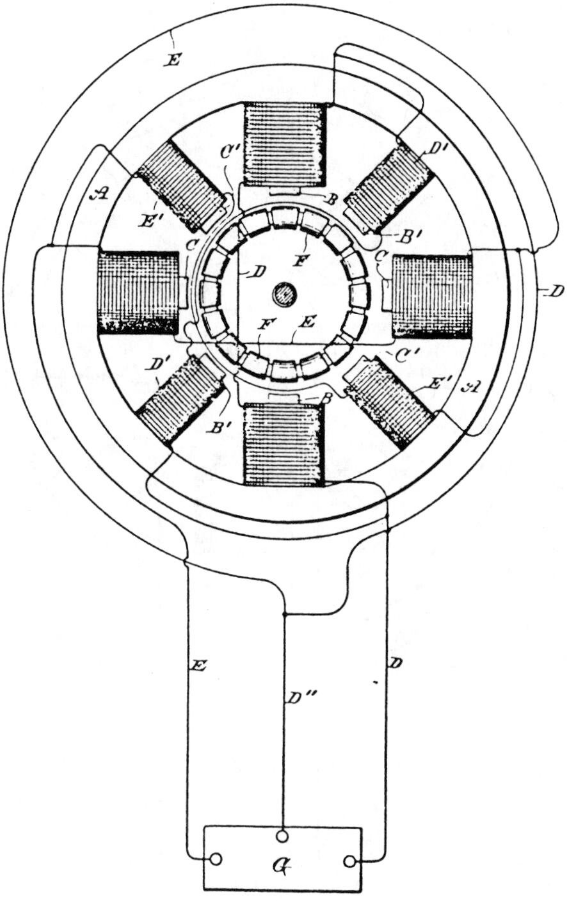

Witnesses:
Raphaël Netter
Robt F. Gaylord

Inventor
Nikola Tesla

By
Duncan, Curtis & Page
Attorneys.

(No Model.) 3 Sheets—Sheet 2.

N. TESLA.

ELECTRO MAGNETIC MOTOR.

No. 416,195. Patented Dec. 3, 1889.

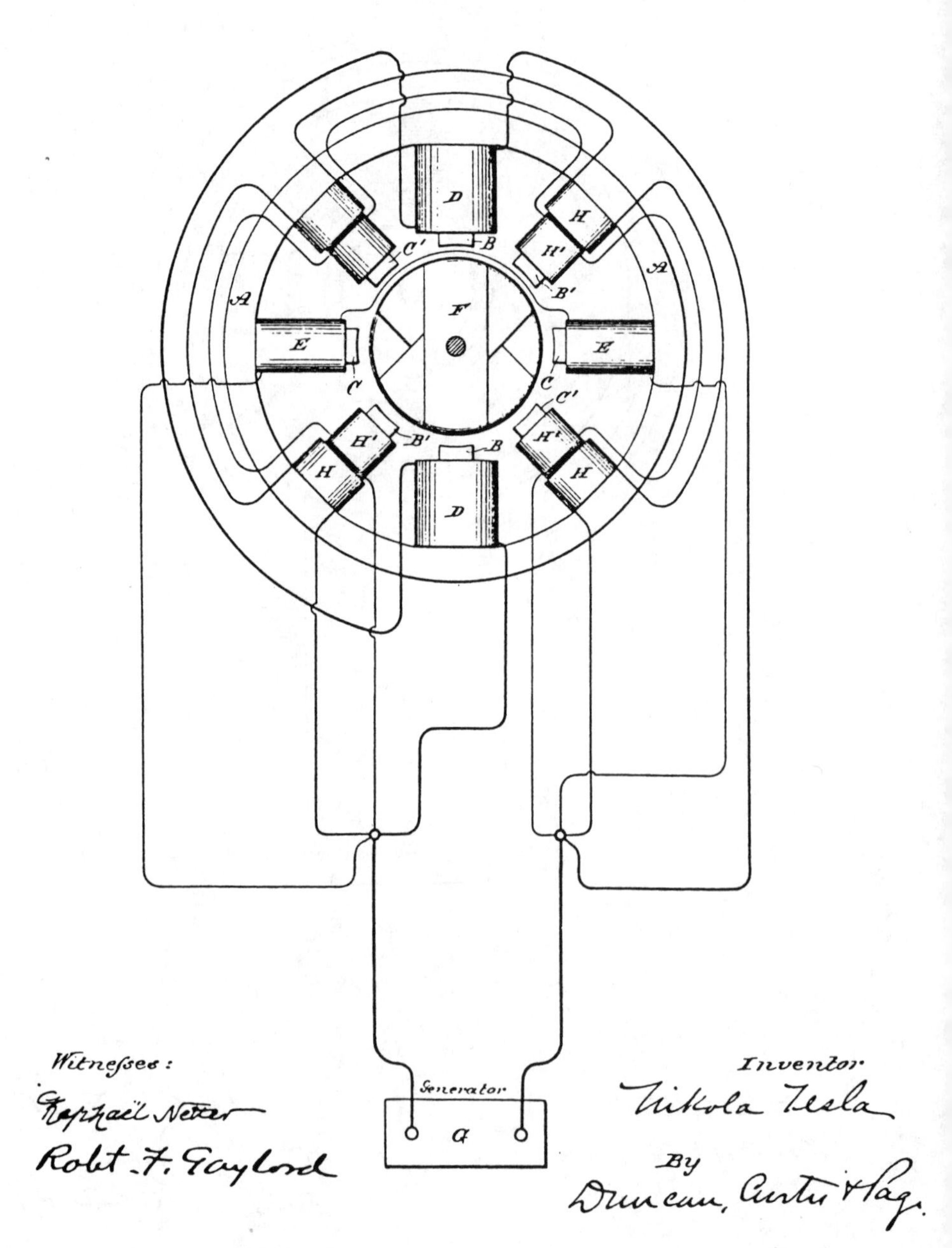

(No Model.) 2 Sheets—Sheet 1.

N. TESLA.

ELECTRO MAGNETIC MOTOR.

No. 424,036. Patented Mar. 25, 1890.

Fig. 1

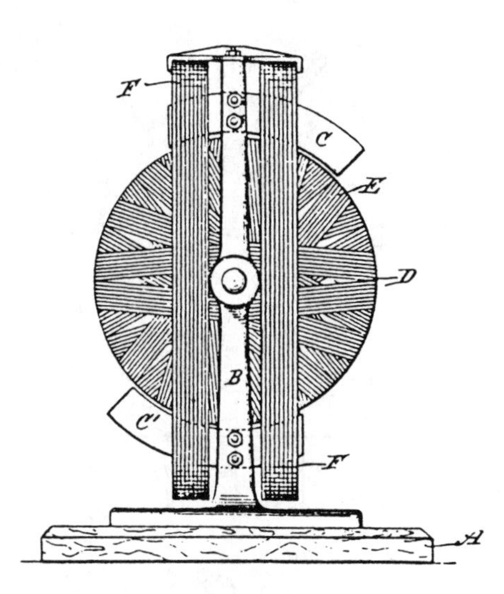

Fig: 2

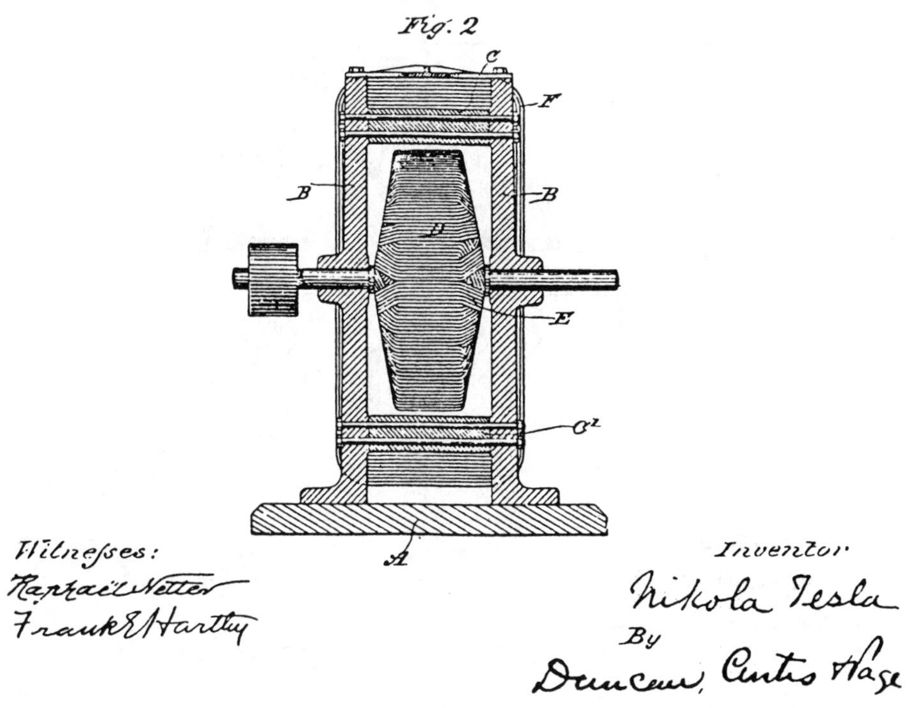

(No Model.)

N. TESLA.

ALTERNATING CURRENT ELECTRO MAGNETIC MOTOR.

No. 433,700. Patented Aug. 5, 1890.

Fig. 1

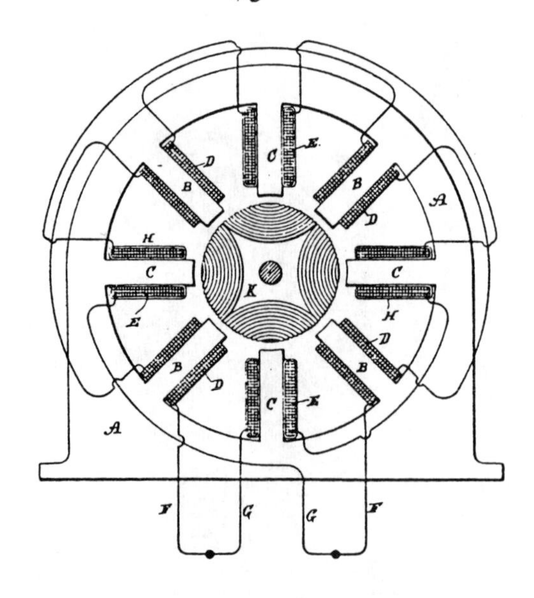

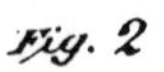

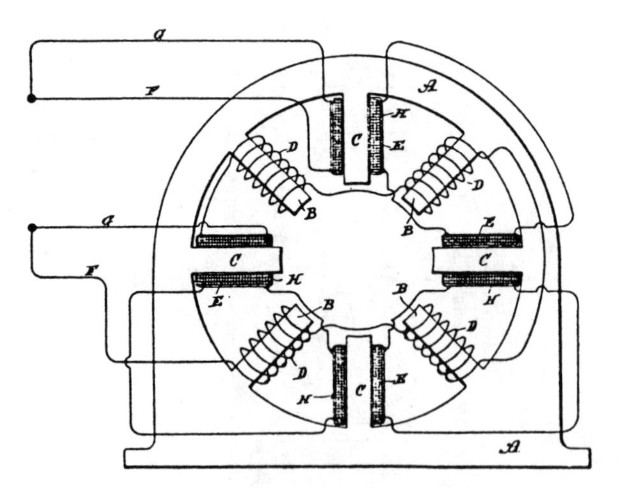

Witnesses:
Raphael Netter
Ernest Hopkinson

Inventor
Nikola Tesla
by
Duncan, Curtis & Page
Attorneys.

N. TESLA.

ELECTROMAGNETIC MOTOR.

No. 445,207. Patented Jan. 27, 1891.

Fig. 1

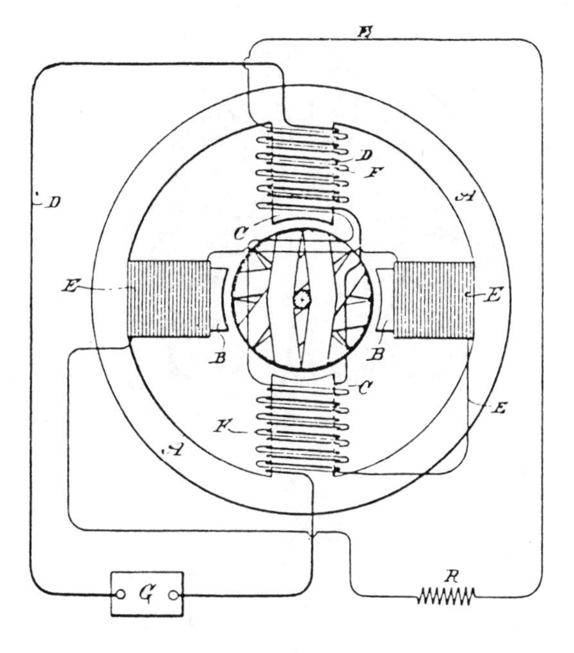

Fig. 2

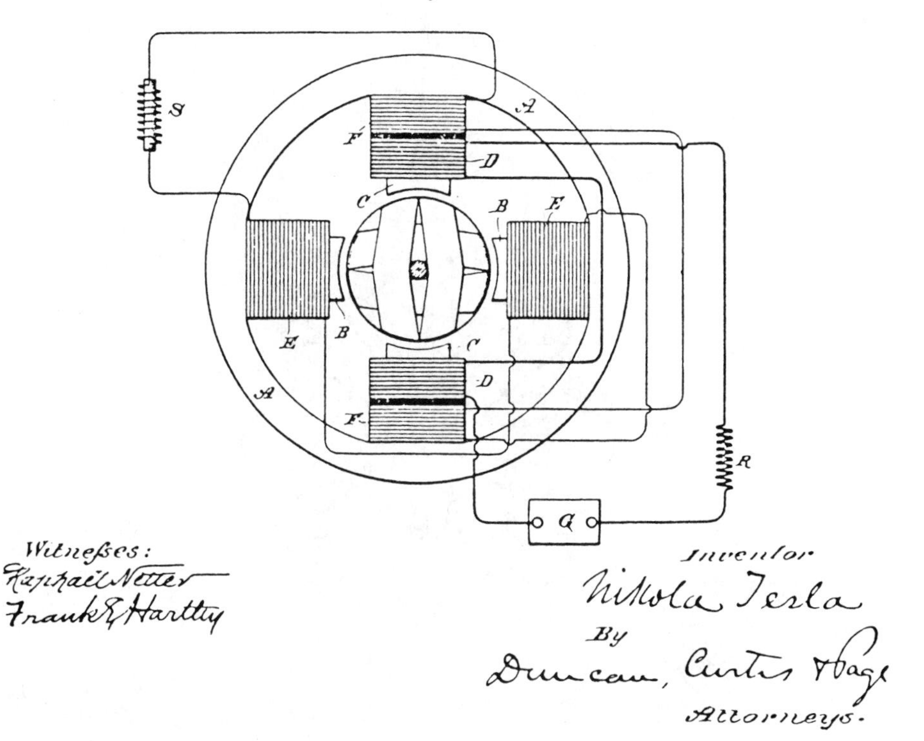

Witnesses:
Raphael Netter
Frank E. Hartley

Inventor
Nikola Tesla
By
Duncan, Curtis & Page
Attorneys.

(No Model.)

3 Sheets—Sheet 2.

N. TESLA.

SYSTEM OF ELECTRICAL TRANSMISSION OF POWER.

No. 487,796.

Patented Dec. 13, 1892.

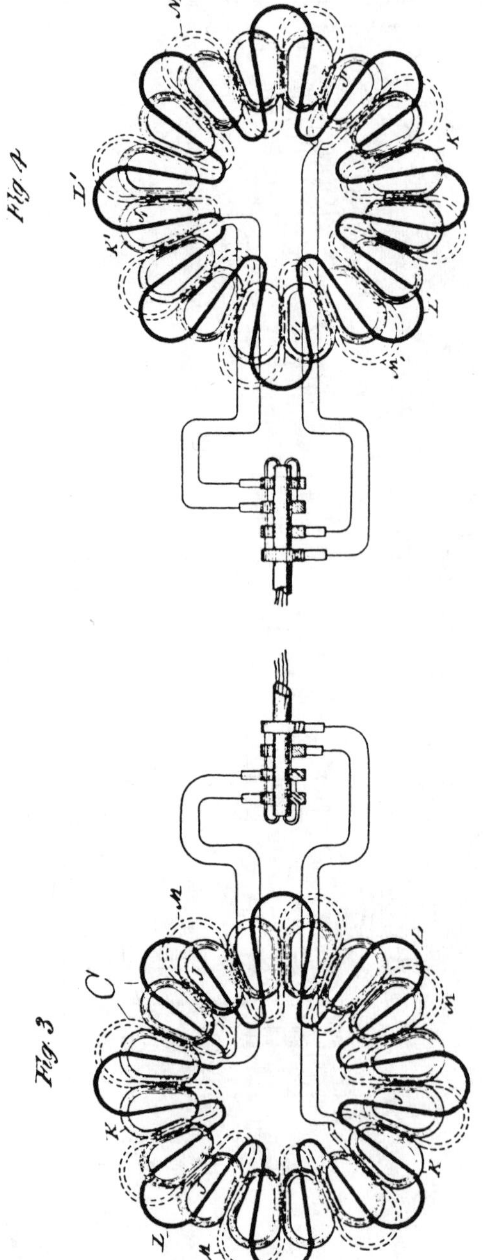

WITNESSES:

Raphael Netter

Allan W. Paige

INVENTOR

Nikola Tesla

BY

Duncan, Curtis & Page

ATTORNEYS.

(No Model.) 3 Sheets—Sheet 1.

N. TESLA.

SYSTEM OF ELECTRICAL POWER TRANSMISSION.

No. 511,560. Patented Dec. 26, 1893.

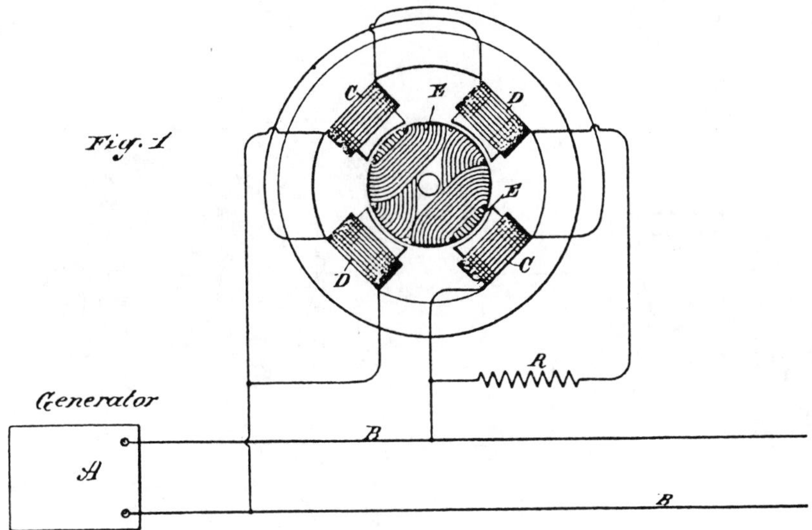

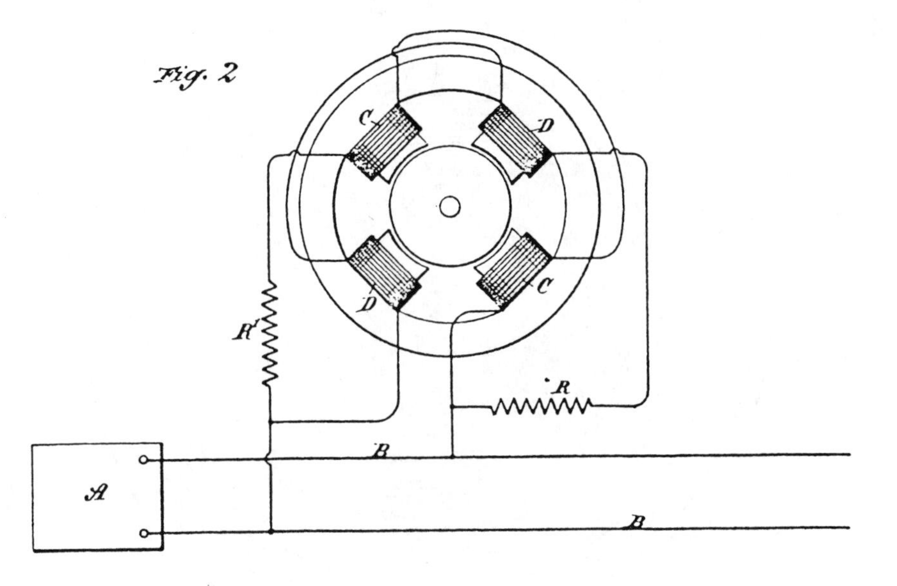

WITNESSES:

Raphaël Netter

Ernest Hopkinson

INVENTOR

Nikola Tesla

BY

Duncan, Curtis & Page

ATTORNEYS.

(No Model.)

N. TESLA.

ELECTRICAL TRANSMISSION OF POWER.

No. 511,915. Patented Jan. 2, 1894.

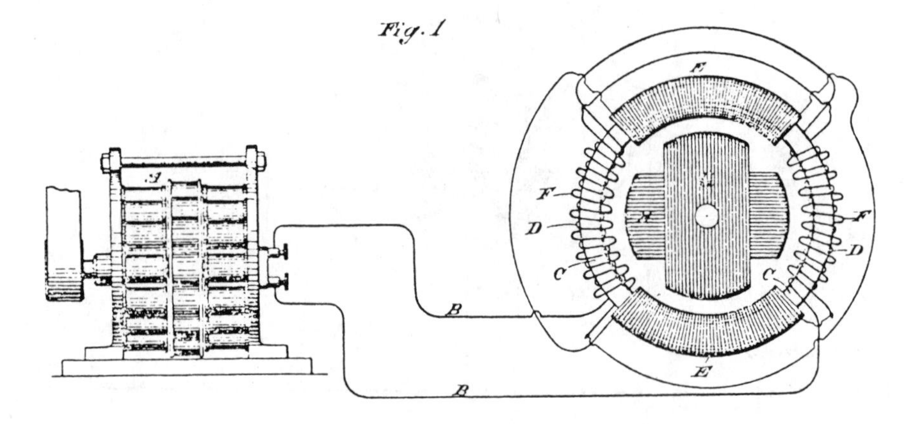

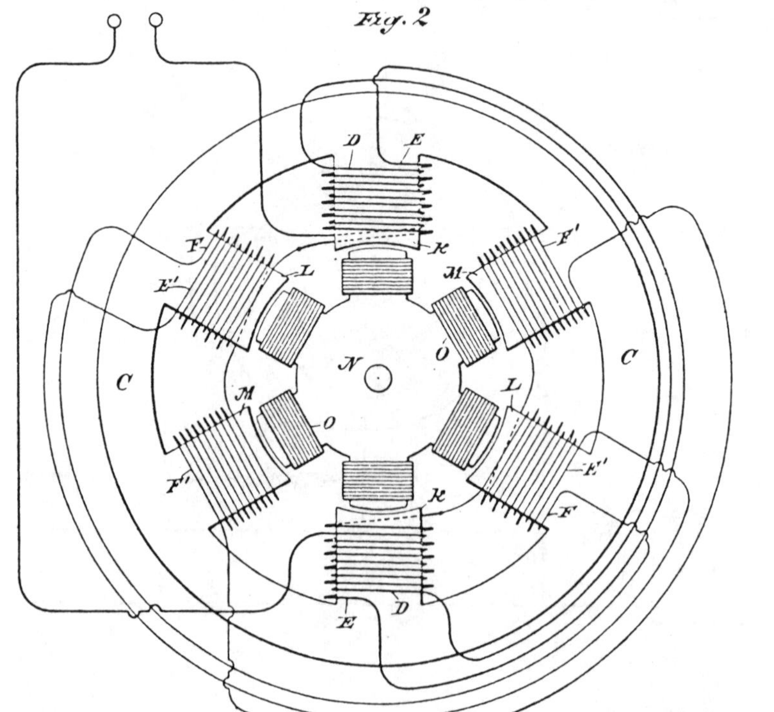

WITNESSES:

Raphaël Netter

R. F. Newbury

INVENTOR

Nikola Tesla

BY

Duncan, Curtis & Page

ATTORNEYS.

(No Model.)

N. TESLA.

COIL FOR ELECTRO MAGNETS.

No. 512,340. Patented Jan. 9, 1894.

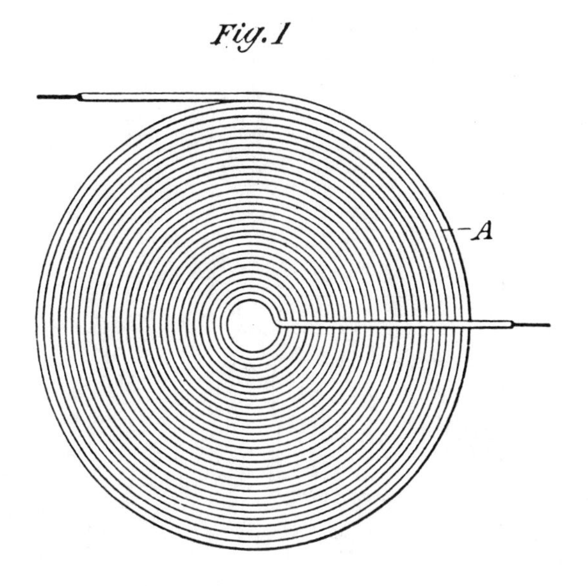

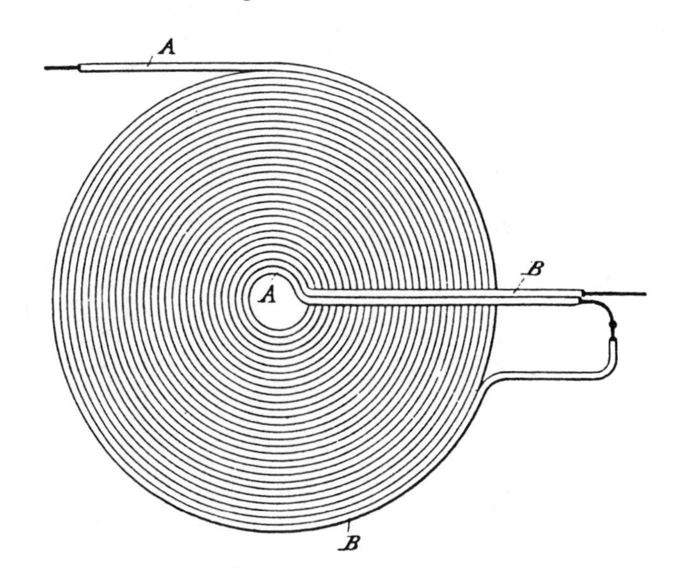

Witnesses
Raphaël Netter
James N. Catlow

Inventor
Nikola Tesla
By his Attorneys
Duncan & Page.

(No Model.)

N. TESLA.

MEANS FOR GENERATING ELECTRIC CURRENTS.

No. 514,168. Patented Feb. 6, 1894.

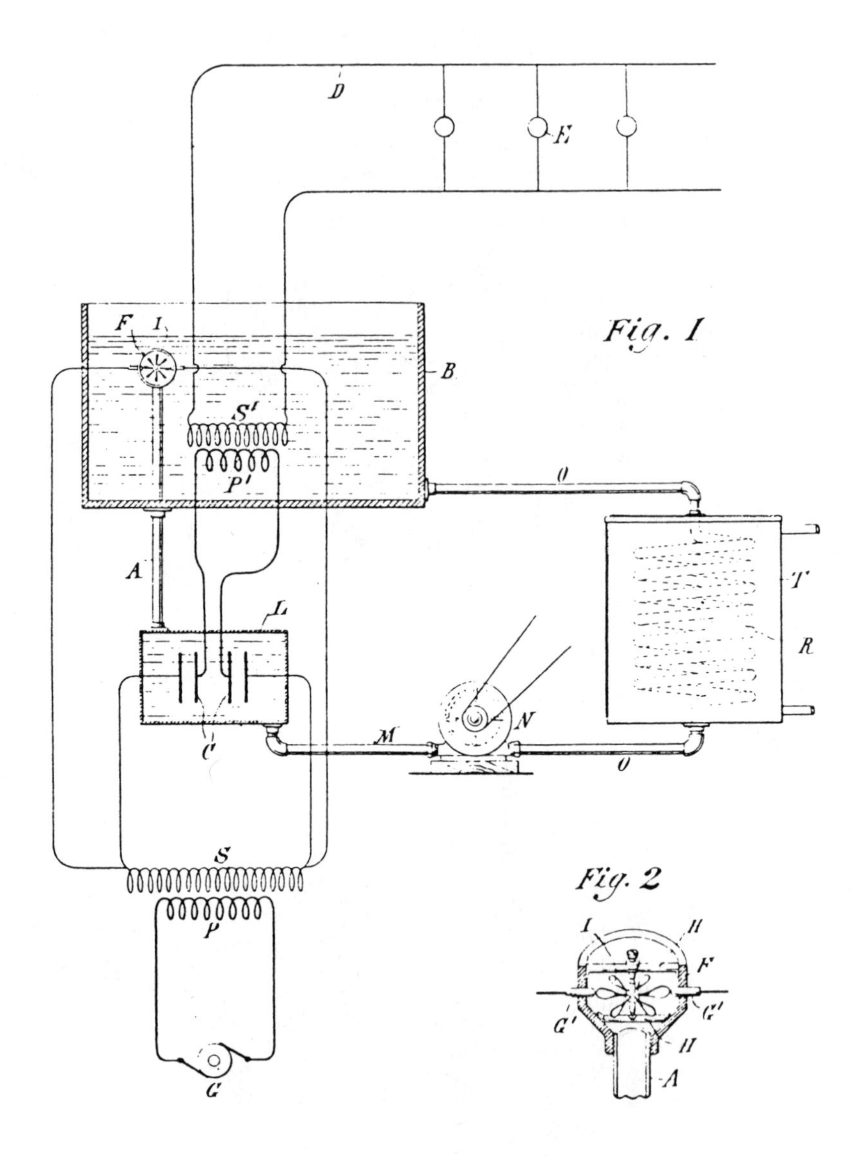

Witnesses
Raphaël Netter
James N. Catlow

Inventor
Nikola Tesla
By his Attorneys
Duncan & Page

(No Model.) 2 Sheets—Sheet 1.

N. TESLA.

ELECTRIC GENERATOR.

No. 511,916. Patented Jan. 2, 1894.

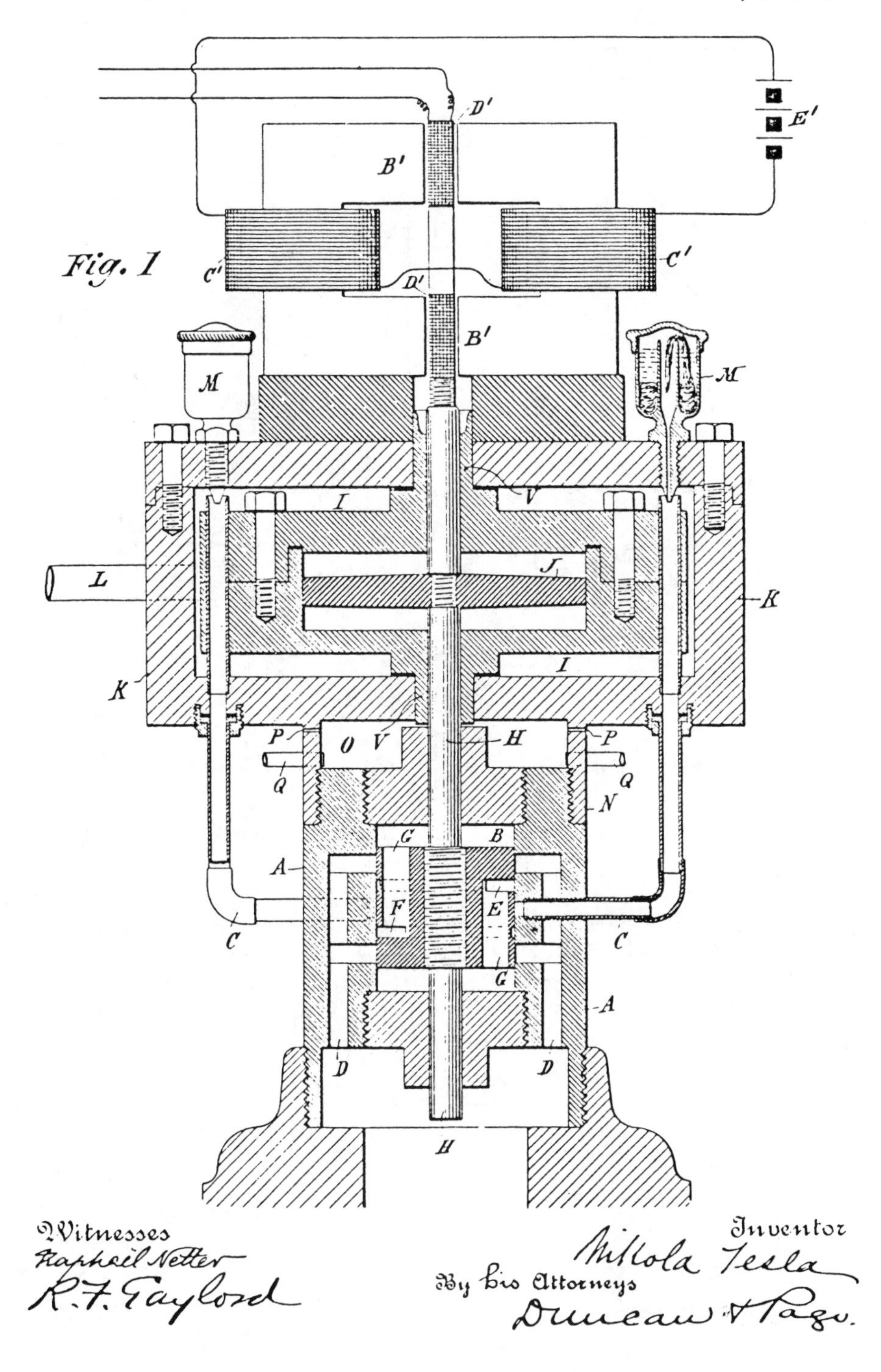

(No Model.) 2 Sheets—Sheet 2.

N. TESLA.

ELECTRIC GENERATOR.

No. 511,916. Patented Jan. 2, 1894.

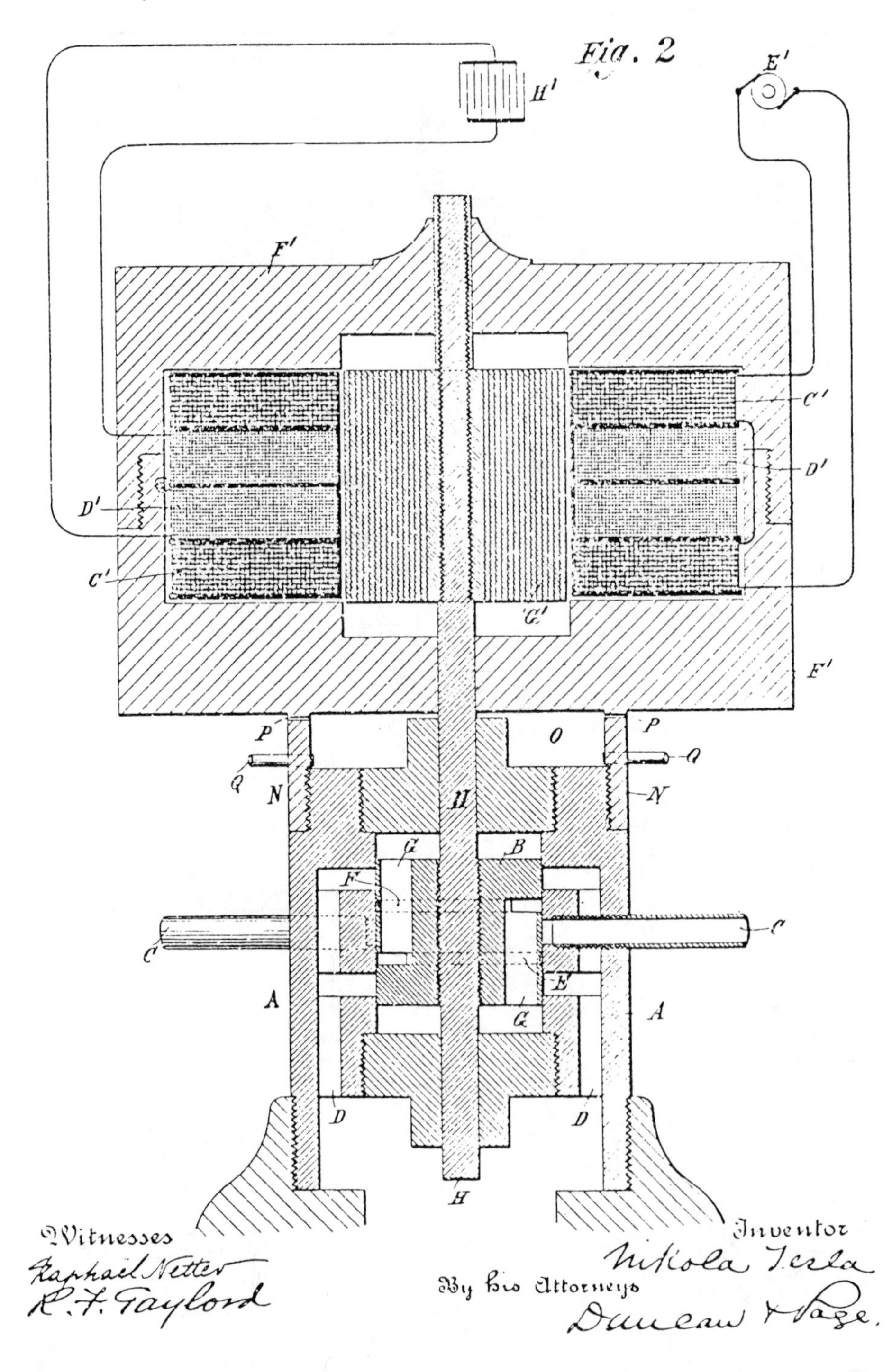

(No Model.)

N. TESLA.

ALTERNATING MOTOR.

No. 555,190. Patented Feb. 25, 1896.

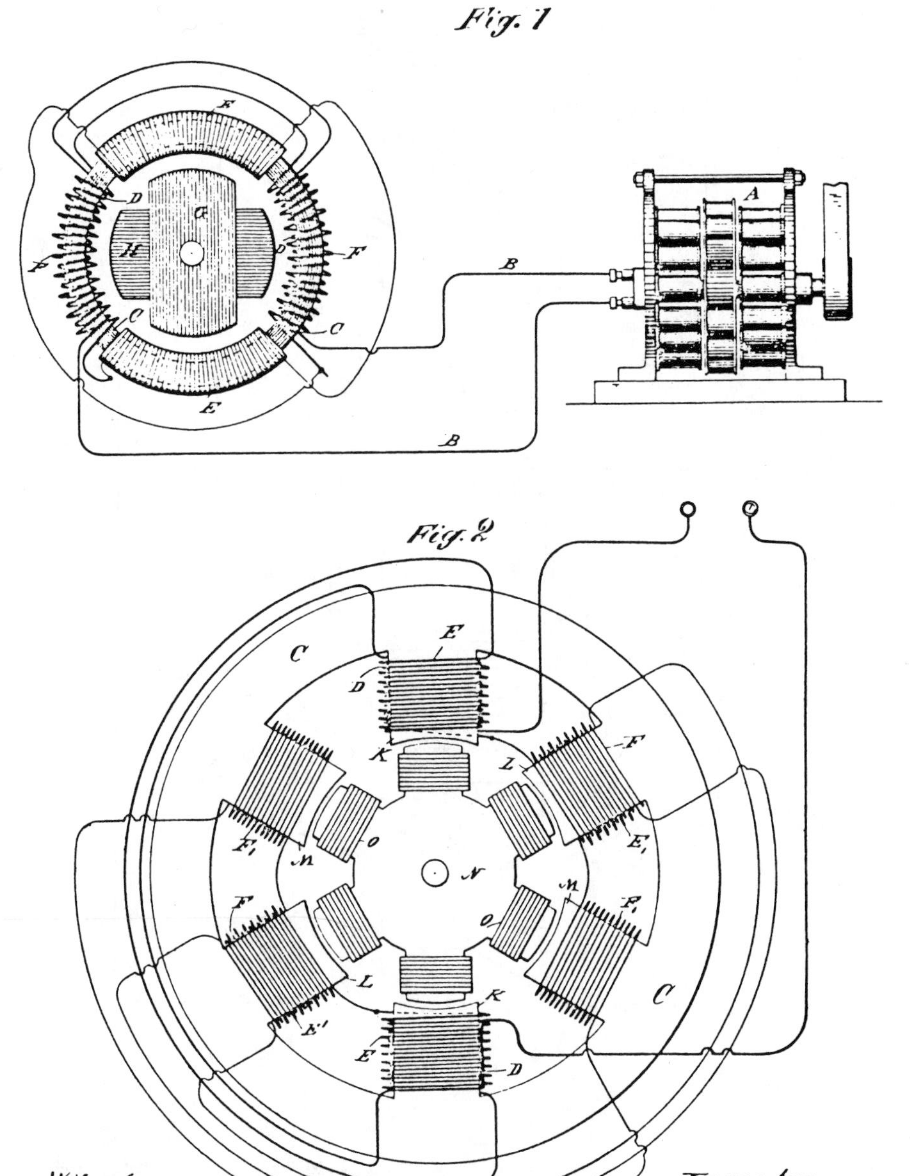

Witnesses:
Raphaël Netter
Robt. F. Gaylord

Inventor
Nikola Tesla
by
Duncan, Curtis & Page
Attorneys.

(No Model.)

2 Sheets—Sheet 1.

N. TESLA.

APPARATUS FOR PRODUCING ELECTRIC CURRENTS OF HIGH FREQUENCY AND POTENTIAL.

No. 568,176. Patented Sept. 22, 1896.

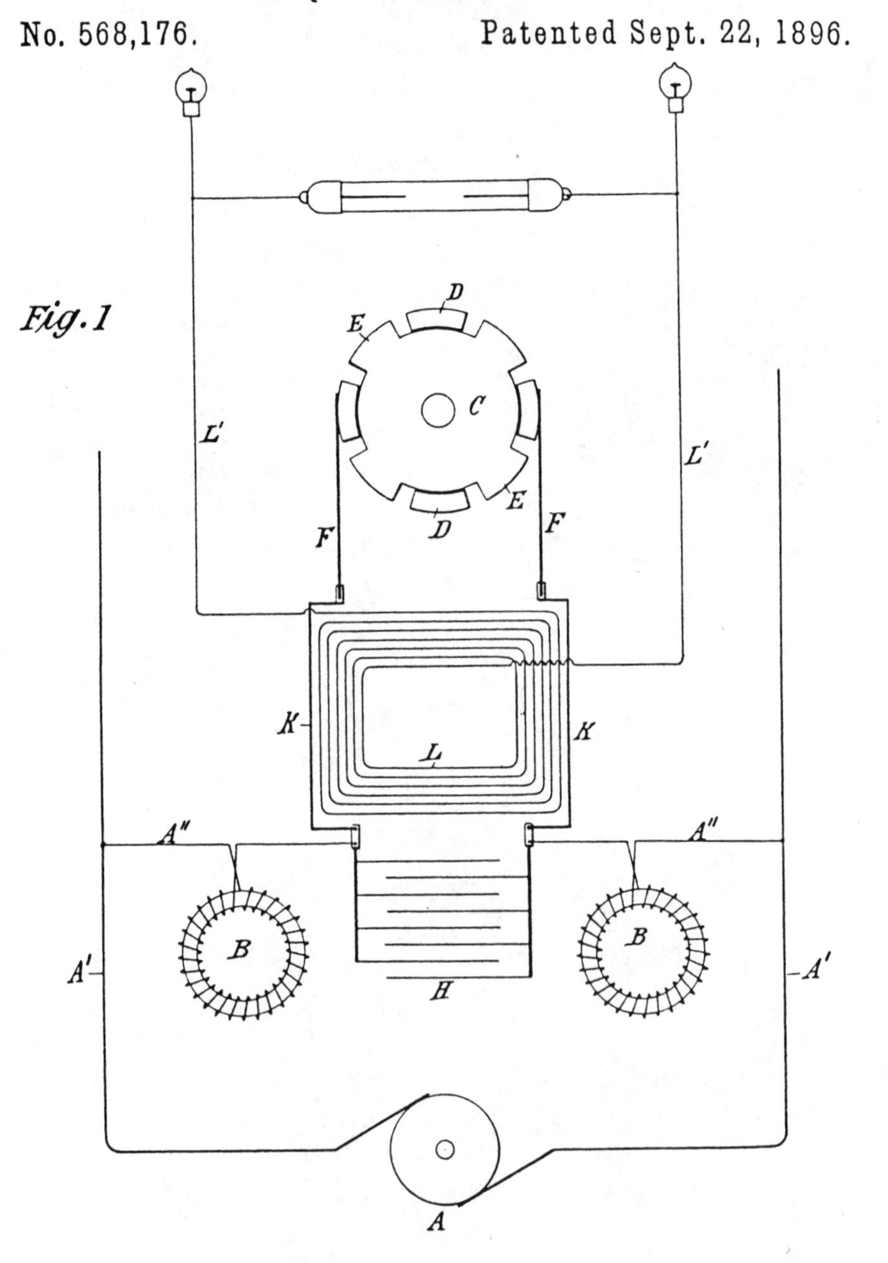

Witnesses:

Raphaël Netter

Drury W. Cooper

Nikola Tesla, Inventor

by Kerr, Curtis & Page.

Att'ys.

(No Model.) 2 Sheets—Sheet 2.

N. TESLA.

APPARATUS FOR PRODUCING ELECTRIC CURRENTS OF HIGH FREQUENCY AND POTENTIAL.

No. 568,176. Patented Sept. 22, 1896.

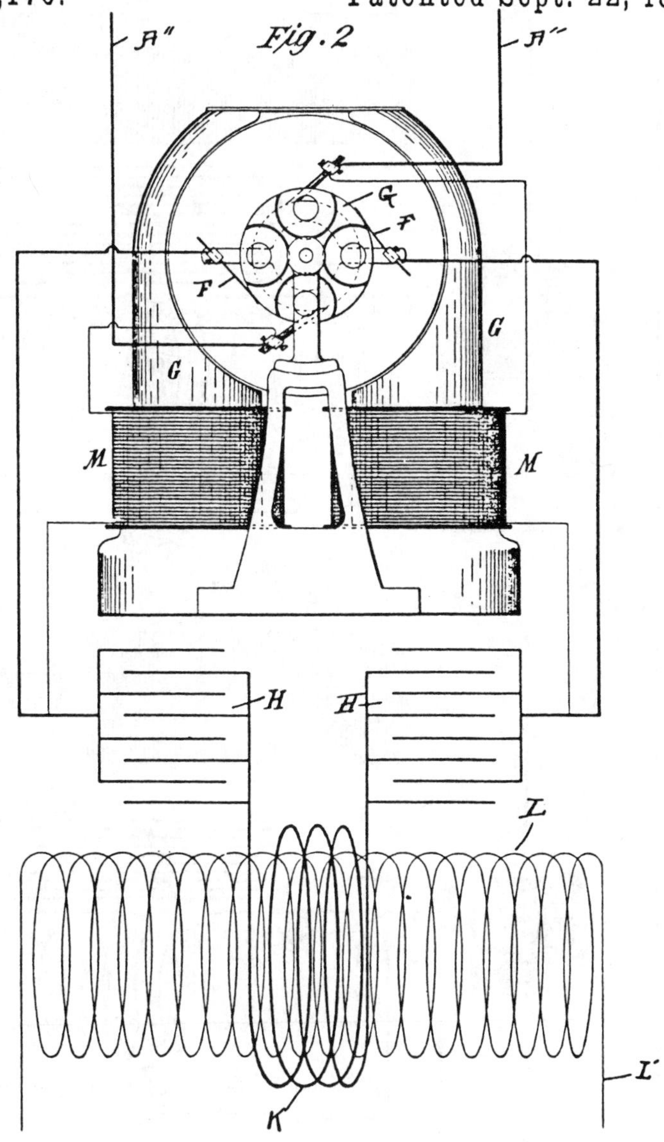

WITNESSES:

M. Lawson Dyer

Edwin B. Hopkinson.

Nikola Tesla INVENTOR

BY

Kerr, Curtis & Page

ATTORNEYS

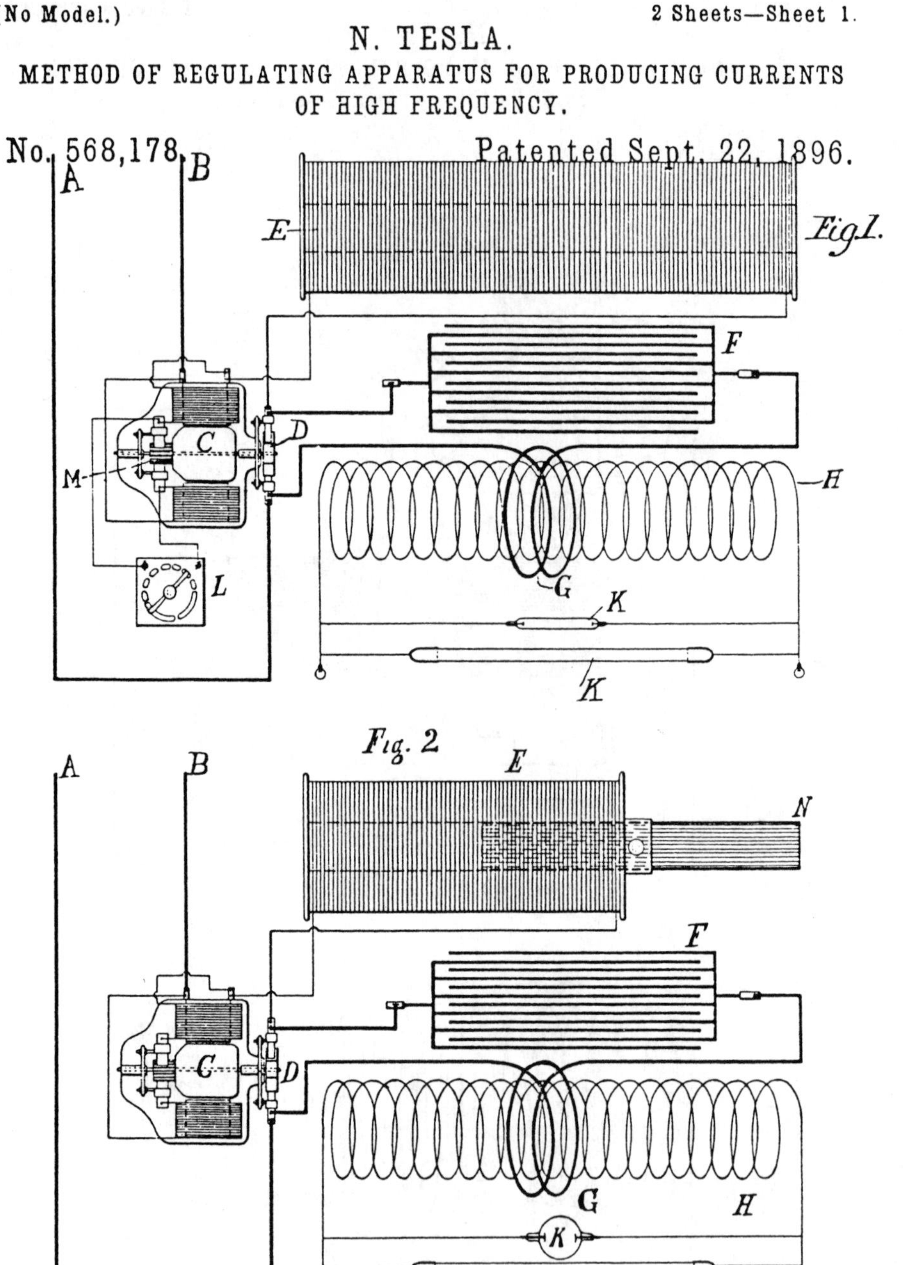

WITNESSES

Edwin B. Hopkinson
M. Lawson Dyer

INVENTOR

Nikola Tesla

BY

Kerr, Curtis & Page

ATTORNEYS

(No Model.) 2 Sheets—Sheet 1.

N. TESLA.

METHOD OF AND APPARATUS FOR PRODUCING CURRENTS OF HIGH FREQUENCY.

No. 568,179. Patented Sept. 22, 1896.

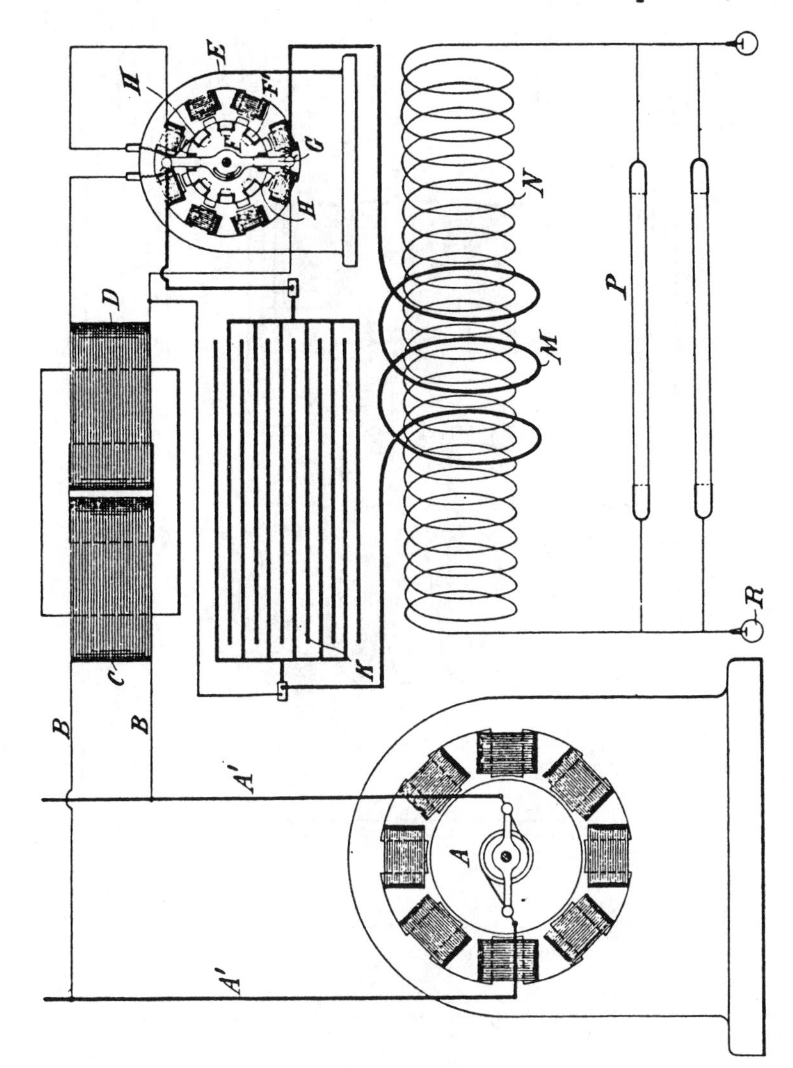

WITNESSES

Drury W. Cooper

Edwin B. Hopkinson.

INVENTOR

Nikola Tesla

BY

Kerr. Curtis & Page.

ATTORNEYS

(No Model.)

2 Sheets—Sheet 2.

N. TESLA.

METHOD OF AND APPARATUS FOR PRODUCING CURRENTS OF HIGH FREQUENCY.

No. 568,179.

Patented Sept. 22, 1896.

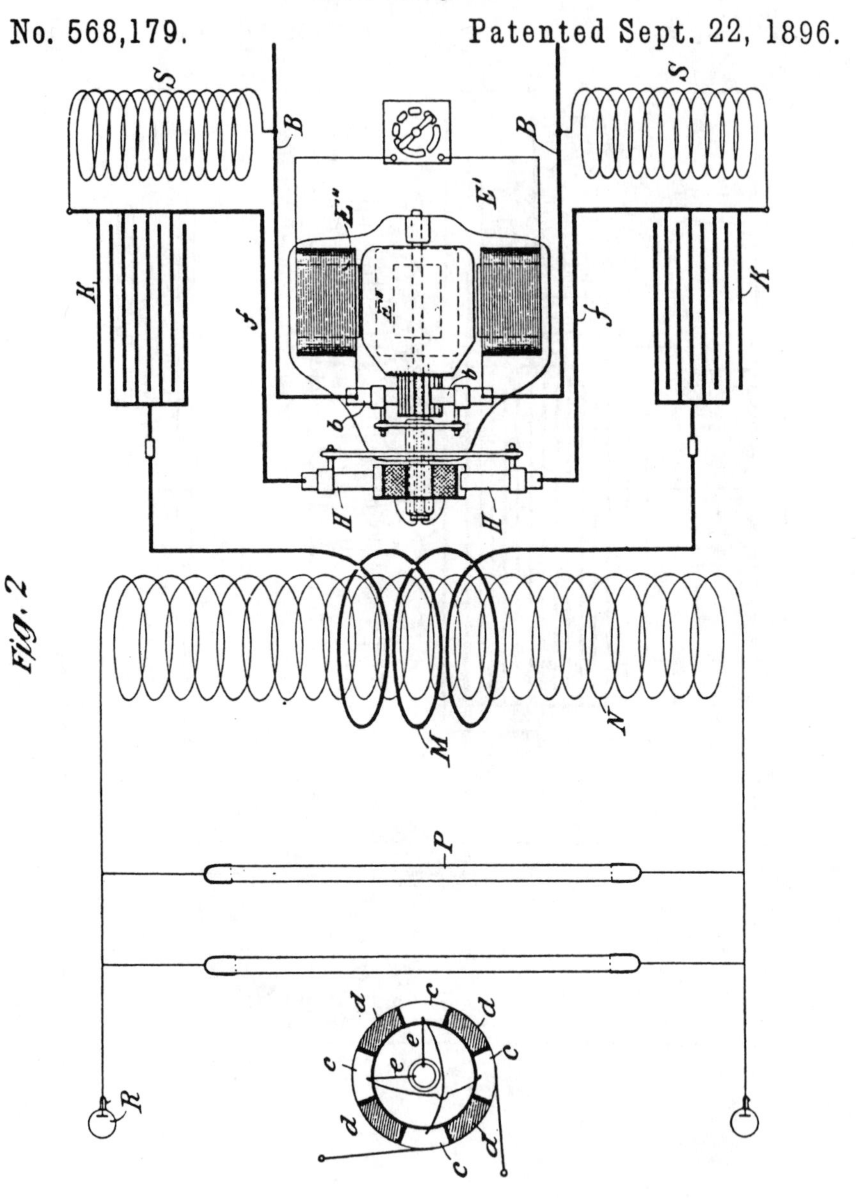

WITNESSES

Drury W. Cooper

Edwin B. Hopkinson.

INVENTOR

Nikola Tesla

BY

Kerr, Curtis & Page.

ATTORNEYS

(No Model.) 2 Sheets—Sheet 2

N. TESLA.

APPARATUS FOR PRODUCING ELECTRICAL CURRENTS OF HIGH FREQUENCY.

No. 568,180. Patented Sept. 22, 1896.

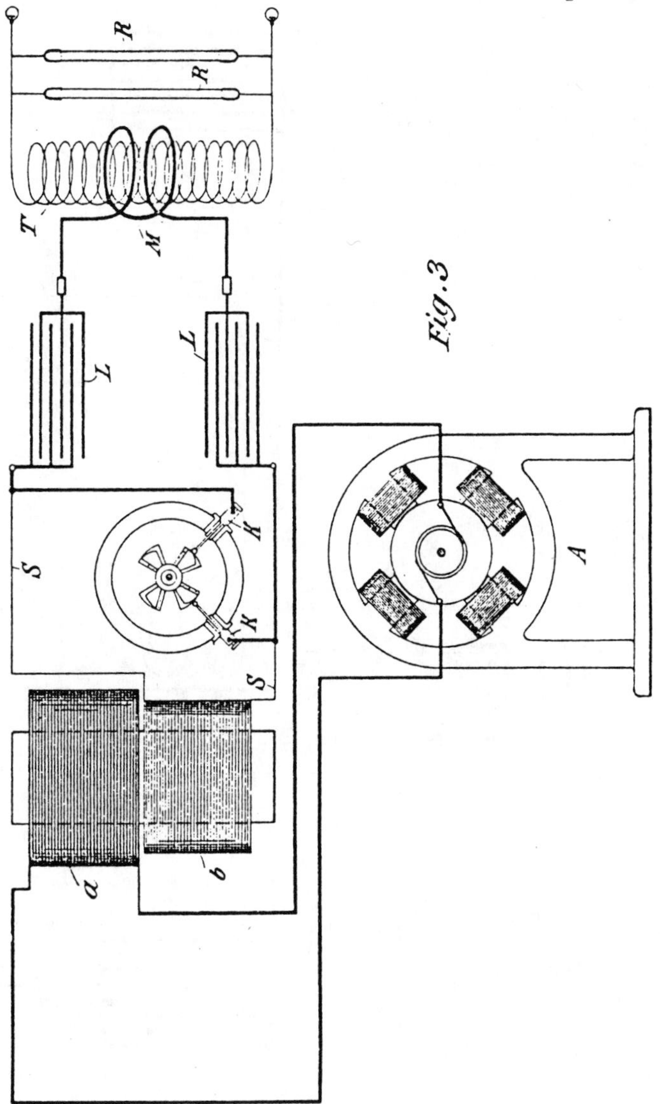

WITNESSES:

Edwin B. Hopkinson.

Benjamin Garstorf

Nikola Tesla INVENTOR

BY

Kerr, Curtis & Page, ATTORNEYS

(No Model.) 2 Sheets—Sheet 1

N. TESLA.

APPARATUS FOR PRODUCING ELECTRIC CURRENTS OF HIGH FREQUENCY.

No. 577,670. Patented Feb. 23, 1897.

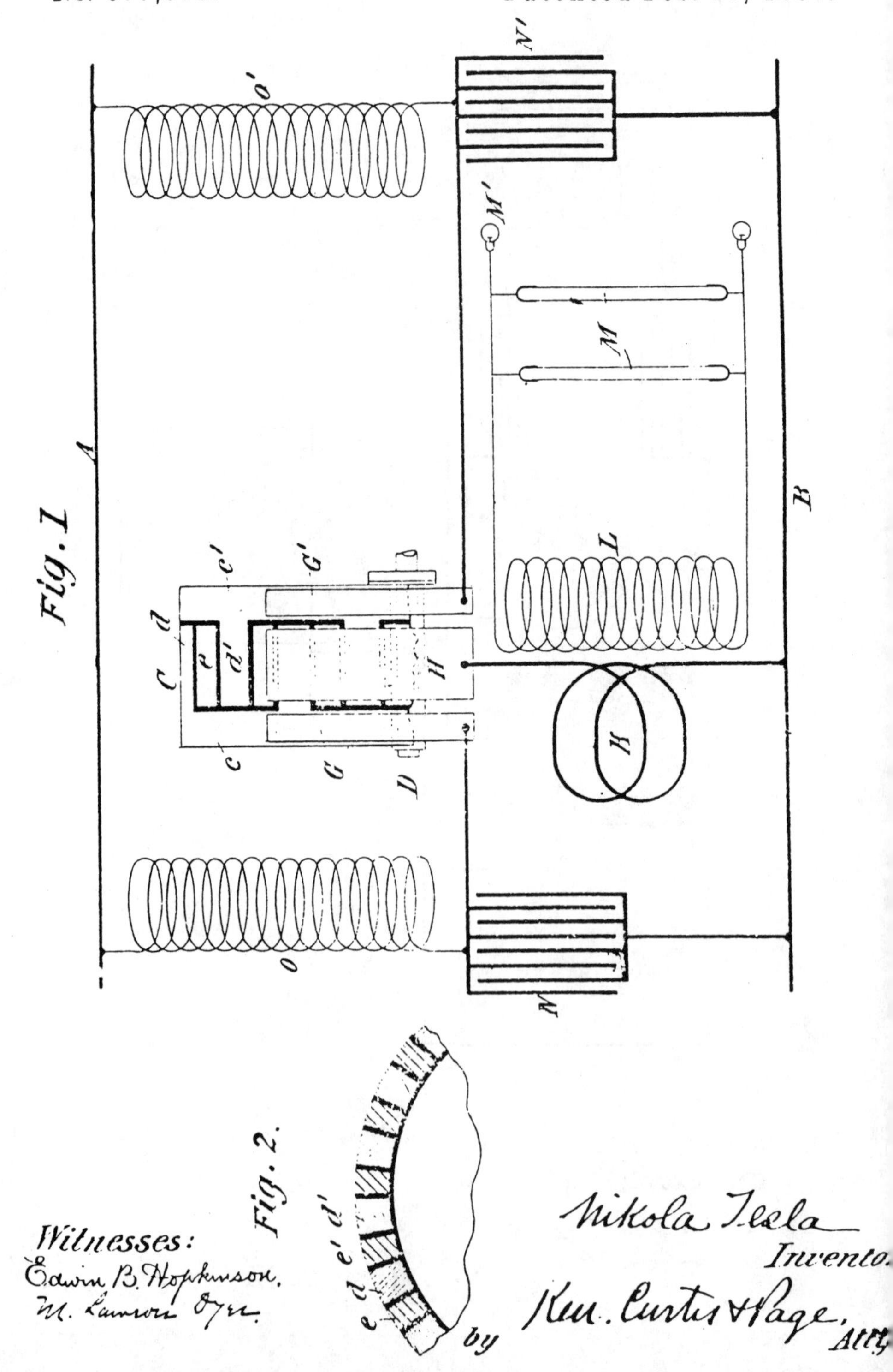

(No Model.) 2 Sheets—Sheet 2.

N. TESLA.

APPARATUS FOR PRODUCING ELECTRIC CURRENTS OF HIGH FREQUENCY.

No. 577,670. Patented Feb. 23, 1897.

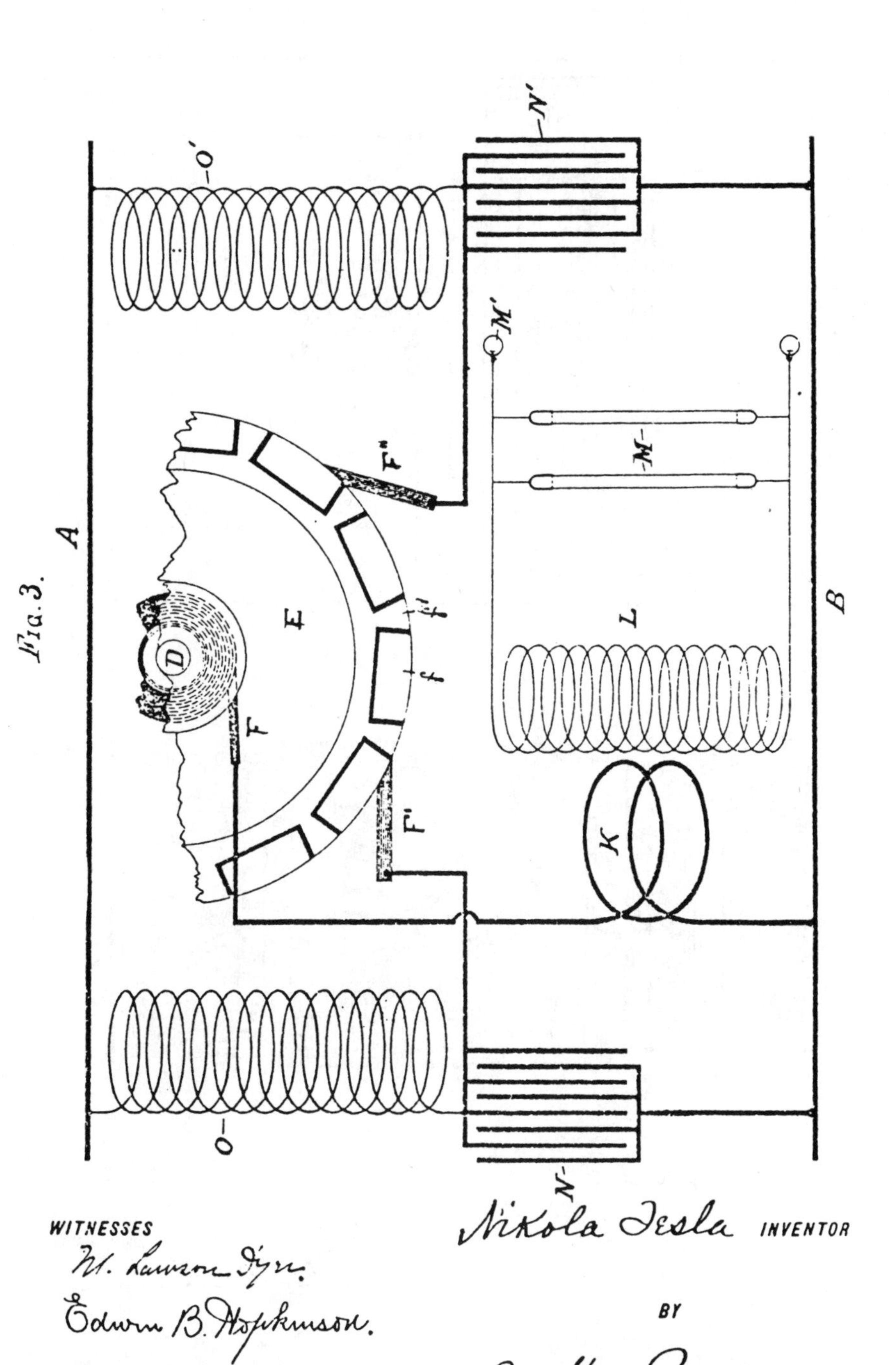

(No Model)

N. TESLA.

APPARATUS FOR PRODUCING CURRENTS OF HIGH FREQUENCY.

No. 583,953. Patented June 8, 1897

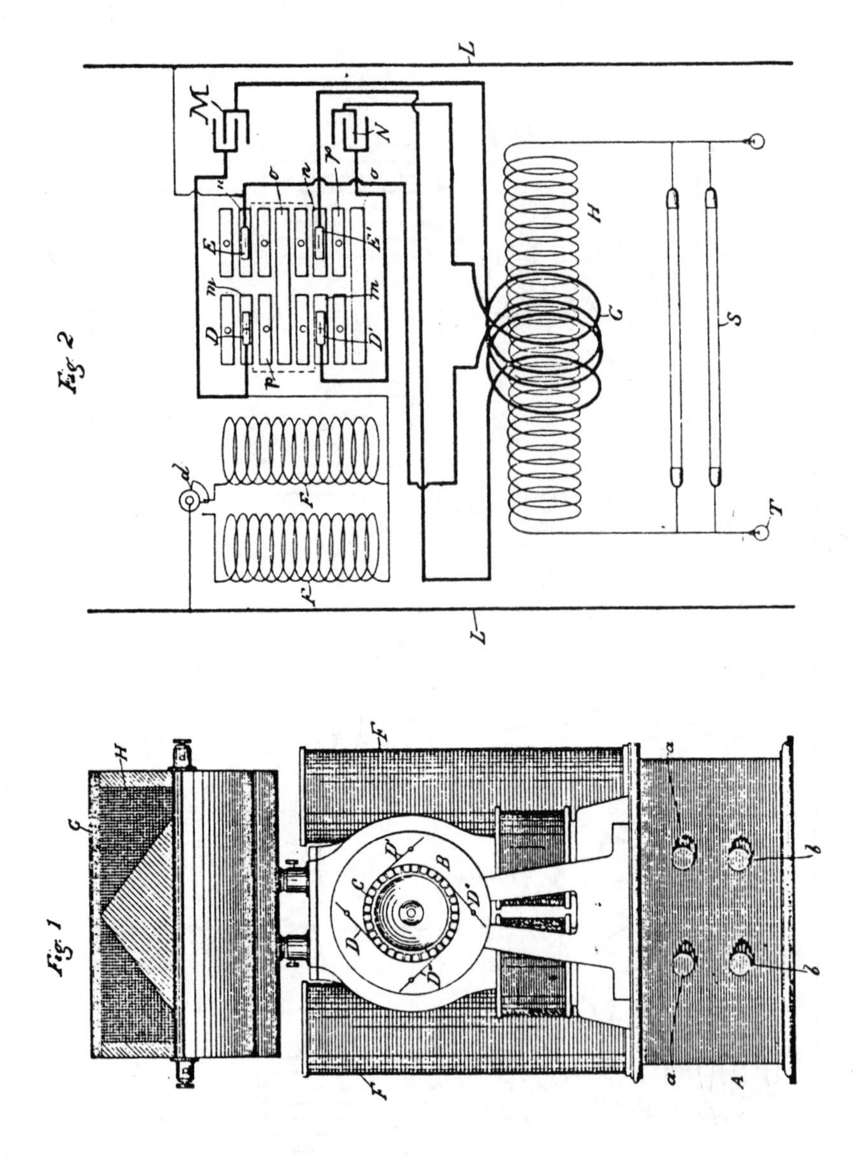

WITNESSES

G. B. Lewis.

Edwin B. Hopkinson.

INVENTOR

Nikola Tesla

BY

Kerr. Curtis & Page

ATTORNEYS.

No. 609,245. Patented Aug. 16, 1898.

N. TESLA.

ELECTRICAL CIRCUIT CONTROLLER.

(Application filed Dec. 2, 1897.)

(No Model.)

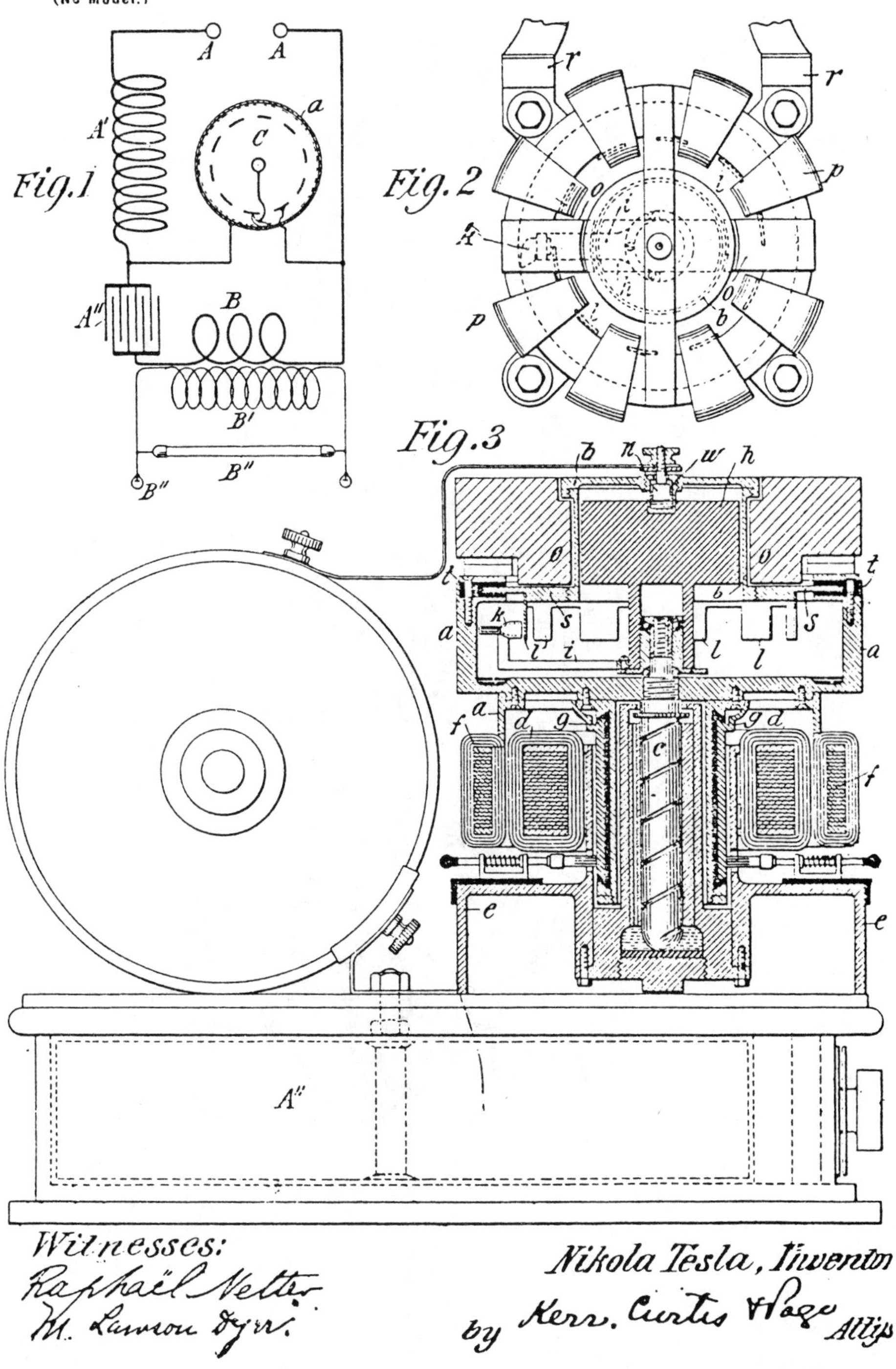

Witnesses:
Raphaël Netter
M. Lawson Dyer.

Nikola Tesla, Inventor
by Kerr, Curtis & Page Atty's

No. 613,809. Patented Nov. 8, 1898.

N. TESLA.

METHOD OF AND APPARATUS FOR CONTROLLING MECHANISM OF MOVING VESSELS OR VEHICLES.

(No Model.) 5 Sheets—Sheet I.

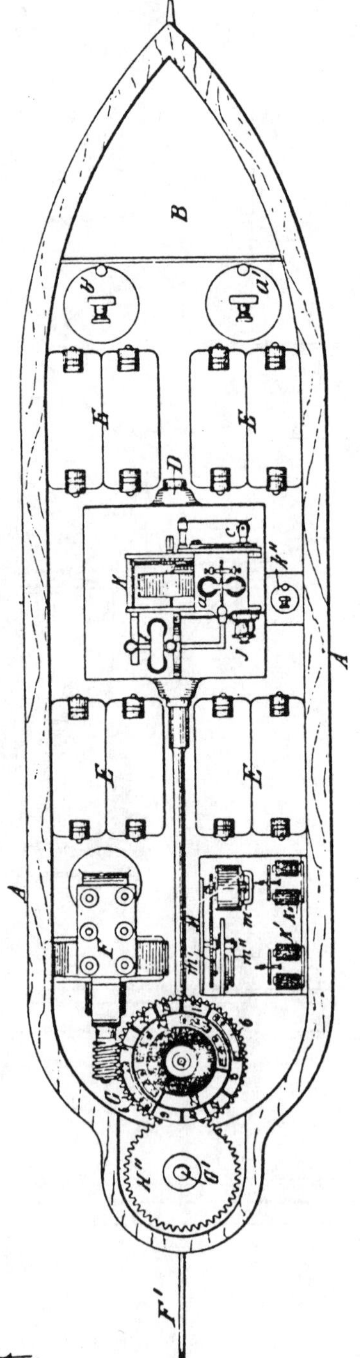

Fig. 1

Witnesses:
Raphaël Netter
George Scherff.

Inventor
Nikola Tesla

Chapter 5

TRANSMISSION OF ELECTRICAL ENERGY WITHOUT WIRES

(Communicated to the Thirtieth Anniversary Number of the *Electrical World and Engineer*, March 5, 1904.)

BY NIKOLA TESLA.

IT is impossible to resist your courteous request extended on an occasion of such moment in the life of your journal. Your letter has vivified the memory of our beginning friendship, of the first imperfect attempts and undeserved successes, of kindnesses and misunderstandings. It has brought painfully to my mind the greatness of early expectations, the quick flight of time, and alas! the smallness of realizations. The following lines which, but for your initiative, might not have been given to the world for a long time yet, are an offering in the friendly spirit of old, and my best wishes for your future success accompany them.

Towards the close of 1898 a systematic research, carried on for a number of years with the object of perfecting a method of transmission of electrical energy through the natural medium, led me to recognize three important necessities: First, to develop a transmitter of great power; second, to perfect means for individualizing and isolating the energy transmitted; and, third, to ascertain the laws of propagation

of currents through the earth and the atmosphere. Various reasons, not the least of which was the help proffered by my friend Leonard E. Curtis and the Colorado Springs Electric Company, determined me to select for my experimental investigations the large plateau, two thousand meters above sea-level, in the vicinity of that delightful resort, which I reached late in May, 1899. I had not been there but a few

Experimental Laboratory, Colorado Springs.

days when I congratulated myself on the happy choice and I began the task, for which I had long trained myself, with a grateful sense and full of inspiring hope. The perfect purity of the air, the unequaled beauty of the sky, the imposing sight of a high mountain range, the quiet and restfulness of the place—all around contributed to make the conditions for scientific observation ideal. To this was added the exhilarating influence of a glorious climate and a singular sharpening of the senses. In those regions the organs undergo perceptible physical changes. The eyes assume an extraordinary limpidity, improving vision; the ears dry out and become more susceptible to sound. Objects can be clearly distinguished there at distances such that I prefer to have them told by someone else, and I have heard—this I can venture to vouch for—the claps of thunder seven and eight hundred kilometers away. I might have done better still, had it not been tedious to wait for the sounds to arrive, in definite intervals, as heralded precisely by an electrical indicating apparatus—nearly an hour before.

In the middle of June, while preparations for other work were going on, I arranged one of my receiving transformers with the view of determining in a novel manner, experimentally, the electric potential of the globe and studying its periodic and casual fluctuations. This formed part of a plan carefully mapped out in advance. A highly sensitive, self-restorative device, controlling a recording instrument, was included in the secondary circuit, while the primary was connected to the ground and an elevated terminal of adjustable capacity. The variations of potential gave rise to electric surgings in the primary; these generated secondary currents, which in turn affected the sensitive device and recorder in proportion to their intensity. The earth was found to be, literally, alive with electrical vibrations, and soon I was deeply absorbed in this interesting investigation. No better opportunities for such observations as I intended to make could be found anywhere. Colorado is a country famous for the natural displays of electric force. In that dry and rarefied atmosphere the sun's rays beat the objects with fierce intensity. I raised steam, to a dangerous pressure, in barrels filled with concentrated salt solution, and the tin-foil coatings of some of my elevated terminals shriveled up in the fiery blaze. An experimental high-tension trans-

Above: Tesla's tower at Wardenclyffe for sending messages across the Atlantic and electricity into the atmosphere as it appeared in 1904. Left: The letterhead for his stationary, promising "ten million Horsepower" of "Electrical oscillator activity."

former, carelessly exposed to the rays of the setting sun, had most of its insulating compound melted out and was rendered useless. Aided by the dryness and rarefaction of the air, the water evaporates as in a boiler, and static electricity is developed in abundance. Lightning discharges are, accordingly, very frequent and sometimes of inconceivable violence. On one occasion approximately twelve thousand discharges occurred in two hours, and all in a radius of certainly less than fifty kilometers from the laboratory. Many of them resembled gigantic trees of fire with the trunks up or down. I never saw fire balls, but as a compensation for my disappointment I succeeded later in determining the mode of their formation and producing them artificially.

In the latter part of the same month I noticed several times that my instruments were affected stronger by discharges taking place at great distances than by those near by. This puzzled me very much. What was the cause? A number of observations proved that it could not be due to the differences in the intensity of the individual discharges, and I readily ascertained that the phenomenon was not the result of a varying relation between the periods of my receiving circuits and those of the terrestrial disturbances. One night, as I was walking home with an assistant, meditating over these experiences, I was suddenly staggered by a thought. Years ago, when I wrote a chapter of my lecture before the Franklin Institute and the National Electric Light Association, it had presented itself to me, but I had dismissed it as absurd and impossible. I banished it again. Nevertheless, my instinct was aroused and somehow I felt that I was nearing a great revelation.

It was on the third of July—the date I shall never forget—when I obtained the first decisive experimental evidence of a truth of overwhelming importance for the advancement of humanity. A dense mass of strongly charged clouds gathered in the west and towards the evening a violent storm broke loose which, after spending much of its fury in the mountains, was driven away with great velocity over the plains. Heavy and long persisting arcs formed almost in regular time intervals. My observations were now greatly facilitated and rendered more accurate by the experiences already gained. I was able to handle my instruments quickly and I was prepared. The recording apparatus being properly adjusted, its indications became fainter and fainter with the increasing distance of the storm, until they ceased altogether. I was watching in eager expectation. Surely enough, in a little while the indications again began, grew stronger and stronger and, after passing through a maximum, gradually decreased and ceased once more. Many times, in regularly recurring intervals, the same actions were repeated until the storm which, as evident from simple computations, was moving with nearly constant speed, had retreated to a distance of about three hundred kilometers. Nor did these strange actions stop then, but continued to manifest themselves with undiminished force. Subsequently, similar observations were also made by my assistant, Mr. Fritz Lowenstein, and shortly afterward several admirable opportunities presented themselves which brought out, still more forcibly, and unmistakably, the true nature of the wonderful phenomenon. No doubt whatever remained: I was observing stationary waves.

As the source of disturbances moved away the receiving circuit came successively upon their nodes and loops. Impossible as it seemed, this planet, despite its vast extent, behaved like a conductor of limited dimensions. The tremendous significance of this fact in the transmission of energy by my system had already become quite clear to me. Not only was it practicable to send telegraphic messages to

Experimental Laboratory, Colorado Springs.

any distance without wires, as I recognized long ago, but also to impress upon the entire globe the faint modulations of the human voice, far more still, to transmit power, in unlimited amounts, to any terrestrial distance and almost without any loss.

With these stupendous possibilities in sight, with the experimental evidence before me that their realization was henceforth merely a question of expert knowledge, patience and skill, I attacked vigorously the development of my magnifying transmitter, now, however, not so much with the original intention of producing one of great power, as with the object of learning how to construct the best one. This is, essentially, a circuit of very high self-induction and small resistance which in its arrangement, mode of excitation and action, may be said to be the diametrical opposite of a transmitting circuit typical of telegraphy by Hertzian or electromagnetic radiations. It is difficult to form an adequate idea of the marvelous power of this unique appliance, by the aid of which the globe will be transformed. The electromagnetic radiations being reduced to an insignificant quantity, and proper conditions of resonance maintained, the circuit acts like an immense pendulum, storing indefinitely the energy of the primary exciting impulses and impressions upon the earth and its conducting atmosphere uniform harmonic oscillations of intensities which, as actual tests have shown, may be pushed so far as to surpass those attained in the natural displays of static electricity.

Simultaneously with these endeavors, the means of individualization and isolation were gradually improved. Great importance was attached to this, for it was found that simple tuning was not sufficient to meet the vigorous practical requirements. The fundamental idea of employing a number of distinctive elements, co-operatively associated, for the purpose of isolating energy transmitted, I trace directly to my perusal of Spencer's clear and suggestive exposition of

N. TESLA.

APPARATUS FOR TRANSMITTING ELECTRICAL ENERGY.

APPLICATION FILED JAN. 18, 1902. RENEWED MAY 4, 1907.

1,119,732. Patented Dec. 1, 1914.

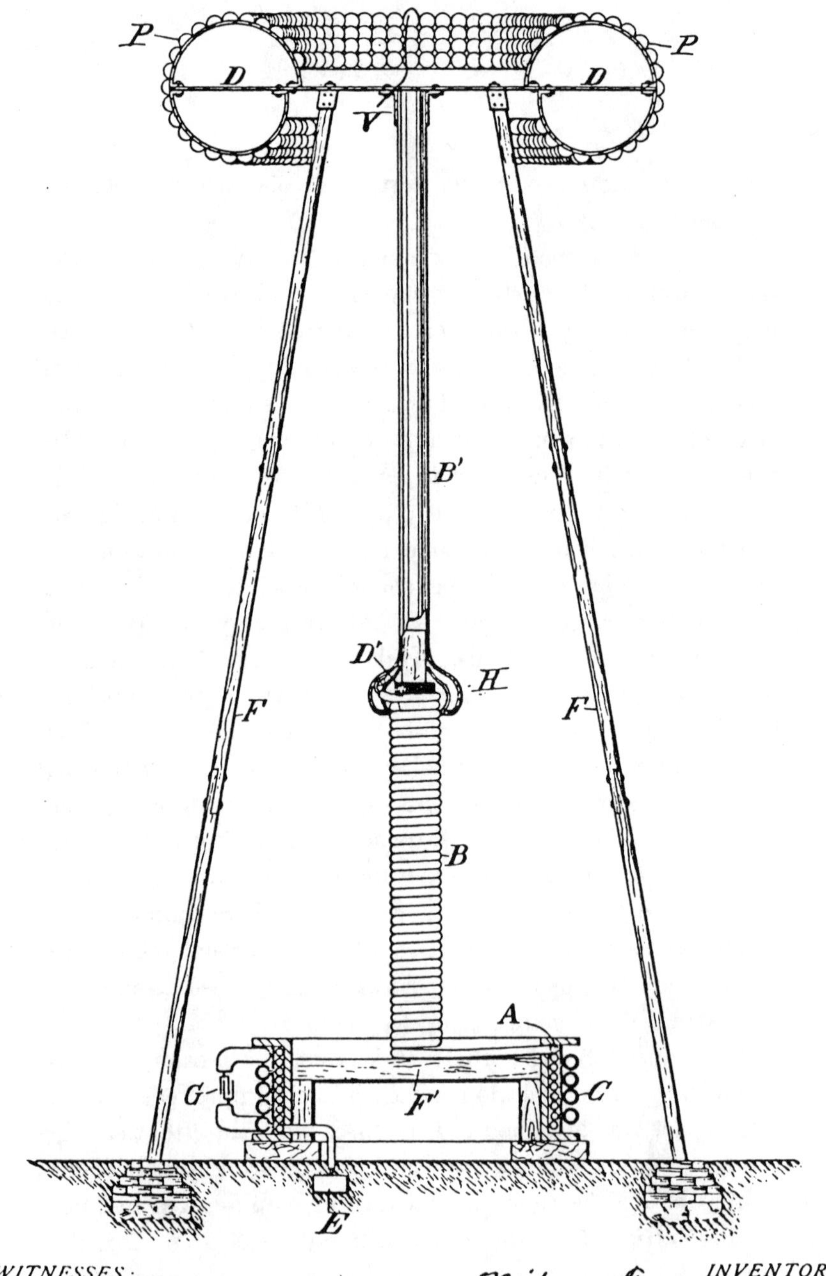

WITNESSES:

M. Lawson Dyer

Benjamin Miller.

Nikola Tesla, INVENTOR,

BY Kerr, Page & Cooper, his ATTORNEYS.

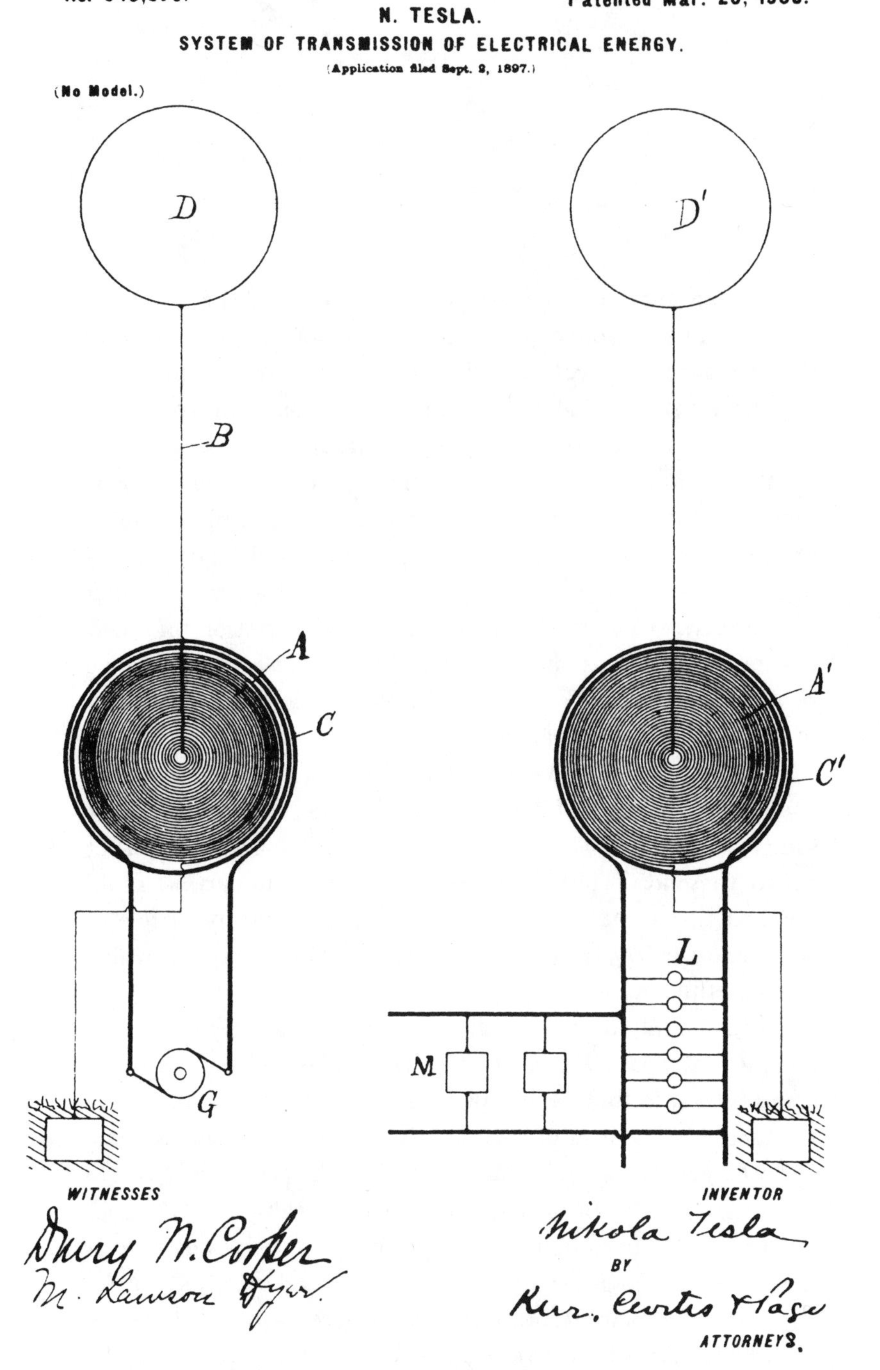

Tesla's perfected system of wireless transmission with four tuned circuits was described in U.S. Patent numbers 645,576 (March 20, 1900) and 649,621 (May15, 1900) . The applications were filed on Sept. 2, 1897.

the human nerve mechanism. The influence of this principle on the transmission of intelligence, and electrical energy in general, cannot as yet be estimated, for the art is still in the embryonic stage; but many thousands of simultaneous telegraphic and telephonic messages, through one single conducting channel, natural or artificial, and without serious mutual interference, are certainly practicable, while millions are possible. On the other hand, any desired degree of individualization may be secured by the use of a great number of co-operative elements and arbitrary variation of their distinctive features and order of succession. For obvious reasons, the principle will also be valuable in the extension of the distance of transmission.

Progress though of necessity slow was steady and sure, for the objects aimed at were in a direction of my constant study and exercise. It is, therefore, not astonishing that before the end of 1899 I completed the task undertaken and reached the results which I have announced in my article in the *Century Magazine* of June, 1900, every word of which was carefully weighed.

Much has already been done towards making my system commercially available, in the transmission of energy in small amounts for specific purposes, as well as on an industrial scale. The results attained by me have made my scheme of intelligence transmission, for which the name of "World Telegraphy" has been suggested, easily realizable. It constitutes, I believe, in its principle of operation, means employed and capacities of application, a radical and fruitful departure from what has been done heretofore. I have no doubt that it will prove very efficient in enlightening the masses, particularly in still uncivilized countries and less accessible regions, and that it will add materially to general

he experimental station at Colorado Springs showing the structure used to determine he rate of incremental capacity with reference to the earth.

safety, comfort and convenience, and maintenance of peaceful relations. It involves the employment of a number of plants, all of which are capable of transmitting individualized signals to the uttermost confines of the earth. Each of them will be preferably located near some important center of civilization and the news it receives through any channel will be flashed to all points of the globe. A cheap and simple device, which might be carried in one's pocket, may then be set up somewhere on sea or land, and it will record the world's news or such special messages as may be intended for it. Thus the entire earth will be converted into a huge brain, as it were, capable of response in every one of its parts. Since a single plant of but one hundred horse-power can operate hundreds of millions of instruments, the system will have a virtually infinite working capacity, and it must needs immensely facilitate and cheapen the transmission of intelligence.

The first of these central plants would have been already completed had it not been for unforeseen delays which, fortunately, have nothing to do with its purely technical features. But this loss of time, while vexatious, may, after all, prove to be a blessing in disguise. The best design of which I know has been adopted, and the transmitter will emit a wave complex of a total maximum activity of ten million horse-power, one per cent. of which is amply sufficient to "girdle the globe." This enormous rate of energy delivery, approximately twice that of the combined falls of Niagara, is obtainable only by the use of certain artifices, which I shall make known in due course.

For a large part of the work which I have done so far I am indebted to the noble generosity of Mr. J. Pierpont Morgan, which was all the more welcome and stimulating, as it was

No. 609,245. Patented Aug. 16, 1898.

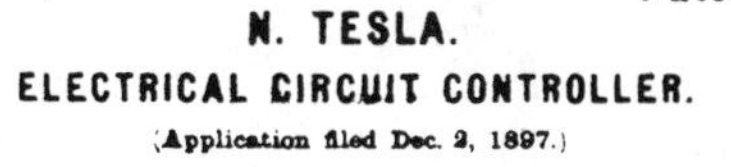

(No Model.)

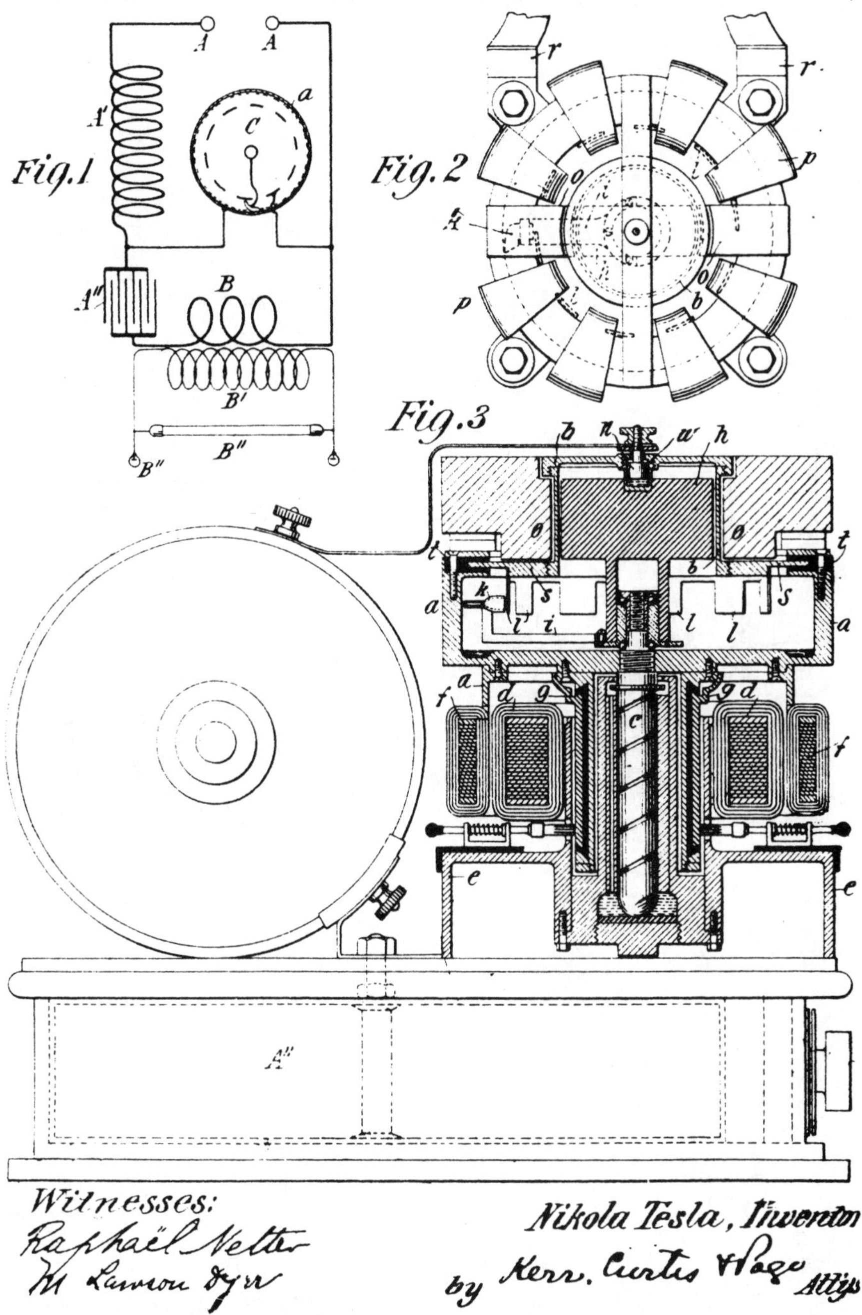

Diagramatic illustrations of the hermetically enclosed mercury break that is described in U.S. Patent No. 609,245 of August 16, 1898.

No. 685,955. Patented Nov. 5, 1901.

N. TESLA.

APPARATUS FOR UTILIZING EFFECTS TRANSMITTED FROM A DISTANCE TO A RECEIVING DEVICE THROUGH NATURAL MEDIA.

(Application filed Sept. 8, 1899. Renewed May 29, 1901.)

(No Model.)

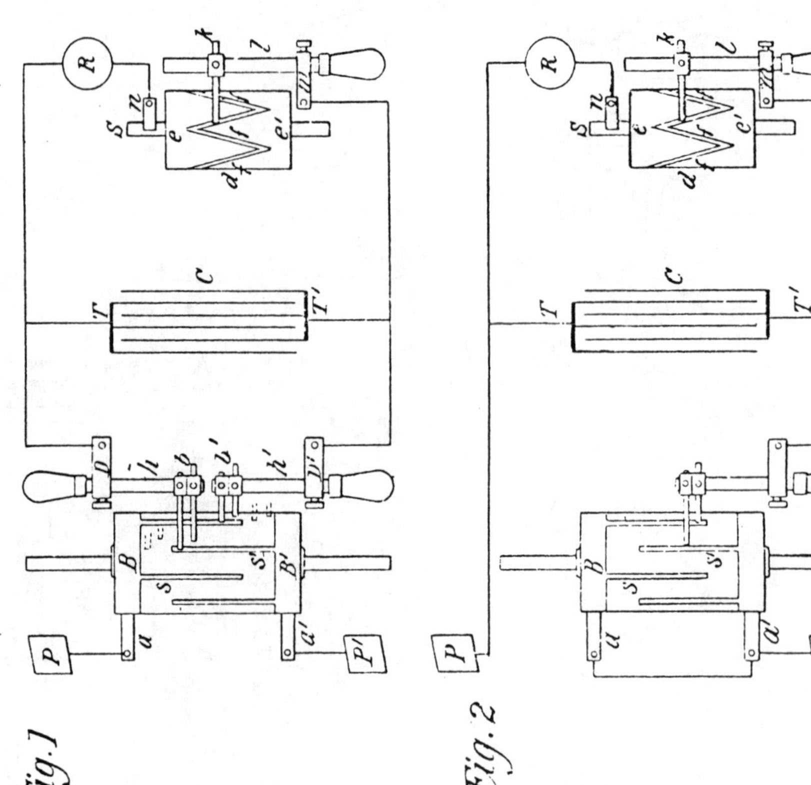

Fig. 1

Fig. 2

(G)

(G)

Witnesses:
G. B. Lewis.
Hillary C. Messimer

Nikola Tesla, Inventor
by Kerr. Page & Cooper.
Att'ys

No. 685,954 Patented Nov. 5, 1901.

N. TESLA.

METHOD OF UTILIZING EFFECTS TRANSMITTED THROUGH NATURAL MEDIA.

(Application filed Aug. 1, 1899. Renewed May 29, 1901.)

(No Model.) 2 Sheets—Sheet 1.

Fig 1.

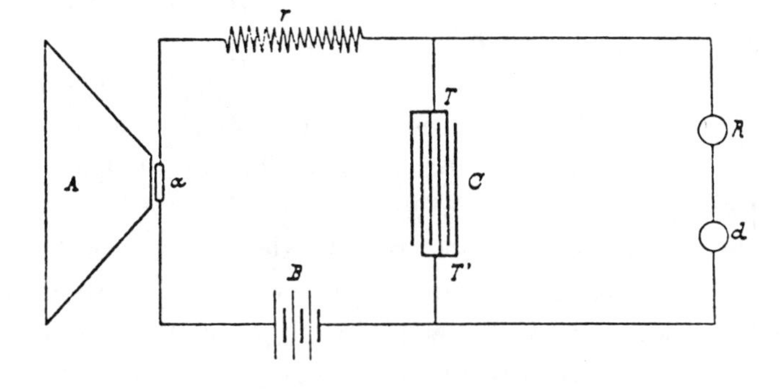

Fig 2.

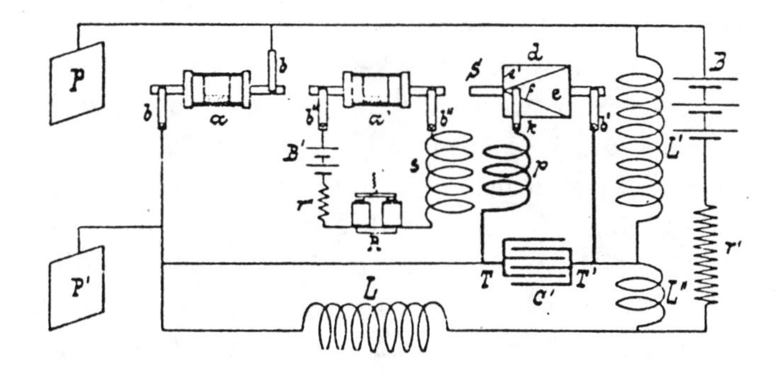

WITNESSES

Benjamin Miller.
M. Lawson Dyer.

INVENTOR

Nikola Tesla

BY

Kerr, Page & Cooper

ATTORNEYS.

extended at a time when those, who have since promised most, were the greatest of doubters. I have also to thank my friend, Stanford White, for much unselfish and valuable assistance. This work is now far advanced, and though the results may be tardy, they are sure to come.

Meanwhile, the transmission of energy on an industrial scale is not being neglected. The Canadian Niagara Power Company have offered me a splendid inducement, and next to achieving success for the sake of the art, it will give me the greatest satisfaction to make their concession financially profitable to them. In this first power plant, which I have been designing for a long time, I propose to distribute ten thousand horse-power under a tension of one hundred million volts, which I am now able to produce and handle with safety.

This energy will be collected all over the globe preferably in small amounts, ranging from a fraction of one to a few horse-power. One of its chief uses will be the illumination of isolated homes. It takes very little power to light a dwelling with vacuum tubes operated by high-frequency currents and in each instance a terminal a little above the roof will be sufficient. Another valuable application will be the driving of clocks and other such apparatus. These clocks will be exceedingly simple, will require absolutely no attention and will indicate rigorously correct time. The idea of impressing upon the earth American time is fascinating and very likely to become popular. There are innumerable devices of all kinds which are either now employed or can be supplied, and by operating them in this manner I may be able to offer a great convenience to the whole world with a plant of no more than ten thousand horse-power. The introduction of this system will give opportunities for invention

No. 685,954. Patented Nov. 5, 1901.

N. TESLA.

METHOD OF UTILIZING EFFECTS TRANSMITTED THROUGH NATURAL MEDIA.

(Application filed Aug. 1, 1899. Renewed May 29, 1901.)

(No Model.) 2 Sheets—Sheet 2.

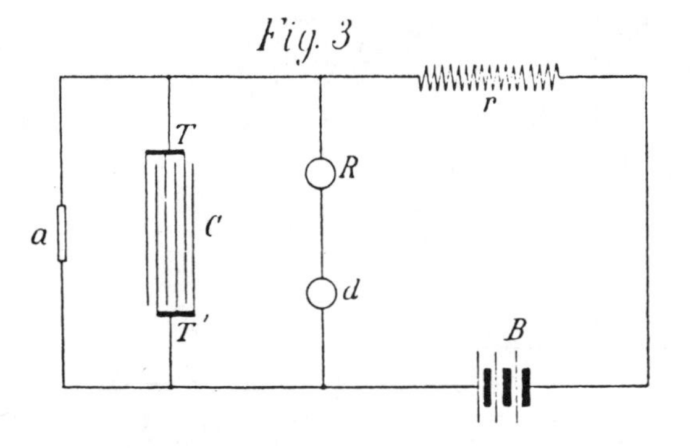

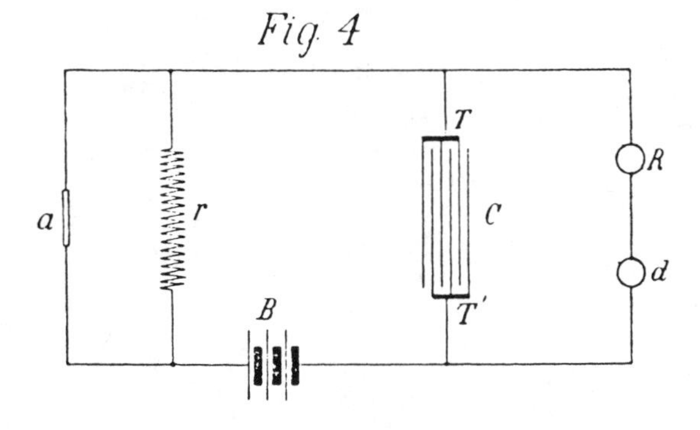

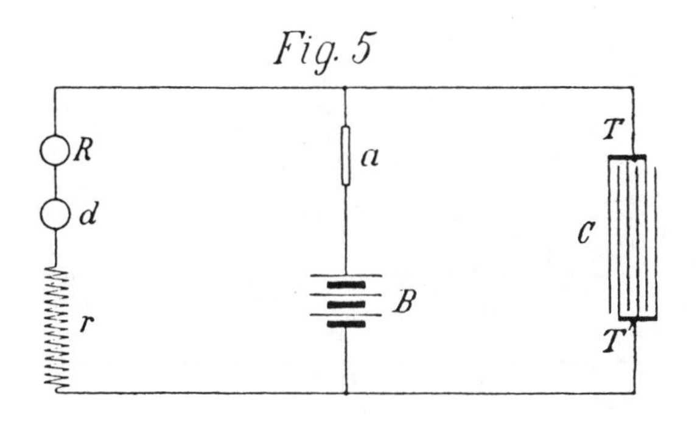

Witnesses:
Raphaël Netter
Benjamin Miller

Nikola Tesla Inventor

by Kerr, Page & Cooper Attys

No. 685,956. Patented Nov. 5, 1901.

N. TESLA.

APPARATUS FOR UTILIZING EFFECTS TRANSMITTED THROUGH NATURAL MEDIA.

(Application filed Nov. 2, 1899. Renewed May 29, 1901.)

(No Model.)

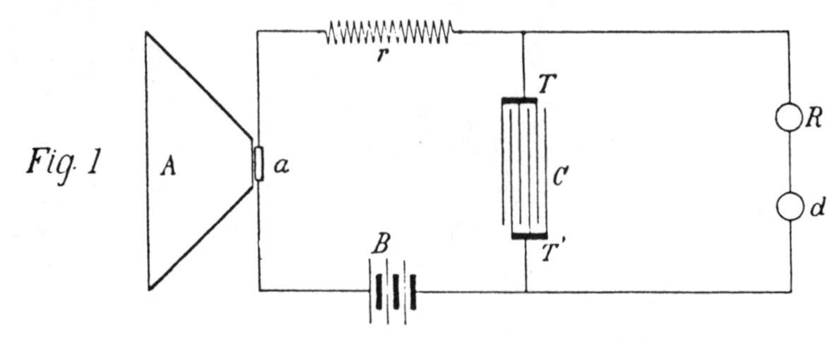

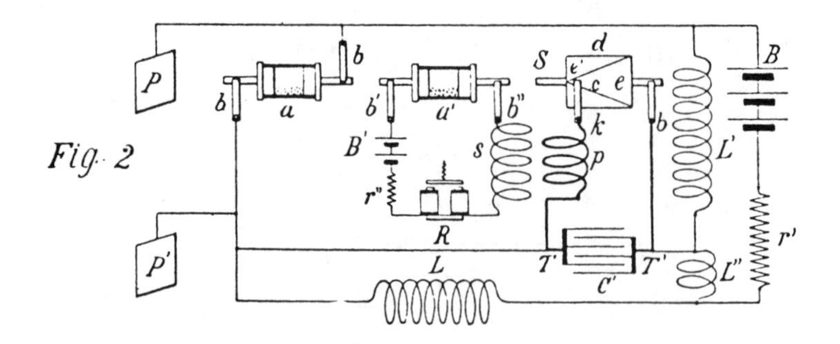

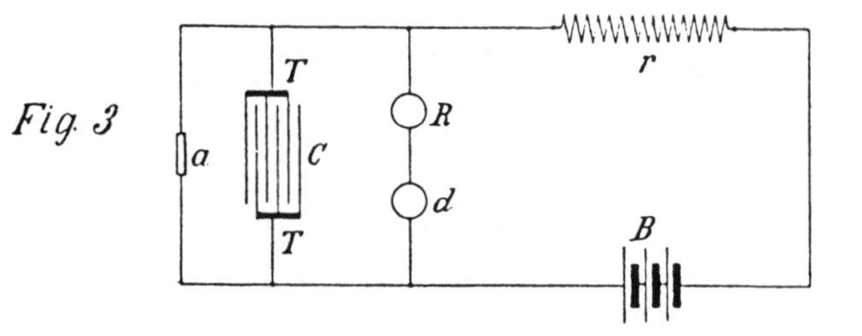

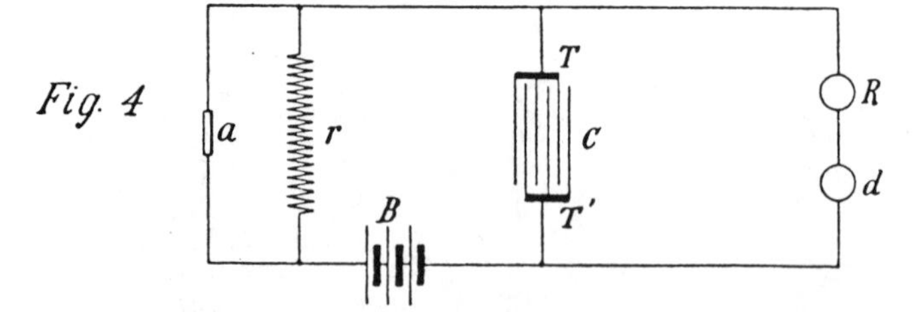

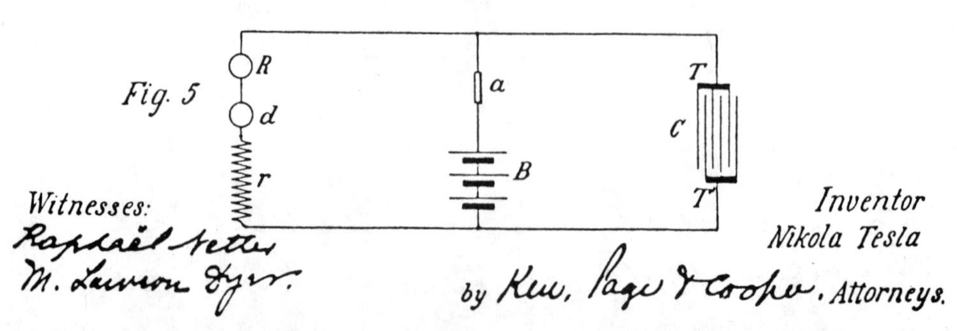

The original 1904 caption to this photo read: Tesla Central Power Plant, Transmitting Tower, and Laboratory for "World Telegraphy," Wardenclyffe, Long Island.

and manufacture such as have never presented themselves before.

Knowing the far-reaching importance of this first attempt and its effect upon future development, I shall proceed slowly and carefully. Experience has taught me not to assign a term to enterprises the consummation of which is not wholly dependent on my own abilities and exertions. But I am hopeful that these great realizations are not far off, and I know that when this first work is completed they will follow with mathematical certitude.

When the great truth accidentally revealed and experimentally confirmed is fully recognized, that this planet, with all its appalling immensity, is to electric currents virtually no more than a small metal ball and that by this fact many possibilities, each baffling imagination and of incalculable consequence, are rendered absolutely sure of accomplishment; when the first plant is inaugurated and it is shown that a telegraphic message, almost as secret and non-interferable as a thought, can be transmitted to any terrestrial distance, the sound of the human voice, with all its intonations and inflections, faithfully and instantly reproduced at any other point of the globe, the energy of a waterfall made available for supplying light, heat or motive power, anywhere—on sea, or land, or high in the air—humanity will be like an ant heap stirred up with a stick: See the excitement coming!

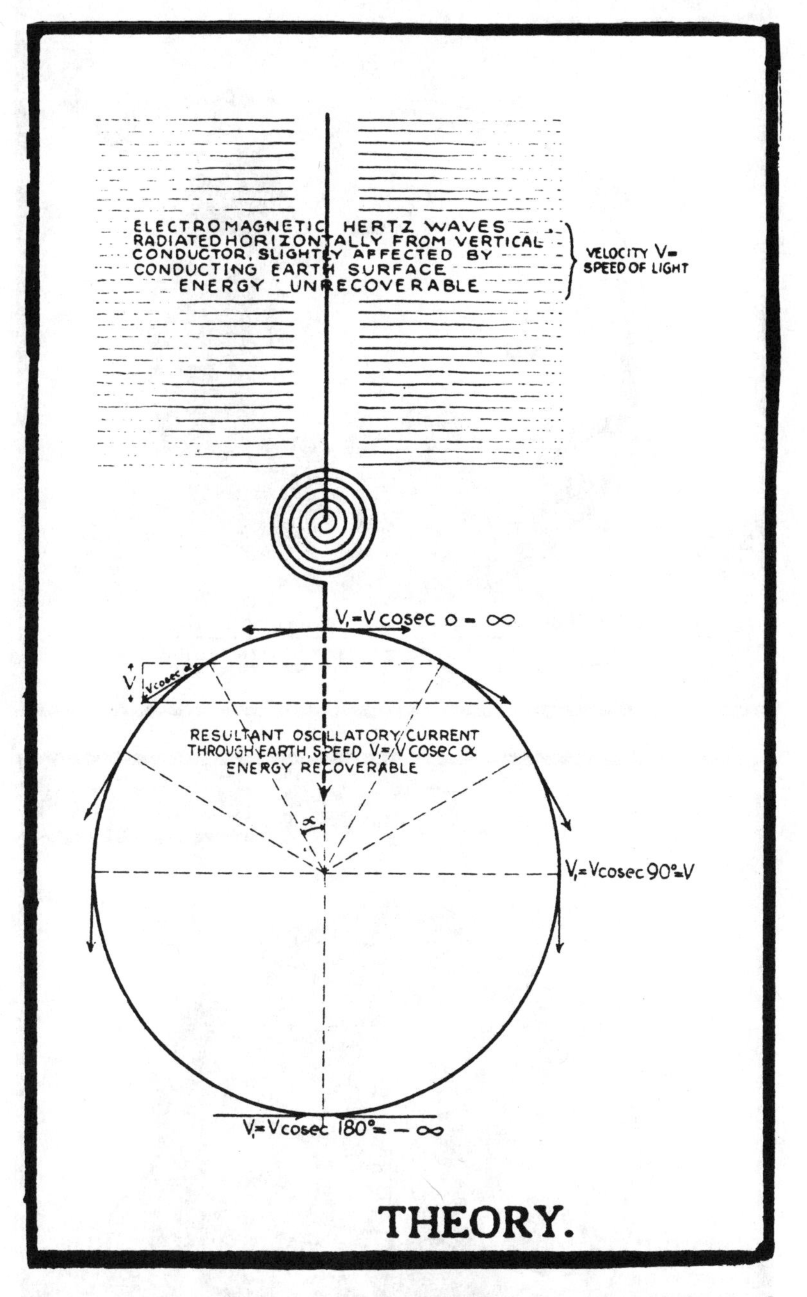

In a 1929 article on "World-Wide Transmission of Electrical Signals, Tesla explained his theory to the general public. The article used the following diagrams to show the "Theory, Analogy, and Realization" of the transmission of Electrical Signals world-wide.

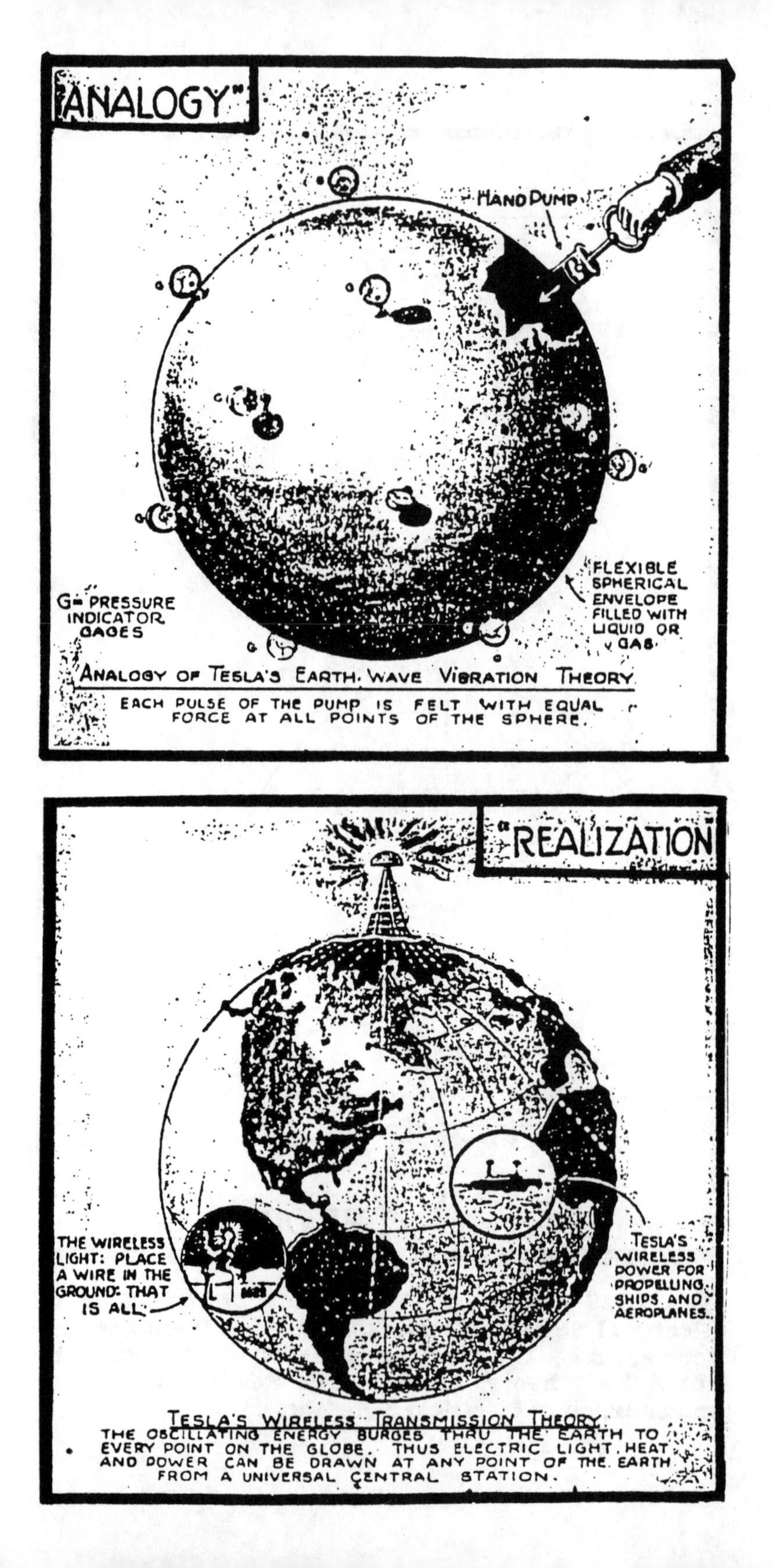
"ANALOGY"
HAND PUMP
G= PRESSURE INDICATOR GAGES
FLEXIBLE SPHERICAL ENVELOPE FILLED WITH LIQUID OR GAS
ANALOGY OF TESLA'S EARTH WAVE VIBRATION THEORY
EACH PULSE OF THE PUMP IS FELT WITH EQUAL FORCE AT ALL POINTS OF THE SPHERE.
"REALIZATION"
THE WIRELESS LIGHT: PLACE A WIRE IN THE GROUND: THAT IS ALL
TESLA'S WIRELESS POWER FOR PROPELLING SHIPS AND AEROPLANES.
TESLA'S WIRELESS TRANSMISSION THEORY
THE OSCILLATING ENERGY SURGES THRU THE EARTH TO EVERY POINT ON THE GLOBE. THUS ELECTRIC LIGHT, HEAT AND POWER CAN BE DRAWN AT ANY POINT OF THE EARTH FROM A UNIVERSAL CENTRAL STATION.

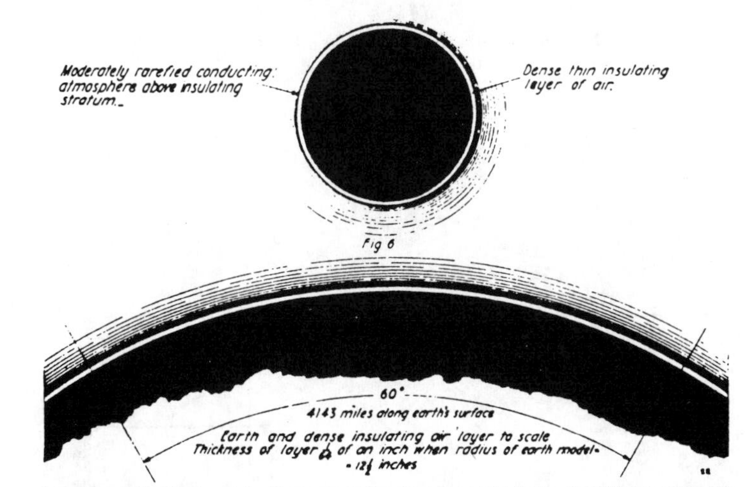

A Section of the Earth and Its Atmospheric Envelope Drawn to Scale. It Is Obvious That the Hertzian Rays Cannot Traverse So Thin a Crack Between Two Conducting Surfaces For Any Considerable Distance, Without Being Absorbed, Says Dr. Tesla, in Discussing the Ether Space Wave Theory.

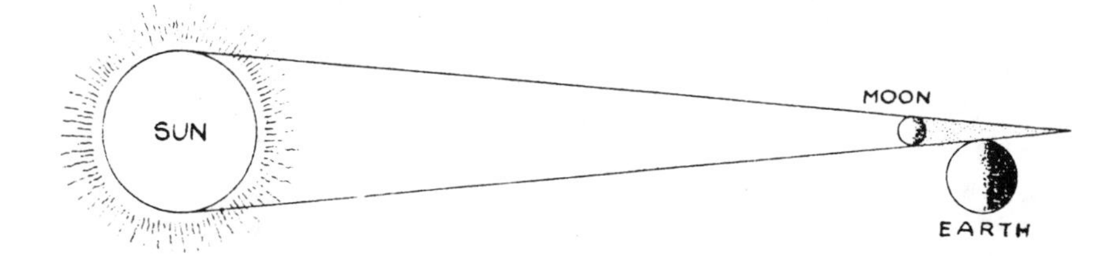

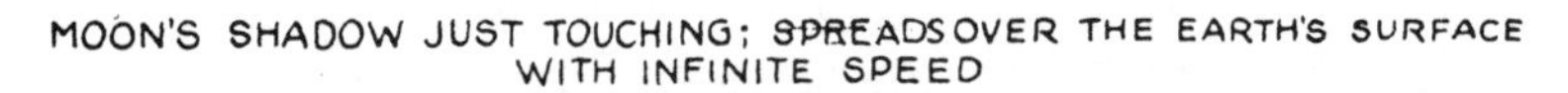

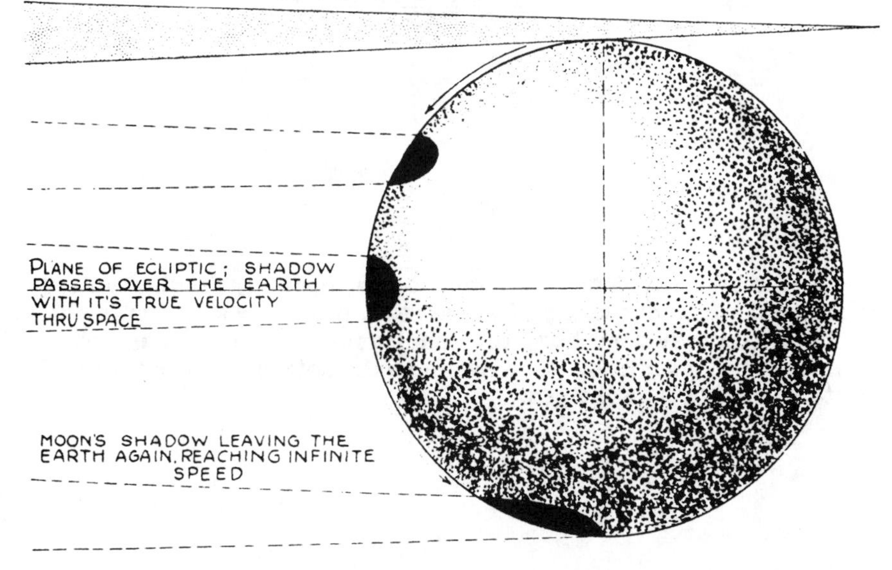

Above: Tesla's drawing of the earth and its atmospheric envelope drawn to scale from the February, 1919 issue of *Electrical Experimenter*, which discussed Tesla's "Ether Space Wave Theory." Below: Another drawing from 1929 demonstrating Tesla's Ether theory and how he theorized he would recover the energy from electromagnetic Hertz waves as oscillatory energy.

No. 685,958. Patented Nov. 5, 1901.

N. TESLA.

METHOD OF UTILIZING RADIANT ENERGY.

(Application filed Mar. 21, 1901.)

(No Model.)

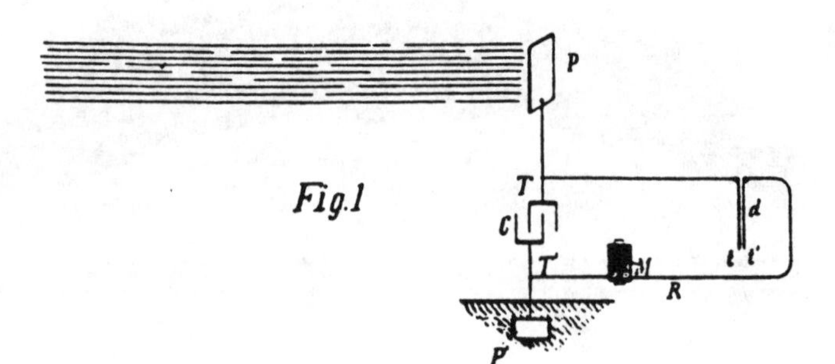

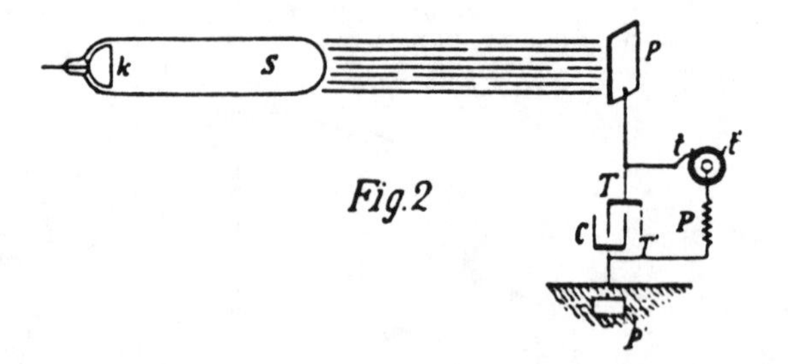

Tesla's X-Ray tube, part of his method for utilizing "radiant energy," which operated from the top of a Tesla coil, providing a means to charge the "elevated insulated body of capacitance" C, with armatures T-T'. "Whenever the circuit is closed owing to the rotation of the terminal t', the stored energy is discharged..."

No. 787,412. PATENTED APR. 18, 1905.

N. TESLA.

ART OF TRANSMITTING ELECTRICAL ENERGY THROUGH THE NATURAL MEDIUMS.

APPLICATION FILED MAY 16, 1900. RENEWED JUNE 17, 1902.

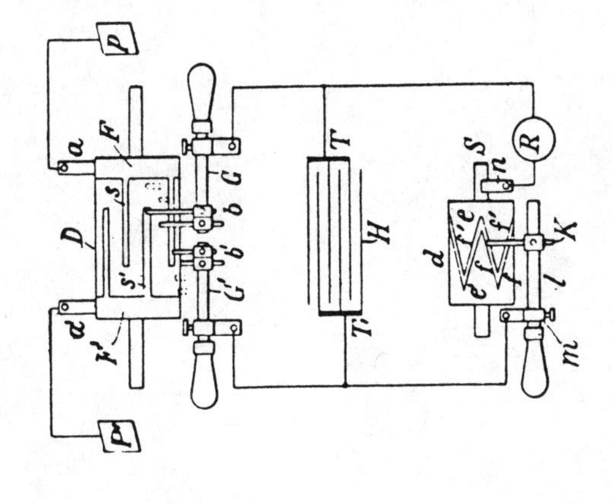

Fig. 2

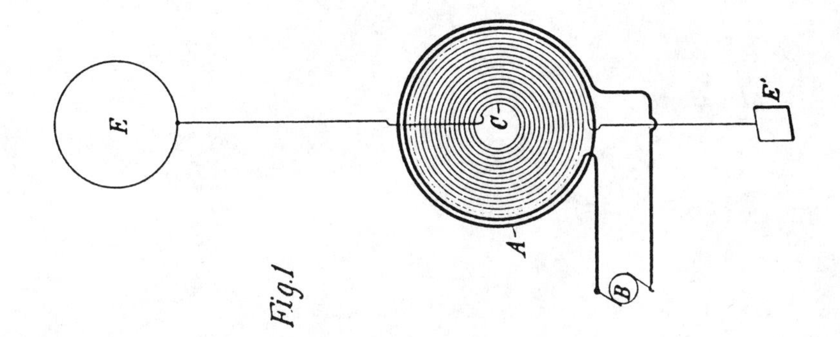

Fig. 1

Witnesses:
Raphaël Netter
M. Lawson Dyer.

Nikola Tesla Inventor
by Ken: Page & Cooper Attys

Interior views of Tesla's experimental Colorado Springs tower, showing the cage generatoring voltage and the banks of batteries.

Chapter 6

TESLA'S DEATH RAYS

Tesla's *Death Ray* was instantly a controversial and popular topic. In his later years, after the Wardencliff Tower project had been stopped by J.P. Morgan and dismantled under F.B.I. supervision, Tesla made little money from his projects and in many cases did not even bother to seek patents. He was more interested in publicity and became a favorite with newspaper reporters for his flamboyant demonstrations, controversial predictions and incredible new inventions.

On July 11, 1934 , the New York Times ran a story which was headlined:

> **TESLA AT 78 BARES**
> **NEW "DEATH-BEAM,"**
> **Invention Powerful Enough to Destroy**
> **10,000 Planes 250 Miles Away,**
> **He Asserts.**
> **DEFENSIVE WEAPON ONLY.**
> **Scientist, in Interview, Tells of Apparatus**
> **That He Says Will Kill Without Trace.**

Tesla's death ray device was a kind of radio-wave-scalar weapon or what might be called an ultra-sound gun. Tesla and death ray made quite a media splash at the time. In the 1930's several Death Ray weapon films came out, including *The Death Ray* (1938) with Boris Karloff, and such serials as *Flash Gordon* and *Radar Men From the Moon.*

In fact, the very first of the Max Fletcher *Superman* cartoons of the 1940's featured Tesla in *The Mad Scientist* (Sept. 1941) in which a crazed, eccentric scientist, obviously patterned after Tesla, battles Superman while he terrorizes New York with his "electrothenasia death ray." In the next cartoon, *The*

Mechanical Monsters (Nov. 1941) Superman again battles Tesla, the mad scientist, who this time unleashes an army of robots on Manhattan. Superman battled Tesla and his Death Ray one last time in *Magnetic Telescope* (April 1942), where Tesla is using a special magneto-gravitic ray that pulls asteroids out of orbit and sends them crashing to earth. With *Japateurs* in September of 1942, the Superman cartoons turned toward War themes, featuring Japanese spies and, to a lesser part, Nazi agents.

It is interesting to think of Tesla as the model for all the "mad scientists" of comic book and cinematic fiction.

In the Spring of 1924 "death rays," were the subject of many newspapers around the world. Harry Grindell-Matthews of London lead the contenders in this early Star Wars race. The New York Times of May 21st had this report:

> **Paris, May 20—If confidence of Grindell Mathew (sic), inventor of the so-called 'diabolical ray,' in his discovery is justified it may become possible to put the whole of an enemy army out of action, destroy any force of airplanes attacking a city or paralyze any fleet venturing within a certain distance of the coast by invisible rays.**

Grindell-Matthews stated that his destructive rays would operate over a distance of four miles and that the maximum distance for this type of weapon would be seven or eight miles. "Tests have been reported where the ray has been used to stop the operation of automobiles by arresting the action of the magnetos, and an quantity of gunpowder is said to have been exploded by playing the beams on it from a distance of thirty-six feet." Grindell-Matthews was able, also, to electrocute mice, shrivel plants, and light the wick of an oil lamp from the same distance away.

Sensing something of importance the *New York Times* copyrighted a story on May 28th of 1924 on a ray-weapon developed by the Soviets. The story opened:

> **"News has leaked out from the Communist circles in Moscow that behind Trotsky's recent war-like utterance lies an electromagnetic invention, by a Russian engineer named Grammachikoff for destroying airplanes.**
> **Tests of the destructive ray, the Times continued, had began the previous August with the aid of German technical experts.**

> **A large, scale demonstration at Podosinsky Aerodome near Moscow was so successful that the revolutionary Military Council and the Political Bureau decided to fund enough electronic anti-aircraft stations to protect sensitive areas of Russia. Similar, but more powerful, stations were to be constructed to disable the electrical mechanisms of warships.**
> **The Commander of the Soviet Air Services, Rosenholtz, was so overwhelmed by the ray weapon demonstration that he proposed "to curtail the activity of the air fleet, because the invention rendered a large air fleet unnecessary for the purpose of defense."**

Tesla appears to have been the renegade scientist, eccentric and brilliant. However, after his finances had been destroyed by Morgan, and indirectly by Westinghouse, Tesla was constantly broke. In lieu of money on rent, in the early 1930's, Tesla gave the management of the Governor Clinton Hotel a supposed invention of his to be used for collateral. He said that the device was very dangerous and worth $10,000. In 1943, an MIT scientist, working for the National Defense Research Committee (NDRC) and accompanied by the office of Naval Intelligence, John O. Trump, went to the hotel to retrieve the device, after Tesla's death.

He was told that the invention could "detonate if opened by an unauthorized person." Trump stated that he reflected momentarily upon his life before he opened the container. In his FBI report he stated

> **"Inside was a handsome wooden chest bound with brass... [containing] a multidecade resistance box of the type used for a Wheatstone bridge resistance measurements—a common standard item found in every electric laboratory before the turn of the century!"**

According to Tesla researcher Dr. Marc Seifer, Tesla appears to have told both his pigeon caretaker and an army engineer named Fizgerald, a friend of Tesla's, that he had built a working model of a Death Ray. Dr. Seifer says that a number of people closely associated with Tesla would recount stories, circa 1918, of Tesla bouncing electronic beams off the moon. Seifer says that this is not a Death Ray, but it certainly supports the hypothesis that the inventor created working models along those lines.

According to Dr. Seifer, Tesla drew up "artist conceptions" in the mid-1930s that were "made of a building with a tower in the form of a cylinder 16.5 feet in diameter, 115 feet tall. The structure was capped at the top by a 10 meter diameter sphere (covered with hemispheric shells as in the 1914 patent)." The inventor had also contacted people at Alcoa Aluminum throughout 1935 who were "ready to start as soon as Tesla advanced the funds."

Two years later, at the 81, the inventor stated at a luncheon attended by ministers of Yugoslavia and Czechoslovakia that he had constructed a number of beam transmission devices including the death ray for protecting a country from incoming invasions and a laser-like machine that could send impulses to the moon and other planets.

According to Dr. Seifer, Tesla also said that he was going to take the death ray to a Geneva conference for world peace.; When pressed by the columnists to "give a full description..., Dr. Tesla said..., "But it is not an experiment... I have built, demonstrated and used it. Only a little time will pass before I can give it to the world."

Another Tesla scholar who believes that Tesla built a "death ray" is Oliver Nichelson, who has written quite a bit on Tesla, including an article entitled "Nikola Tesla's Long Range Weapon" (1989).

Picking up the death ray stories on the wire services on the other side of the world, the Colorado Springs Gazette, ran a local interest item on May 30th. With the headline: "Tesla Discovered 'Death Ray' in Experiments He Made Here," the story recounted, with a feeling of local pride, the inventor's 1899 researches financed by John Jacob Astor.

Tesla's Colorado Springs tests were well remembered by local residents. With a 200 foot pole topped by a large copper sphere rising above his laboratory he generated potentials that discharged lightning bolts up to 135 feet long. Thunder from the released energy could be heard 15 miles away in Cripple Creek. People walking along the streets were amazed to see sparks jumping between their feet and the ground, and flames of electricity would spring from a tap when anyone turned them on for a drink of water. Light bulbs within 100 feet of the experimental tower glowed when they were turned off. Horses at the livery stable received shocks through their metal shoes and bolted from the stalls. Even insects were affected: Butterflies became electrified and "helplessly swirled in circles—their wings spouting blue halos of 'St. Elmo's Fire.'"

The most pronounced effect, and the one that captured the attention of death ray inventors, occurred at the Colorado Springs Electric Company generating station. One day while Tesla was conducting a high power test, the crackling from inside the laboratory suddenly stopped. Bursting into the lab

Tesla demanded to know why his assistant had disconnected the coil. The assistant protested that had not anything. The power from the city's generator, the assistant said, must have quit. When the angry Tesla telephoned the power company he received an equally angry reply that the electric company had not cut the power, but that Tesla's experiment had destroyed the generator!

According to Oliver Nichelson, Tesla explained to *The Electrical Experimenter,* in August of 1917 what had happened. While running his transmitter at a power level of "several hundred kilowatts" high frequency currents were set up in the electric company's generators. These powerful currents "caused heavy sparks to jump thru the winds and destroy the insulation." When the insulation failed, the generator shorted out and was destroyed.

Some years later, 1935, he elaborated on the destructive potential of his transmitter in the February issue of Liberty magazine:

> **"My invention requires a large plant, but once it is established it will be possible to destroy anything, men or machines, approaching within a radius of 200 miles."**

He went on to make a distinction between his invention and those brought forward by others. He claimed that his device did not use any so-called "death rays" because such radiation cannot be produced in large amounts and rapidly becomes weaker over distance. Here, he likely had in mind a Grindell-Matthews type of device which, according to contemporary reports, used a powerful ultra-violet beam to make the air conducting so that high energy current could be directed to the target. The range of an ultra-violet searchlight would be much less than what Tesla was claiming. As he put it: "all the energy of New York City (approximately two million horsepower [1.5 billion watts]) transformed into rays and projected twenty miles, would not kill a human being." On the contrary, he said:

> **"My apparatus projects particles which may be relatively large or of microscopic dimensions, enabling us to convey to a small area at a great distance trillions of times more energy than is possible with rays of any kind. Many thousands of horsepower can be thus transmitted by a stream thinner than a hair, so that nothing can resist."**

According to Oliver Nichelson, what Tesla had in mind with this defensive system was a large scale version of his Colorado Springs lightning bolt machine. As airplanes or ships entered the electric field of his charged tower, they would set up a conducting path for a stream of high energy particles that would destroy the intruder's electrical system.

A drawback to having giant Tesla transmitters poised to shoot bolts of lightning at an enemy approaching the coasts is that they would have to be located in an uninhabited area equal to its circle of protection. Anyone stepping into the defensive zone of the coils would be sensed as an intruder and struck down. Today, with the development of oil drilling platforms, this disadvantage might be overcome by locating the lightning defensive system at sea.

As ominous as death ray and beam weapon technology will be for the future, there is another, more destructive, weapon system alluded to in Tesla's writings.

According to Oliver Nichelson, when Tesla realized, as he pointed out in the 1900 Century article, "The Problem of Increasing Human Energy," that economic forces would not allow the development of a new type of electrical generator able to supply power without burning fuel he "was led to recognize [that] the transmission of electrical energy to any distance through the media as by far the best solution of the great problem of harnessing the sun's energy for the use of man." His idea was that a relatively few generating plants located near waterfalls would supply his very high energy transmitters which, in turn, would send power through the earth to be picked up wherever it was needed.

The plan would require several of his transmitters to rhythmically pump huge amounts of electricity into the earth at pressures on the order of 100 million volts. The earth would become like a huge ball inflated to a great electrical potential, but pulsing to Tesla's imposed beat.

Receiving energy from this high pressure reservoir only would require a person to put a rod into the ground and connect it to a receiver operating in unison with the earth's electrical motion. As Tesla described it, "the entire apparatus for lighting the average country dwelling will contain no moving parts whatever, and could be readily carried about in a small valise."

However, the difference between a current that can be used to run, say, a sewing machine and a current used as a method of destruction, however, is a matter of timing. If the amount of electricity used to run a sewing machine for an hour is released in a millionth of a second, it would have a very different, and negative, effect on the sewing machine.

Tesla said his transmitter could produce 100 million volts of pressure with currents up to 1000 amperes which is a power level of 100 billion watts. If it was resonating at a radio frequency of 2 MHz, then the energy released during one period of its oscillation would be 100,000,000,000,000,000 Joules of energy, or roughly the amount of energy released by the explosion of 10 megatons of TNT.

Such a transmitter, would be capable of projecting the energy of a nuclear warhead by radio. Any location in the world could be vaporized at the speed of light.

Not unexpectedly, many scientists doubted the technical feasibility of Tesla's wireless power transmission scheme whether for commercial or military purposes. The secret of how through- the-earth broadcast power was found not in the theories of electrical engineering, but in the realm of high energy physics.

Dr. Andrija Puharich, in 1976, was the first to point out that Tesla's power transmission system could not be explained by the laws of classical electrodynamics, but, rather, in terms of relativistic transformations in high energy fields. He noted that according to Dirac's theory of the electron, when one of those particles encountered its oppositely charged member, a positron, the two particles would annihilate each other. Because energy can neither be destroyed nor created the energy of the two former particles are transformed into an electromagnetic wave. The opposite, of course, holds true. If there is a strong enough electric field, two opposite charges of electricity are formed where there was originally no charge at all. This type of transformation usually takes place near the intense field near an atomic nucleus, but it can also manifest without the aid of a nuclear catalyst if an electric field has enough energy. Puharich's involved mathematical treatment demonstrated that power levels in a Tesla transmitter were strong enough to cause such pair production.

The mechanism of pair production offers a very attractive explanation for the ground transmission of power. Ordinary electrical currents do not travel far through the earth. Dirt has a high resistance to electricity and quickly turns currents into heat energy that is wasted. With the pair production method electricity can be moved from one point to another without really having to push the physical particle through the earth - the transmitting source would create a strong field, and a particle would be created at the receiver.

If the sending of currents through the earth is possible from the viewpoint of modern physics, the question remains of whether Tesla actually demonstrated the weapons application of his power transmitter or whether it remained an unrealized plan on the part of the inventor. Circumstantial evidence points to there having been a test of this weapon.

The clues are found in the chronology of Tesla's work and financial fortunes between 1900 and 1915.

1900: Tesla returned from Colorado Springs after a series of important tests of wireless power transmission. It was during these tests that his magnifying transmitter sent out waves of energy causing the destruction of the power company's generator.

He received financial backing from J. Pierpont Morgan of $150,000 to build a radio transmitter for signaling Europe. With the first portion of the money he obtained 200 acres of land at Shoreham, Long Island and built an enormous tower 187 feet tall topped with a 55 ton, 68 foot metal dome. He called the research site "Wardenclyffe."

As Tesla was just getting started, investors were rushing to buy stock offered by the Marconi company. Supporters of the Marconi Company include his old adversary Edison.

On December 12th, Marconi sent the first transatlantic signal, the letter "S," from Cornwall, England to Newfoundland. He did this with, as the financiers noted, equipment much less costly than that envisioned by Tesla.

1902: Marconi is being hailed as a hero around the world while Tesla is seen as a shirker by the public for ignoring a call to jury duty in a murder case (he was excused from duty because of his opposition to the death penalty).

1903: When Morgan sent the balance of the $150,000, it would not cover the outstanding balance Tesla owed on the Wardenclyffe construction. To encourage a larger investment in the face of Marconi's success, Tesla revealed to Morgan his real purpose was not to just send radio signals but the wireless transmission of power to any point on the planet. Morgan was uninterested and declined further funding.

A financial panic that Fall put an end to Tesla's hopes of financing by Morgan or other wealthy industrialists. This left Tesla without money even to buy the coal to fire the transmitter's electrical generators.

1904: Tesla writes for the Electrical World, "The Transmission of Electrical Energy Without Wires," noting that the globe, even with its great size, responds to electrical currents like a small metal ball.

Tesla declares to the press the completion of Wardenclyffe.

1904: The Colorado Springs power company sues for electricity used at that experimental station. Tesla's Colorado laboratory is torn down and is sold for lumber to pay the $180 judgement; his electrical equipment is put in storage.

1905: Electrotherapeutic coils are manufactured at Wardenclyffe, for hospitals and researchers to help pay bills.

Tesla is sued by his lawyer for non-payment of a loan. In an article, Tesla comments on Peary's expedition to the North Pole and tells of his, Tesla's, plans for energy transmission to any central point on the ground.

Tesla is sued by C.J. Duffner, a caretaker at the experimental station in Colorado Springs, for wages .

1906: "Left Property Here; Skips; Sheriff's Sale," was the headline in the Colorado Springs Gazette for March 6th. Tesla's electrical equipment is sold to pay judgement of $928.57.

George Westinghouse, who bought Tesla's patents for alternating current motors and generators in the 1880's, turns down the inventor's power transmission proposal.

Workers gradually stop coming to the Wardenclyffe laboratory when there are no funds to pay them.

1907: When commenting on the destruction of the French ship Iena, Tesla noted in a letter to the New York Times that he has built and tested remotely controlled torpedoes, but that electrical waves would be more destructive. "As to projecting wave energy to any particular region of the globe ... this can be done by my devices," he wrote. Further, he claimed that "the spot at which the desired effect is to be produced can be calculated very closely, assuming the accepted terrestrial measurements to be correct."

1908: Tesla repeated the idea of destruction by electrical waves to the newspaper on April 21st. His letter to the editor stated, "When I spoke of future warfare I meant that it should be conducted by direct application of electrical waves without the use of aerial engines or other implements of destruction." He added: "This is not a dream. Even now wireless power plants could be constructed by which any region of the globe might be rendered uninhabitable without subjecting the population of other parts to serious danger or inconvenience."

1915: Again, in another letter to the editor, Tesla stated: "It is perfectly practical to transmit electrical energy without wires and produce destructive effects at a distance. I have already constructed a wireless transmitter which makes this possible... When unavoidable, the [transmitter] may be used to destroy property and life."

Important to this chronology is the state of Tesla's mental health. One researcher, Marc J. Seifer, a psychologist, believes Tesla suffered a nervous breakdown catalyzed by the death of one the partners in the Tesla Electric Company and the shooting of Stanford White, the noted architect, who had designed Wardenclyffe. Seifer places this in 1906 and cites as evidence a letter from George Scherff, Tesla's secretary:

> **Wardenclyffe, 4/10/1906**
> **Dear Mr. Tesla:**
>
> **I have received your letter and am very glad to know you are vanquishing your illness. I have scarcely ever seen you so out of sorts as last Sunday; and I was frightened.**

In the period from 1900 to 1910 Tesla's creative thrust was to establish his plan for wireless transmission of energy. Undercut by Marconi's accomplishment, beset by financial problems, and spurned by the scientific establishment, Tesla was in a desperate situation by mid-decade. The strain became too great by 1906 and he suffered an emotional collapse. In order to make a final effort to have his grand scheme recognized, he may have tried one high power test of his transmitter to show off its destructive potential. This would have been in 1908.

The Tunguska event took place on the morning of June 30th, 1908. An explosion estimated to be equivalent to 10-15 megatons of TNT flattened 500,000 acres of pine forest near the Stony Tunguska River in central Siberia. Whole herds of reindeer were destroyed. The explosion was heard over a radius of 620 miles. When an expedition was made to the area in 1927 to find evidence of the meteorite presumed to have caused the blast, no impact crater was found. When the ground was drilled for pieces of nickel, iron, or stone, the main constituents of meteorites, none were found down to a depth of 118 feet.

Many explanations have been given for the Tunguska event. The officially accepted version is that a 100,000 ton fragment of Encke's Comet, composed mainly of dust and ice, entered the atmosphere at 62,000 m.p.h., heated up, and exploded over the earth's surface creating a fireball and shock wave but no crater. Alternative versions of the disaster see a renegade mini-black hole or an alien space ship crashing into the earth with the resulting release of energy.

According to Oliver Nichelson, the historical facts point to the possibility that this event was caused by a test firing of Tesla's energy weapon.

In 1907 and 1908, Tesla wrote about the destructive effects of his energy transmitter. His Wardenclyffe transmitter was much larger than the Colorado Springs device that destroyed the power station's generator. His new transmitter would be capable of effects many orders of magnitude greater than the Colorado device.

In 1915, he said he had already built a transmitter that "when unavoidable ... may be used to destroy property and life." Finally, a 1934 letter from Tesla to J.P. Morgan, uncovered by Tesla biographer Margaret Cheney, seems to conclusively point to an energy weapon test. In an effort to raise money for his defensive system he wrote:

> **The flying machine has completely demoralized the world, so much so that in some cities, as London and Paris, people are in mortal fear from aerial bombing. The new means I have perfected affords absolute protection against this and other forms of attack... These new discoveries I have carried out experimentally on a limited scale, created a profound impression (emphasis added).**

Again, the evidence is circumstantial but, to use the language of criminal investigation, Tesla had motive and means to be the cause of the Tunguska event. He also seems to confess to such a test having taken place before 1915. His transmitter could generate energy levels and frequencies that would release the destructive force of 10 megatons, or more, of TNT. And the overlooked genius was desperate.

The nature of the Tunguska event, also, is not inconsistent with what would happen during the sudden release of wireless power. No fiery object was reported in the skies at that time by professional or amateur astronomers as would be expected when a 200,000,000 pound object enters the atmosphere. The sky glow in the region, mentioned by some witnesses, just before the explosion may have come from the ground, as geological researchers discovered in the 1970's. Just before an earthquake the stressed rock beneath the ground creates an electrical effect causing the air to illuminate.

According to Oliver Nichelson, if the explosion was caused by wireless energy transmission, either the geological stressing or the current itself would cause an air glow. Finally, there is the absence of an impact crater. Because there is no material object to impact, an explosion caused by broadcast power would not leave a crater.

Given Tesla's general pacifistic nature it is hard to understand why he would carry out a test harmful to both animals and the people who herded the animals even when he was in the grip of financial desperation. The answer is that he probably intended no harm, but was aiming for a publicity coup and, literally, missed his target.

At the end of 1908, the whole world was following the daring attempt of Peary to reach the North Pole. Peary claimed the Pole in the Spring of 1909, but the winter before he had returned to the base at Ellesmere Island, about 700 miles from the Pole. If Tesla wanted the attention of the international press, few things would have been more impressive than the Peary expedition sending out word of a cataclysmic explosion on the ice in the direction of the North Pole. Tesla, then, if he could not be hailed as the master creator that he was, could be seen as the master of a mysterious new force of destruction.

The test, it seems, was not a complete success, says Nichelson. It must have been difficult controlling the vast amount of power in transmitter and guiding it to the exact spot Tesla wanted. Alert, Canada on Ellesmere Island and the Tunguska region are all on the same great circle line from Shoreham, Long Island. Both are on a compass bearing of a little more than 2 degrees along a polar path. The destructive electrical wave overshot its target.

Whoever was privy to Tesla's energy weapon demonstration must have been dismayed either because it missed the intended target and would be a threat to inhabited regions of the planet, or because it worked too well in devastating such a large area at the mere throwing of a switch thousands of miles away. Whichever was the case, Tesla never received the notoriety he sought for his power transmitter.

In 1915, the Wardenclyffe laboratory was deeded over to Waldorf- Astoria, Inc. in lieu of payment for Tesla's hotel bills. In 1917, Wardenclyffe was dynamited on orders of the new owners to recover some money from the scrap.

Oliver Nichelson's exotic theory may be pure fantasy, or perhaps, Nikola Tesla did shake the world in a way that has been kept secret for over 80 years.

Today, Stars Wars threatens to control the entire population of this planet from earth orbit. Tesla's death ray inventions can be utilized in a variety of ways: as scalar wave howitzers, world radar, earthquake contrivances, brain wave manipulation, particle beam weapons, wave-train impulses, hand-held phasers, and an infinite variety of more devices.

On the good side of this technology, there is free energy and the use of Tesla Shields, the forming of an energy shell around a city, community or installation that is impenetrable. Blasts from a Tesla Howitzer could destroy the communications network of any major city with a well placed jolt of many millions volts, and airstrikes can be called in from space. The military applications for many of Tesla's inventions are myriad, and so the need for a cover-up of Tesla and his inventions would behoove the military industrial complex.

Above & Below: Tesla's Death Ray popularized in a drawing from "Diabolical Rays" in the November, 1915 issue of *Popular Radio* magazine. The fear of these "diabolical death rays," was one of the reasons given for the dismantling of Tesla's Wardencliff Tower.

THE RAY IN OPERATION EXPLODING GUNPOWDER.

TESLA, AT 78, BARES NEW 'DEATH-BEAM'

Invention Powerful Enough to Destroy 10,000 Planes 250 Miles Away, He Asserts.

DEFENSIVE WEAPON ONLY

Scientist, in Interview, Tells of Apparatus That He Says Will Kill Without Trace.

Nikola Tesla, father of modern methods of generation and distribution of electrical energy, who was 78 years old yesterday, announced a new invention, or inventions, which he said, he considered the most important of the 700 made by him so far.

He has perfected a method and apparatus, Dr. Tesla said yesterday in an interview at the Hotel New Yorker, which will send concentrated beams of particles through the free air, of such tremendous energy that they will bring down a fleet of 10,000 enemy airplanes at a distance of 250 miles from a defending nation's border and will cause armies of millions to drop dead in their tracks.

Times Wide World Photo.

NOTED INVENTOR 78.

Nikola Tesla.

-A- Fig. 3 -B-

Two Optional Forms of Wireless Antennae Formed of Searchlight Beams—Ionized Atmospheric Streams.

Above: The *New York Times* article on Tesla's Death Ray of July 11, 1934. Below: Two illustrations from an article in the March, 1920 issue of Electrical Experimenter entitled Wireless Transmission of Power Now Possible. The illustrations show his prototype devices for "directed ionized beam transmissions," a "death-ray—searchlight" device. Curiously, powerful searchlight-beams have frequently been reported as part of unidentified discoid and cigar-shaped craft since the late 1800s.

Section 2 — GENERAL NEWS — OBITUARIES — SHIPPING AND MAILS

The New York Times.

SUNDAY, JULY 11, 1937.

Copyright, 1937, by The New York Times Company.

7,000 BOY SCOUTS SEE SIGHTS HERE

Khaki-Clad Youngsters From All Parts of Nation Break Trips Home From Jamboree

ARRIVE BY TRAIN AND BUS

Chartered Motors and River Boats Share Honors as Boys Try to See Everything

HEAT IS NO DETERRENT

Which May Be Partly Explained by Fact Many Come From the South and Southwest

Sending of Messages to Planets Predicted by Dr. Tesla on Birthday

LAST OF THE DINOSAURS

Inventor, 81, Talks of Key to Interstellar Transmission and Tube to Produce Radium Copiously and Cheaply—Decorated by Yugoslavia and Czechoslovakia

CALEDONIA'S CREW RISES ABOVE HEAT

Captain and Aides Seek a Refuge in Tower of the Empire State Building

THEN GO TO SWIM POOL

Reason is Britishers Brought Only Heavy Uniforms and Rather Warm Tweeds

'FREEWAY' IS URGED ON MOTOR PARKWAY

Breaking Up of Long Island Thoroughfare Is Opposed by Regional Plan Group

EXPRESS ARTERY FAVORED

'Priceless Chance' Is Seen for Direct Route From World's Fair to Farmingdale

Dr. Nicoll Urges State Alienists To Fight Abuse of Insanity Plea

Permanent Board Proposed by Health Official to Prevent Killers' Escape From Death Penalty—Would Take Decision From Juries and Confine Psychopaths

SECURITY ACT HELD FAR TOO LIMITED

Wider and Larger Benefits Required, 20th Century Fund Says, Suggesting Reforms

DELAY CALLED NEEDLESS

Beginning of Payments in 1939 Is Urged—Levy on Employe Found Too High

RESERVE FUND CRITICIZED

Pay-as-You-Go System Proposed as Substitute—More Federal Aid to States Backed

26,923 CARS A DAY CROSS TRIBORO...

Above: The New York Times for Sunday, July 11, 1937 calling Tesla a "Dinosaur." Tesla, a man living far ahead of his time, rather than behind the times, speaks of sending messages to Mars on his 81st birthday. Marconi and his scientists were already preparing to journey to Mars with their electro-gravitic spacecraft. Right: A recent article on Tesla's advanced science by Oliver Nicholson in the January, 1990 issue of FATE magazine.

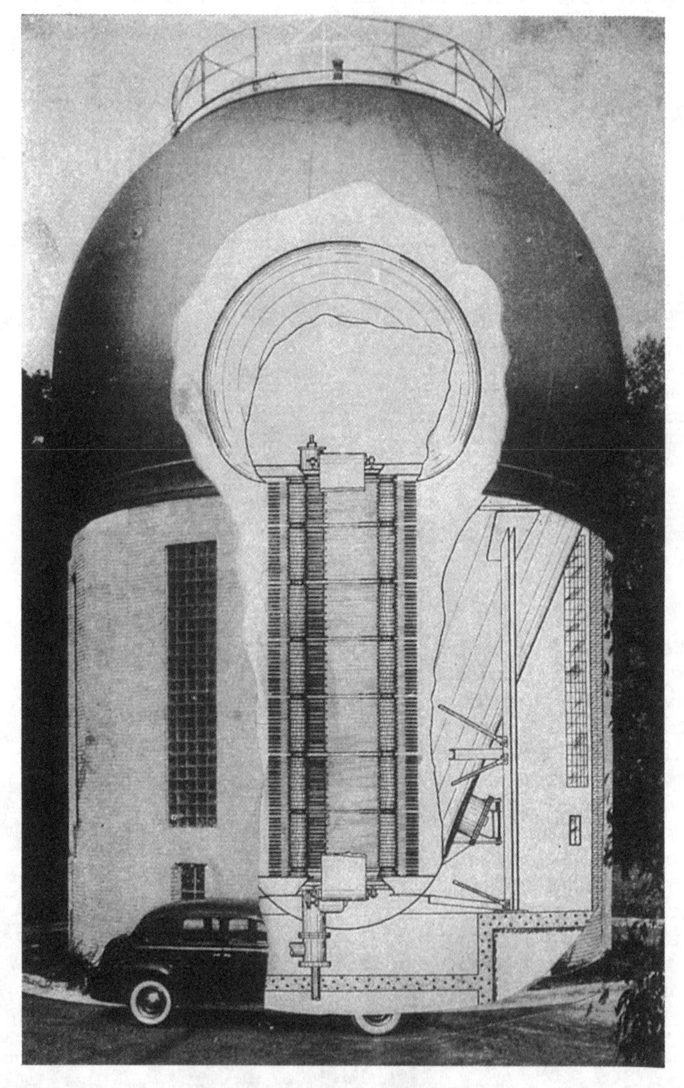

This electrostatic atom-smasher was built at the Carnegie Institution in Washington D.C., and used between 1920 and 1940. The cross-section shows a spherical conductor, its insulating supports, and tube in which particles are accelerated. The charging belt is shown cut-off near the top and bottom. This structure was also the talk of "death-rays."

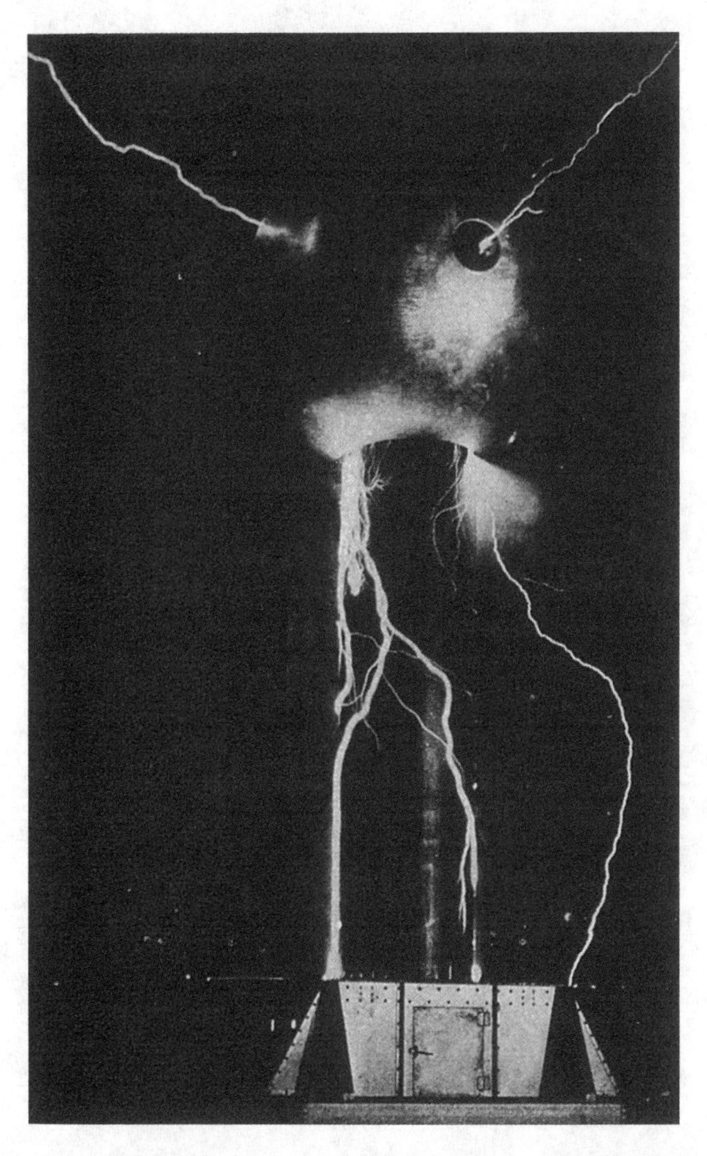

The Van de Graff electrostatic generator of the Carnegie Institution in Washington D.C. in action. Note the man-sized door at the bottom of the building. This gives a good idea of how Tesla's Wardenclyffe tower might have appeared when operational.

The amazing Wardencliff Tower of Long Island in full action as Tesla envisioned it. The tower is broadcasting power to anti-gravity airships and electric airplanes that hover around it. Note the powerful searchlight-beams on the airships. These were a combination of searchlight and death-ray, as commonly spoken of by Tesla.

TESLA SCALAR WAVE SYSTEMS: THE EARTH AS A CAPACITOR

by Richard L. Clark

Nikola Tesla engineered his communications and power broadcast systems based on the Earth as a spherical capacitor plate with the ionosphere as the other plate. The frequencies that work best with this system are 12 Hz and its harmonics and the "storm" frequency around 500 KHz. The basic Earth electrostatic system and the basic Tesla designs are shown in the figure below. All lengths or circuits must be one-quarter wavelength or some odd multiple of it.

The elevated capacitor has really two functions, Capacity to Ground (Cg) and Capacity to Ionosphere (Ci). The bottom plate only to ground is Cg, and both plates are Ci. L2 and C3 are a resonant stepdown air core coupling system at the desired frequency. Simple calculations will allow resonant frequency values to be determined from the Tesla Equivalent Circuit diagram. Be extemely careful of the high voltages in this system.

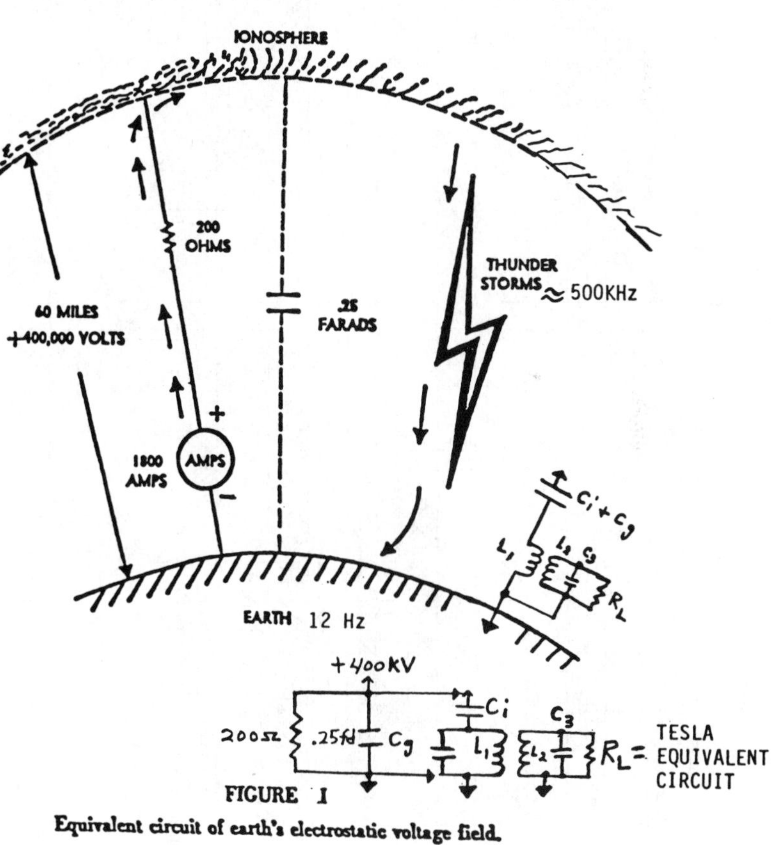

FIGURE I

Equivalent circuit of earth's electrostatic voltage field.

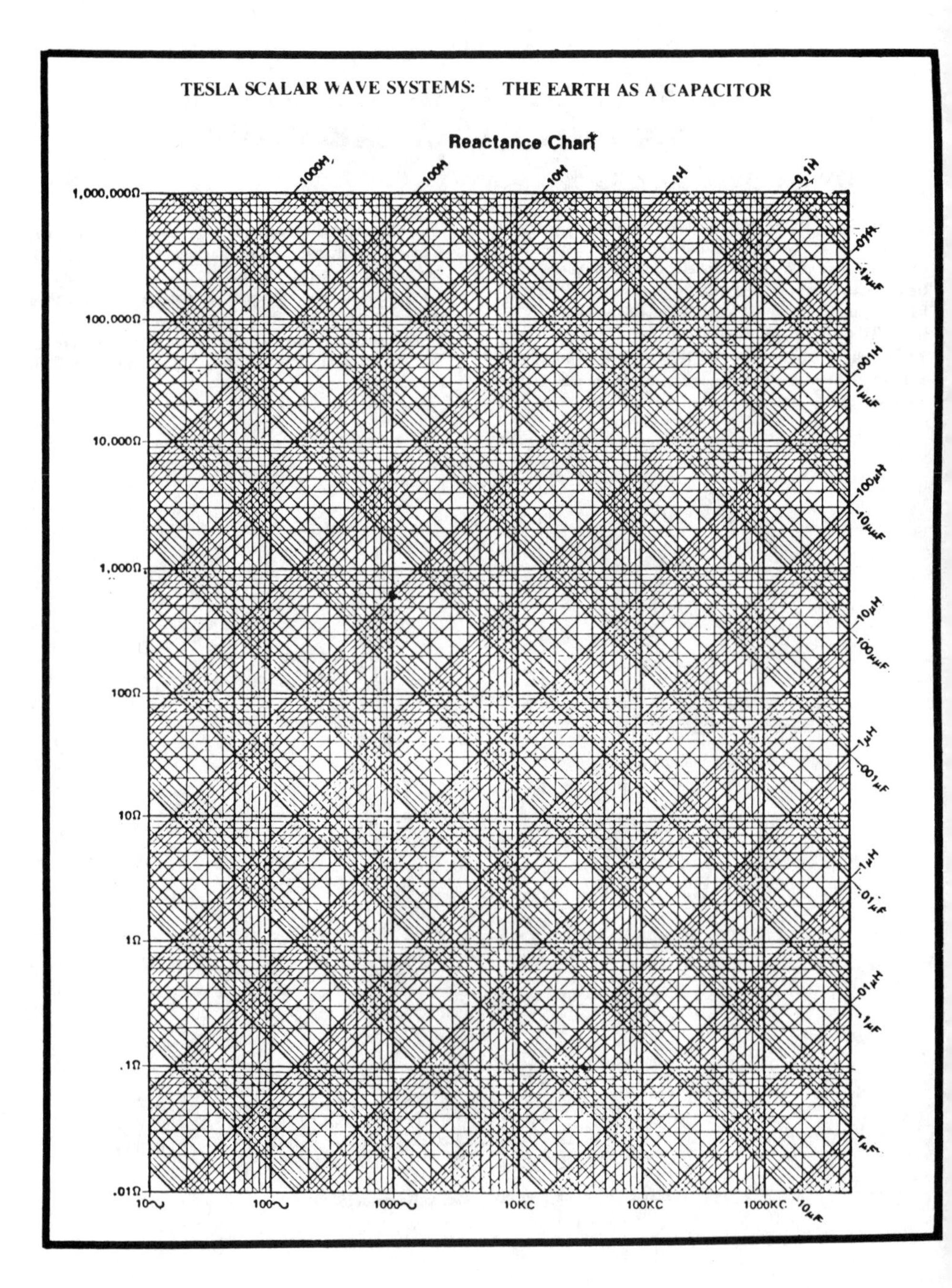
Reactance Chart
1000H
100H
10H
1H
1,000,000Ω
100,000Ω
10,000Ω
1,000Ω
100Ω
10Ω
1Ω
.1Ω
.01Ω
10~
100~
1000~
10KC
100KC
1000KC
100μH
10μF
10μH
100μF
1μH
.001μF
.01μH
1μF
10μF

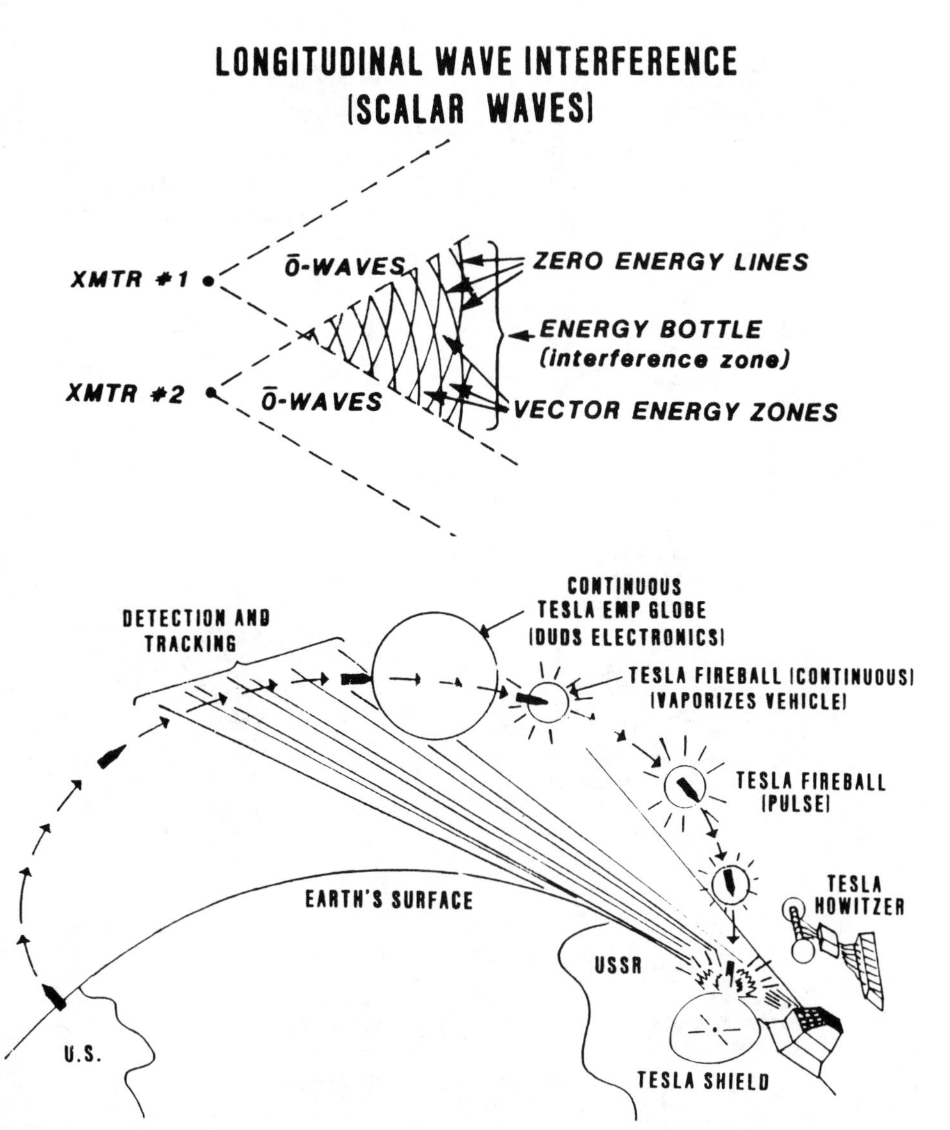

Bearden's Scalar Wave weapons in action. Tomarrow's science-fiction weaponery was yesterday's reality. Yet science has apparently not moved forward with this technology for eighty years—or has it?

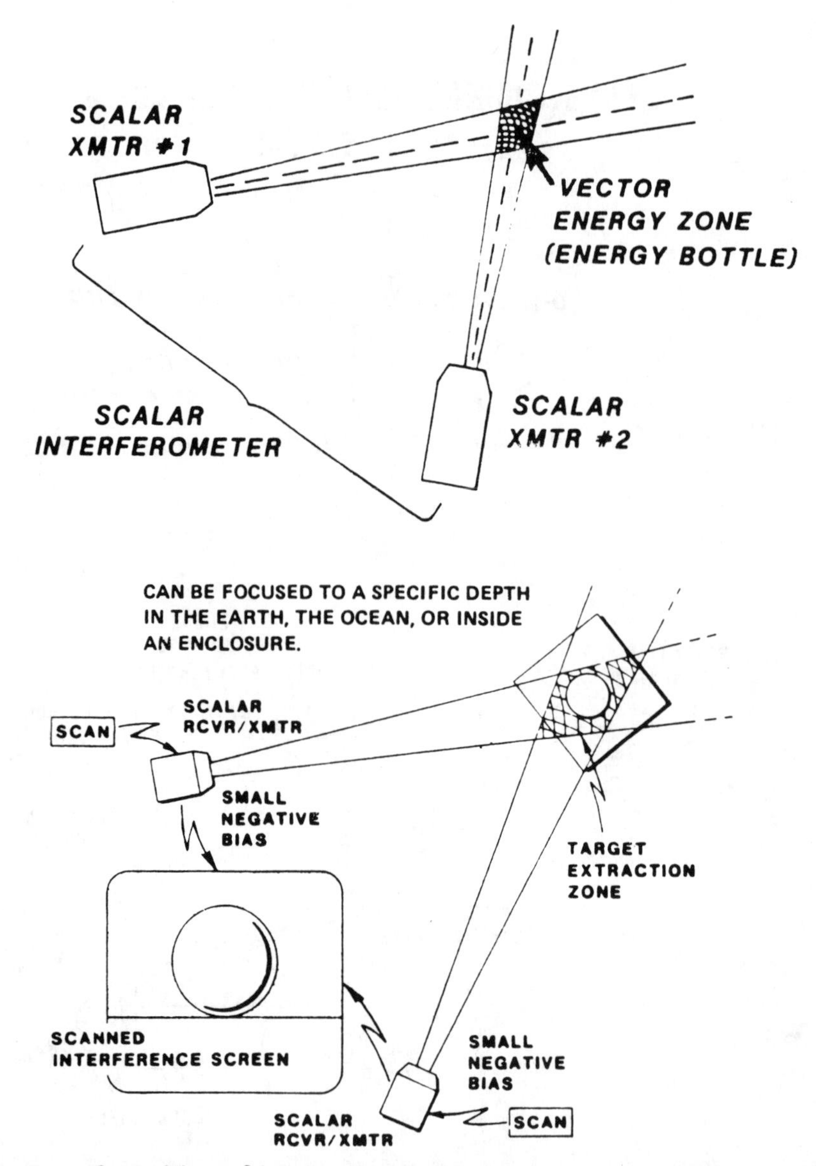

Col. Tom Bearden's idea of how a "Tesla Howizter" system using current scalar wave technology might work. Compare to Tesla's 1920 illustration for his "directed ionized beam transmissions."

Chicago Tribune, Sunday, August 10, 1986

Was Edison adversary father of 'Star Wars'?

By James Coates
Chicago Tribune

COLORADO SPRINGS, Colo.—Giants have trod the ground here. Zebulon Pike, legendary explorer of the unknown West, gave his name to the majestic white- capped peak just outside of town.

President Dwight Eisenhower came here to carve America's ultimate nuclear war command center, the awesome North American Aerospace Defense Command [NORAD] bunker, into the granite underneath Pike's Peak's neighboring summit, Cheyenne Mountain.

Most impressive of all, the man who invented radio and who discovered the way that the world transmits its electrical power did much of his creative work here.

But, wait. Weren't we taught that radio was invented by an Italian named Guglielmo Marconi? And that the legendary Thomas Alva Edison devised today's electrical power system in his New Jersey laboratories?

"We were taught wrong," said Toby Grotz, president of the International Tesla Society based here in honor of a little-known flamboyant genius named Nikola Tesla.

Two years before Marconi demonstrated his wireless radio transmission, Tesla, a naturalized Yugoslavian immigrant, performed an identical feat at the 1893 World's Fair in Chicago.

On June 21, 1943, in the case of Marconi Wireless Telegraph Co. vs. the United States the Supreme Court ruled that that Tesla's radio patents had predated those of the Italian genius.

To be sure, Edison invented the incandesent light bulb. But he powered it and all of his other projects with inefficient direct current [DC] electricity.

It was Tesla who discovered how to use the far more powerful phased form of alternating current [AC] electricity that is virtually the universal type of electricity employed by modern civilization.

And now, there are indications that Tesla also discovered many of the devices which the United States military-industrial complex is seeking to develop and build for the Pentagon's controversial Star Wars antimissile defense system.

Grotz and other Tesla experts speculate that recent puzzling reports of immense clouds forming within minutes over Soviet arctic territory are indications that the Soviet Union is testing devices for transmitting energy over large distances developed nearly a century ago by Tesla.

Of particular interest to Tesla researchers, said Grotz, is a widely reported April 9, 1984, event in which at least four airline pilots reported seeing an eruption near Japan that appeared to be a nuclear explosion cloud that billowed to a height of 60,000 feet and a width of 200 miles within just two minutes and enveloped their aircraft.

In late July the Cox News Service reported that all four of these planes had been examined by the U.S. Air Force at Anchorage, Alaska, and were found to be free of radiation despite the fact they had flown through the mysterious cloud in question.

Grotz said that such clouds could form if someone were attempting to implement Tesla's plans for broadcasting energy by "creating resonances inside the earth's ionospheric cavity" calculated in Colorado Springs during 1899 experiments by the electrical genius.

Nikola Tesla: Is his research helping the Soviet Union build the ultimate weapon?

Each year about 400 members of the Tesla Society, sanctioned by the prestigous International Institute of Electric Engineering [IIEE], meet here where the wizard of electricity carried out his most startling lightning-crackling experiments to discuss one of the strangest stories in the annals of American science.

It is a story of tormented genius. It also is the story of a little-known but intensely bitter feud that pitted Edison and the fabulously wealthy financier J.P. Morgan on one side and Tesla and his ally, the equally powerful George Westinghouse on the other. And, finally, it is a spy story.

Many in the Tesla Society are convinced

that foolish U.S. bureaucrats shipped the secrets needed to build Star Wars that Tesla discovered to communist-controlled Yugoslavia shortly after World War II, thereby allowing the Soviets an enormous head start in the quest for a particle beam weapon that is deemed essential to building any missile shield.

In an interview between sessions at this August's Tesla symposium, Grotz explained that Tesla was drawn to Colorado Springs because he needed both the dry climate and the furiously powerful lightning storms that so often come tumbling down the sides of Pikes Peak and Cheyenne Mountain.

"Tesla dreamed of supplying limitless amounts of power freely and equally available to all persons on Earth," said Grotz.

And he was convinced he could do so by broadcasting electrical power across large distances just as radio transmits far smaller amounts of energy, explained Grotz.

The same energy beams, of course, could be directed at the speed of light to destroy enemy planes and missiles as well as to supply electricity, he noted.

Such investigations take one into the realm of the most complicated question facing science today, the so-called Unified Field Theory that Albert Einstein himself confessed was beyond his abilities, acknowledged Grotz, an engineer for the Martin Marietta Aerospace company in Denver.

Tesla believed that he could broadcast power by producing vibrations in the atmosphere that were perfectly in phase with the natural vibrations that exist in thunderstorms, said Grotz.

Then, anyone with a receiver could simply tap into broadcasts and acquire elecricity just as they receive radio or TV broadcasts.

On a hilltop just where the prairies sweep up to the foot of the Rockies, Tesla erected a gigantic version of what is known as the Tesla Coil, a device that produces dramatic arcs of electricity by rapidly changing its resistance.

Nearly every natural history museum and high school physics lab in the world sports a Tesla Coil capable of making delighted students' hair stand on end or of arcing dramatic sparks from the fingertips of someone who, standing firmly on a rubber mat, holds the other hand over the coil's top.

At the corner of Foote and Kowia streets in Colorado Springs, Tesla erected a coil 122 feet high. Tapping into the entire city electric system, the electricial genius sent millions of volts of current into the structure and bolts of man-made lightning leaped as much as 135 feet into the brooding sky to mingle with other bolts created in nature.

The first time he threw the switch, the entire city was blacked, tests created artificial clouds around his installation and caused lights to burn as much as 26 miles away, according to news reports of the time.

The Colorado Springs artificial lightning bolts created during the single year that Tesla lived here, 1899-1900, have never been duplicated, said Grotz.

The experiments established that lightning storms as they swooped down the Rockies and then rumbled across the plains into Kansas were resonating at a frequency of 7.68 cycles per second.

This natural phenomenon was rediscovered in the 1960s by researcher W.O. Schumann while working for the Navy on ways to broadcast nuclear war orders to submerged submarines, said Grotz.

A paper widely circulated at the Tesla symposium called "Star Wars Now! The Bohm-Aharonov Effect, Scalar Interferometry and Soviet Weaponization" speculates that the mysterious clouds that frightened airline pilots were created when energy was drained from one area and transmitted to another using Tesla principles.

The paper's author, T.E. Beaden, a retired Pentagon war games expert and active consulting engineer to the Defense Department, said the result of such energy transmissions is a "cold explosion" that could be enormously destructive.

Noting that the cloud covered 150 miles, Beaden wrote, "A single shot of such a weapon could almost instantly freeze every NATO soldier in that area into a block of ice."

Grotz acknowledged that much of the world's mainstream scientific community doubts the claims made by Tesla fans like himself and Beaden.

"But," he added, "Tesla always was rejected by the establishment."

After Tesla began building AC dynamos, motors and other devices with financial backing from Westinghouse, Edison and his General Electric Company waged a campaign to discredit AC by emphasizing its dangers, according to Tesla biographer Margaret Cheney in her "Tesla, Man Out of Time."

Edison would force dogs and cats to stand on steel plates energized by AC current and then throw a switch, electrocuting them. He called the process "Westinghousing," Cheney wrote.

Ultimately Tesla lost out to Edison and other foes, even though his AC power system prevailed.

The visionary died in 1943 in a New York hotel room he shared with several pigeons that he considered his only friends, the biographer said.

After the war, Tesla's relatives in Yugoslavia petitioned Washington to receive 17 trunks of papers and laboratory equipment that he had stored in a New York garage.

In 1952 these items were sent to Belgrade where they are housed in a Tesla museum.

But, said Grotz, "What do you suppose are the chances that everything was first copied by the KGB?"

"In the USA we don't even give him credit for inventing the radio and the Soviet bloc is building Tesla museums," said the engineer.

"Why do they respect him so much?"

Our Future Motive Power

By

Nikola

Tesla

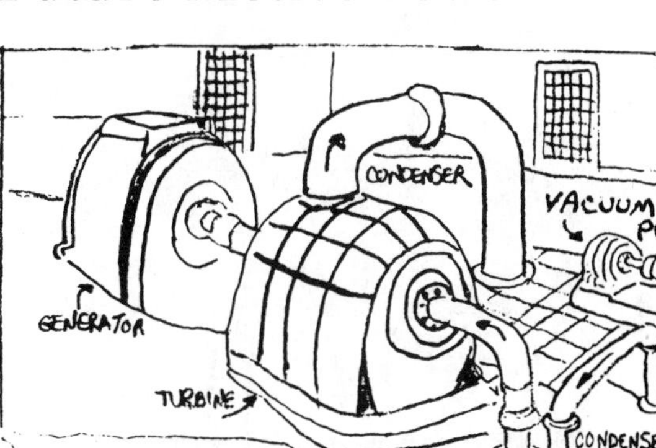

December 1931

Above and to the right, the arrangement of one of the great terrestrial-heat power plants of the future. Water is circulated to the bottom of the shaft, returning as steam to drive the turbines, and then returned to liquid form in the condenser, in an unending cycle.

Internal heat of the earth is great and in comparison with the demands which man can make upon it, is practically inexhaustible: since the heated contents of the earth are sex-trillions of tons.

This drawing illustrates the essential parts comprising a boiler at a great depth, a condenser, cooled by river or other water available, on the ground, a turbine coupled to a generator, and a motor-driven high vacuum pump. The steam or vapor generated in the boiler is conveyed to the turbine and condenser through a insulated central pipe while another smaller pipe, likewise provided with a thermal covering serves to feed the condensate into the boiler by gravity. All that is necessary to open up unlimited resources of power throughout the world is to find some economic and speedy way of sinking deep shafts.

CONDENSED STEAM RETURNED TO PIT

STEAM

WATER

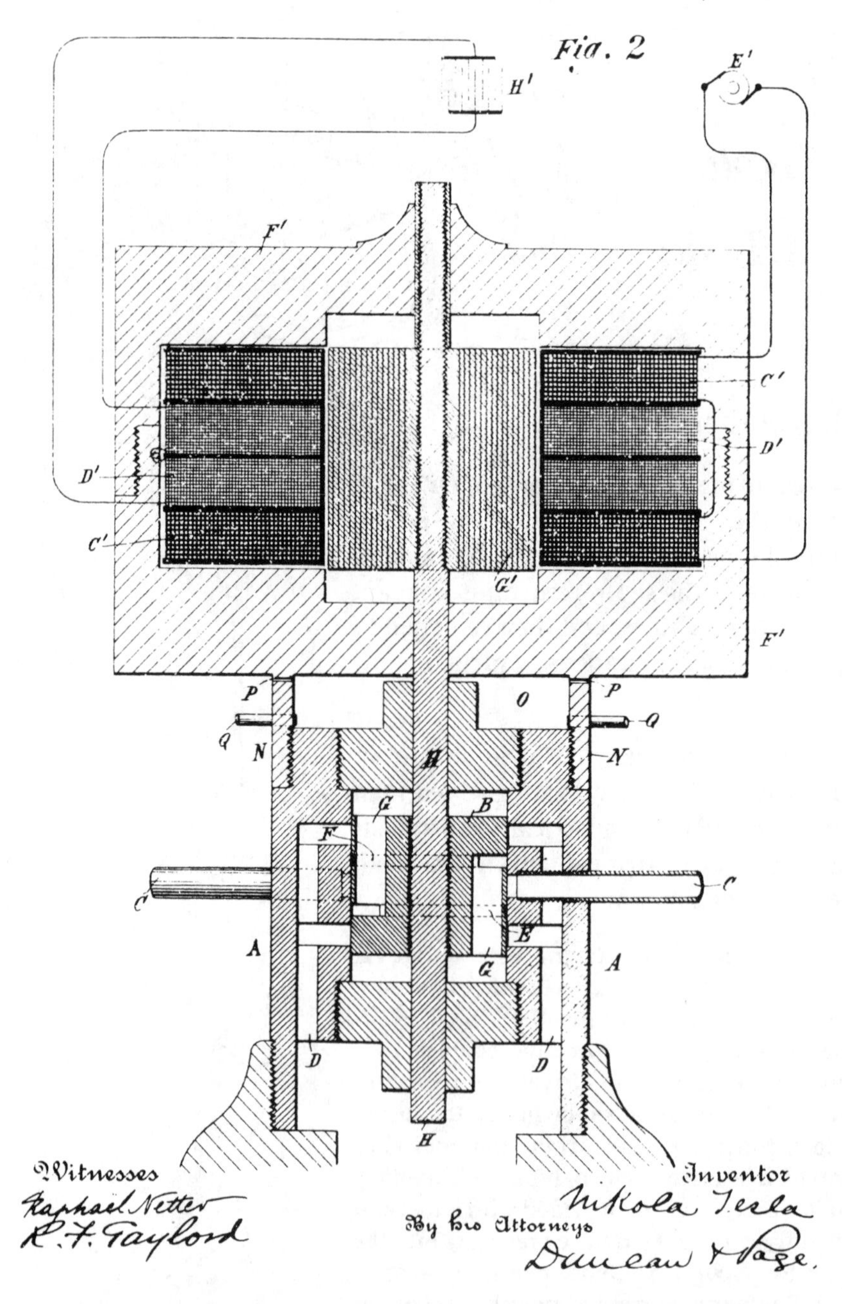

Tesla's fascinating patent of Jan. 2, 1894 is for a mechanical oscillator with a controlling electromagnetic system.

Chapter 7

THE MOST UNUSUAL INVENTIONS

Many of Tesla's inventions, now 90 to 100 years old, still seem like science-fiction to us today.

Much of Tesla's inventions and overall power system in development, does indeed seem to be a recreation of an advanced and ancient system of transmitting power to parts of the globe. As early as 1899, in the Colorado Springs tests, Tesla sent electrical beams through the earth and lighted up light bulbs five miles away.

Tesla is credited with so many inventions, that one might begin to think that he invented much of our modern technology. Tesla's battle for recognition as the inventor of the radio, a device still usually credited to Marconi, is familiar to most Tesla enthusiasts.

Among the incredible inventions that Tesla actually conceived, frequently patented, were:

The Electric Submarine

In 1898, Tesla patented the "Teleautomaton Boat," (#613,809) an electrically powered submarine. This submarine would pick-up electricity that was being broadcast to it by a receiver. Power could also be stored in batteries and the electric submarine could be operated by remote control.

Tesla's VTOL

His design for a vertical take-off and landing (VTOL) aircraft received its patent on January 3, 1928. This was to become Tesla's last patented invention. After this, he no longer sought patents on any of his inventions.

No. 613,809. Patented Nov. 8, 1898.

N. TESLA.

METHOD OF AND APPARATUS FOR CONTROLLING MECHANISM OF MOVING VESSELS OR VEHICLES.

(No Model.) 5 Sheets—Sheet 2.

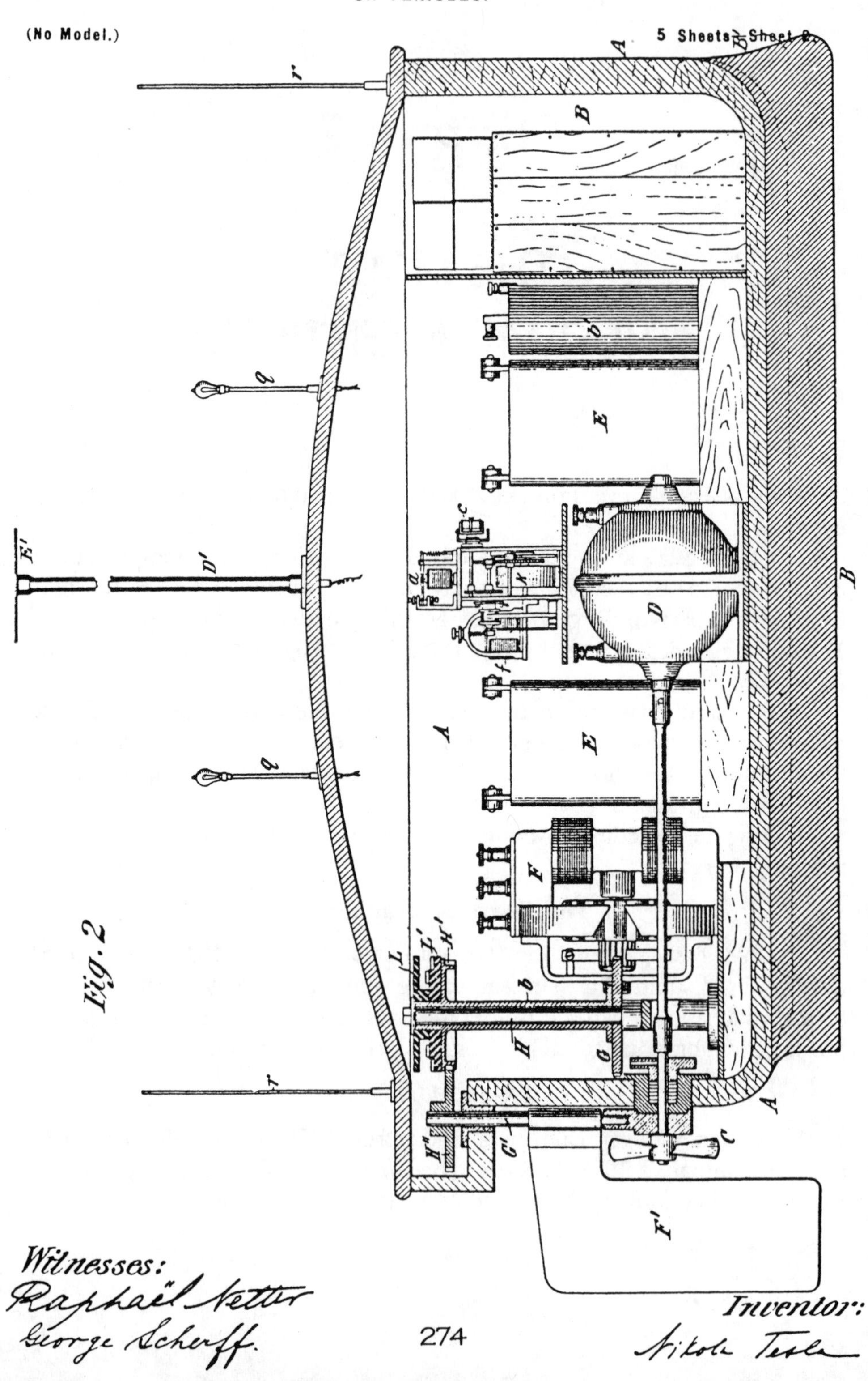

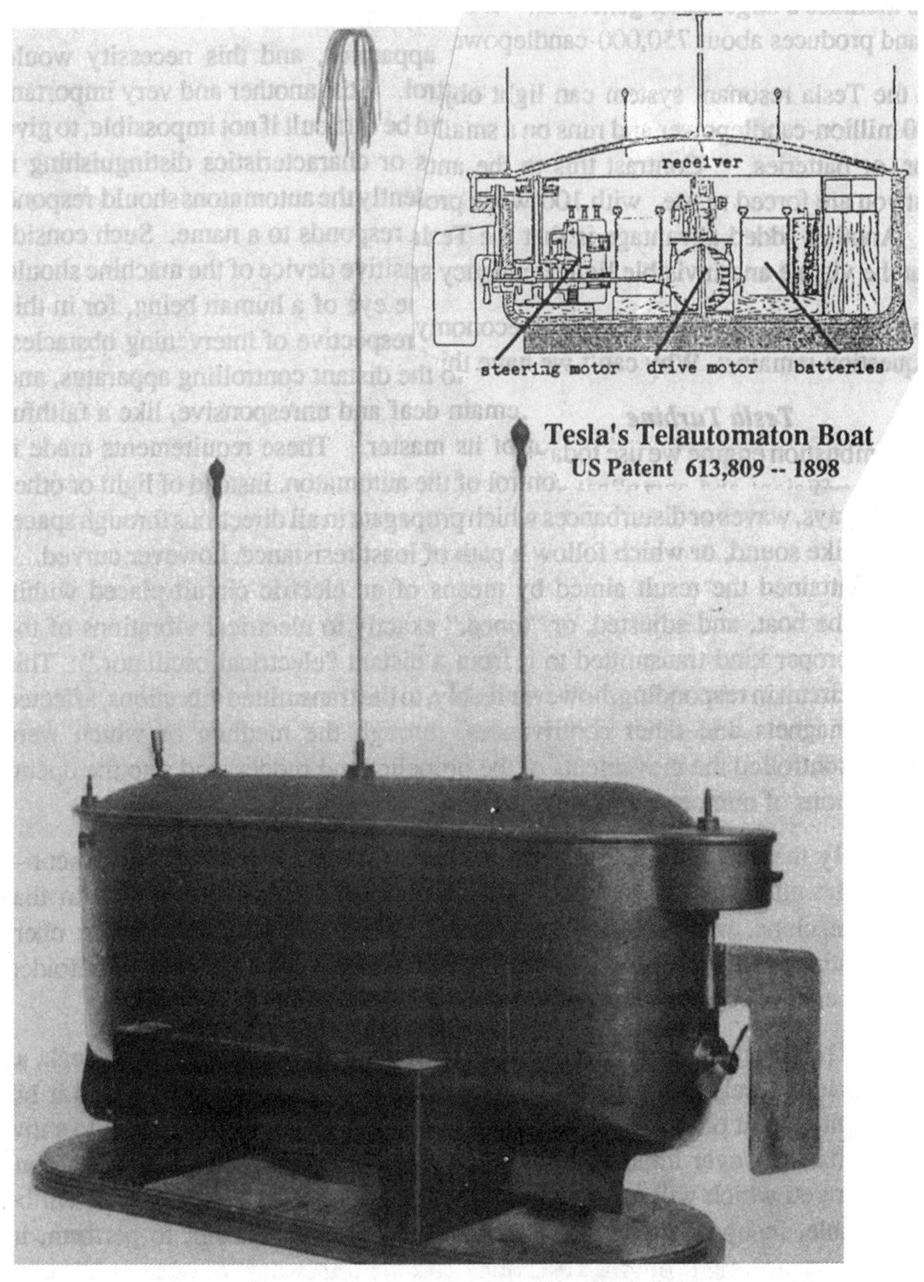

THE FIRST PRACTICAL TELAUTOMATON.

A machine having all its bodily or translatory movements and the operation of the interior mechanism controlled from a distance without wires. The crewless boat shown in the photograph contains its own motive power, propelling--and steering-machinery, and numerous other accessories, all of which are controlled by transmitting from a distance, without wires, electrical oscillations to a circuit carried by the boat and adjusted to respond only to these oscillations.

The Electro Dynamic Induction Lamp

In 1894, Tesla's Electro-Dynamic Induction Lamp was given U.S. patent 514-170. The Electro-Dynamic Induction Lamp is a type of light bulb that is said to be far in advance than those currently available.

The Bladeless Tesla Turbine

This bladeless turbine, patented in 1916 (#1,329,559), uses a series of rotating discs to pump liquids or gases through a turbine engine. Hovercraft, speed boats, or simple pumps can utilize the Bladeless Turbine. It is said to be the world's most efficient engine, and is 20 times more efficient than a conventional turbine, yet, it is still not in use today!

Improved Unipolar Generator

In 1891, Tesla published an article in The Electrical Engineer (New York, Sept. 2, 1891 about his improved version of the Unipolar Generator. His rotating disc and opposing magnets has been copied by many "magic-motor" builders over the years.

Tesla's Mechanical Oscillator

An unusual and little-known device invented by Tesla was the Mechanical Oscillator which compressed air until the oxygen became a liquid. It was built in the form of an air cylinder and contained several chambers, each of which successively cools the air until it becomes liquid. Tesla stated that the device was highly efficient and could be used as a power generating system if magnets were attached to the oscillating pistons. Tesla believed that an "oxygen recycle system" was a vast improvement to gasoline engines and intended to conduct important experiments with LIQUID OXYGEN for new turbine engines capable of developing extraordinary power.

Tesla's Ozone generator

Tesla's ozone generator. US Patent 568,177, issued in 1896. Ozone generator's are currently banned for medical use in the U.S. despite the claims of some doctors that ozone therapy can cure cancer and AIDS.

Tesla's "Thought Photography" Machine

This was perhaps Nikola Tesla's most fantastic invention, a device for photographing thought. Tesla once said in 1933, when 78 years old, "I expect to photograph thoughts... In 1893, while engaged in certain investigations, I became convinced that a definite image formed in thought, must by reflex action, produce a corresponding image on the retina, which might be read by a suitable apparatus. This brought me to my system of television which I announced at that time... My idea was to employ an artificial retina receiving an object of the image seen, an optic nerve and another retina at the place of reproduction... both being fashioned somewhat like a checkerboard, with the

This was perhaps Nikola Tesla's most fantastic invention, a device for photographing thought. Tesla once said in 1933, when 78 years old, "I expect to photograph thoughts... In 1893, while engaged in certain investigations, I became convinced that a definite image formed in thought, must by reflex action, produce a corresponding image on the retina, which might be read by a suitable apparatus. This brought me to my system of television which I announced at that time... My idea was to employ an artificial retina receiving an object of the image seen, an optic nerve and another retina at the place of reproduction... both being fashioned somewhat like a checkerboard, with the optic nerve being a part of earth." Tesla made his transition before revealing too many details of his invention. The above were taken from a newspaper interview that was released to wire services on September 10, 1933.

optic nerve being a part of earth." Tesla made his transition before revealing too many details of his invention. The above were taken from a newspaper interview that was released to wire services on September 10, 1933.

Anti-Gravity & the Wall of Light

When matter is considered to be linked by electromagnetic wave propagations from the sun, manifesting as sunlight, then a literal *Wall of Light* can be created, and through this wall, all manifestations of time, space, gravity and matter can be manipulated. The mystical Wall of Light, used frequently in Tesla references, and is the title of a book about Tesla, is allegorical to columns of light in the sky, and to the manipulation of energy and matter.

Tesla was unquestionably a visionary and a mystic. Anti-gravity airships were typically depicted in illustrations of his interviews and advanced predictions. He often spoke of the coming world in which anti-gravity aircraft will carry cargo across the continent, drawing power from centrally located power stations along the earth grid.

Tesla has been credited with several space drives, though plans that have been published are dubious. In his mind he had no doubt created an electro-gravitic craft that would draw power from his Wardenclyffe Tower plant. Could such a craft have ever been built in secret? Tesla certainly had the plans for such a craft in his head—all he needed was a wealthy financier. Perhaps a Jules Verne-type character like Captain Nemo.

It is interesting to note here that Tesla's electric submarine could also be the proto-design for the airship, as these cigar-shaped craft can allegedly go underwater, and act as submarines, as well as airships.

Teleportation and Time Travel Devices

Tesla's *Death Ray*, a kind of radio-wave-scalar weapon or ultra-sound gun, was the stepping stone to more important inventions, like teleportation and Time Travel devices. H.G. Wells had already popularized the idea, but Tesla may have actually experimented with such devices.

With such popular time-travel tales as The Philadelphia Experiment and The Montauk Project, it would only seem natural that secret government research on time-travel and teleportation would owe something to the work of Nikola Tesla. If Tesla was truly the genius that some believe he was, he could have made his own time-machine and gone into the future, or maybe teleport himself to Mars. Perhaps he built a flying saucer and flew away, after cleverly faking his death.

Tesla was like an odd-ball hero from the past. A literal *Man Out of Time.* He had visions of his inventions, even as a teenager.

Tesla and Atlantis

According to the *Unarius Academy* of San Diego, California, Nikola Tesla was the reincarnation of an Atlantean engineer and inventor who was responsible for the energy supply first used to power on a now destroyed island in the Atlantic. According to Unarius, from the great central pyramid in Atlantis, power beams would be relayed from reflectors on mountaintops into the different homes where these power beams would be converted into light, heat or even to cool the house.

According to *Unarius*, a round glass globe or sphere about a foot in diameter was filled with certain rare gases that would fluoresce and give off a soft white light, just as does a modern fluorescent light. Heating or cooling was also quite simple: Air being made up of molecules of gases, each molecule composed of a number of atoms. Electrical energy of a certain frequency was then radiated through the air and converted into heat through "hysteresis" in the electromagnetic fields of the atoms.

According to *Unarius*, the same proposition in reverse makes the air become cold. Similarly, the atmosphere on the earth is always converting certain electromagnetic energy into heat. Speaking from the point of absolute zero (495 degrees Fahrenheit), all air on the surface of the earth is comparatively warm, even at the poles.

Cooling or heating the air at any given point means merely to decrease or increase the "electromagnetic hysteresis." As a definition for a Pabst hysteresis-synchronous motor, Unarius says that it is the "inductive principle of cosmic hysteresis, and add that "The reference to "hysteresis" is not the earth-electronics definition, but rather an electromagnetic conversion process wherein cyclic (4th dim.) waveform-structures are transformed into lower (3rd dim.) waveform-structures."

Minoan homes are used as an example, where it is said that a small object a foot or so square sitting on the floor of any room could be both the heater and the cooler. It would, according to the dictates of a thermostat, radiate certain energies into the room which would either slow down hysteresis and make the air cooler or speed up hysteresis and make the air warmer; a far different process than our present-day crude, clumsy, inefficient and enormous heating and cooling systems which must always either heat air in a furnace or cool it by means of refrigeration and, with a fan, blow it into the room through a large duct.

The Atlantean Power System

Tesla's Atlantean power system, according to Unarius, was a huge rotating squirrel-cage generator turned by a motor was linked up to an electronic computer which was housed in a twenty-foot square metal box on the floor just above the generator. This computer automatically made and broke connections—with banks of power collector cells on the outside pyramid surface in such a sequential manner that a tremendous oscillating voltage was built up. On the ten-foot ball which stood atop the metal box, this oscillating electricity discharged more than six hundred feet straight up to a similar metal ball hanging down from the pyramid apex on a long metal rod.

Unarius compares the Atlantean-Tesla system to that of a 1900's scientist named Steinmetz, a friend of Tesla's. Steinmetz hurled thunder-bolts from two large metal spheres one hundred feet apart in a manner which is somehow strangely similar to the process used in the Atlantean Pyramid 16,000 years ago. This discharge across the two metal balls served as a tank-circuit, as it is called, and again a similarity to our modern early-day wireless, a motor turning a rimless rotary wheel from which protruded a number of spokes, actually electrodes. As the wheel rotated about 2,000 rpm (rotations per minute), a sizzling white spark jumped from the spokes to another electrode placed about one-half inch away from the spokes. It was this spark-gap which created the necessary high-intensity voltage.

According to *Unarius,* on top of the Atlantean pyramid was a fifty-foot metal column, something like a thick flagpole, which terminated in a circular bank of what looked like the spokes on a wheel. About ten feet long and sixteen inches in diameter, these spokes protruded at a number of irregular intervals, each one carefully sighted like a rifle, to a near or distant receiver. These spokes were actually composed of an exotic mixture of metals and formed into a homogeneous, crystalline aggregate under extreme pressure and magnetic hysteresis. Each rod or spoke then contained billions of tiny crystals; each one pointed, so to speak, toward the outside flat of the rod. They absorbed energy and like a boy who'd eaten too much watermelon, they reached certain capacity and discharged their energy toward the outside end of the rod.

The net total of these charge and discharge oscillations were on the order of millions of megacycles per second and as they functioned from the end of the rod, a beam of pure coherent energy emerged—and at the rate of more than 186,000 miles per second straight to a receiver, a beam of enormous power. How similar to our present first versions of the laser: A six-inch synthetic ruby rod, one inch in diameter and containing many chromium molecules; these chromium molecules were charged with electricity from an outside source of condenser banks and other associated equipment which generated a high-frequency impulse. As the chromium molecule atoms reached their saturation point, they discharged their energies which began to oscillate ping-pong fashion from each end of the optically-ground and slivered ends of the rod. When this oscillating energy reached a certain point, it discharged through the more lightly silvered end in a single straight coherent beam of great intensity and power.

The power beams which emerged from the Atlantis pyramid were intercepted by similar metallic rods of crystallized metal which, because they oscillated in a similar manner an frequency, presented no resistance to the enormous power of the beam. The beam then traveled straight through the rod or was broken up and separated into separate beams by a crystal prism, which again sent beams pulsating through crystalline rods and on a new tangent to another receiver.

In utilizing these power beams in a dwelling, a metal ball fitted on top of a metal rod, like a small flagpole, contained a crystal of certain prismatic configurations which directed the beam down through the hollow center of the rod to a disburser instrument which energized the entire house by means of induction so that the round milky-white crystal globes would glow with light, motors turn, etc.

The Generator-Oscillator Banks

Unarius's technical description goes on to describe the generator-oscillator banks beneath the pyramid and the generation of the 'flame'. In the subterranean chamber beneath the floor stood a motor-generator combination mounted on a vertical shaft. This piece of machinery "worked exactly similar to our present day Pabst synchronous-hysteresis motor, that is, exactly in reverse to ordinary motors which have a rotor rotating inside fixed stationary field coils. In the Pabst motor, the rotor is stationary and the metal field terminals rotate around it, similar to a squirrel cage.

"The Atlantean motor-generator combination works as follows: a huge externally-powered, (A.C.) alternating current motor rotated the squirrel cage which was actually a large number of extremely powerful high-gauss, high-intensity magnets affixed to the metal frame which rotated around what would normally be the rotor which was made from a high-permeability, soft iron core. Wound around a large number of these poles were almost countless thousands of turns of insulated wire.

"These coils were, in turn, connected up to different banks of cells on the outside skin surface of the pyramid. The sequence of this wiring was such, that when the magnets turned around the rotor, the cells and the magnetic currents so generated were in extremely rapid sequence which built up an extremely high-frequency oscillating voltage which discharged across the two balls which I described previously. The purpose of this gap was to stabilize these oscillations under resistive conditions in open air.

"Increasing the frequency increases the voltage or power which is why a laser beam can pierce a diamond with less energy than would light a small flashlight. The energy from a five-foot long lightning bolt from a Tesla coil (500,000 Cycles per second) is less than two millionths of all ampere and would cause only a mild tingling sensation. A lightning bolt traveling from a cloud to the earth contains only enough energy to light a hundred-watt bulb for about thirty seconds."

According to Unarius, electronic scientists of today "are still a bit mixed up on the proposition of voltage versus frequency. They string 1/2 inch thick laminated cable across the countryside for hundreds of miles from tall steel towers and push electricity through these cables in far-away cities at voltages in excess of 300,000 and at only 60 cycles per second alternating frequency, whereas a small pencil-thin power beam oscillating at hundreds of millions of times per second could be reflected from tower to tower across country; one beam carrying sufficient power to energize the largest city."

Protective Metal Helmets

According to Unarius, and other esoteric groups that expound on ancient science, in ancient Egypt, Mexico, and other lands where there were pyramids, the Egyptians and others tried to duplicate the round spoke-like wheel which glowed with a blue-white corona and which shot beams of intense light in different directions. The Egyptians topped their stone pyramid with a large ball-like contrivance covered with small plates of pure polished gold in a scale-like manner; and as the earth turned, shafts of light were reflected in all directions.

Several thousands of years later, these metal balls with scales of gold had disappeared, so had the alabaster white coating except for small sections near the top, in order to use the smaller surface stones in nearby cities for building purposes.

The modern Egyptians wore in their temples and palaces a metallic headdress and woven metal scafes interwoven with threads of gold which hung down over their shoulders just as they did in the ancient Atlantis when, after the scientists had gone, the Atlanteans started to worship the flame in the temple pyramid.

Unarius mentions that the metallic headdress plus a metallic robe was necessary to protect them from the strong electromagnetic field in the pyramid and through various priesthoods the metallic headdress has arrived in our present modern time in the form of a scarf worn by women in a Catholic church, or the uraeus worn by the priest.

Here we see how the Egyptian gold headdress may have originated from the ancient Atlantean power station engineers, and it is fascinating to note that the celebrated *Face On Mars* is also wearing a similar protective helmet! Are the pyramids of Mars part of a similar Atlantean Power system as Tesla was planning to build on earth? This brings us to the final mystery of Nikola Tesla: his involvement with Guglielmo Marconi and the Pyramids of Mars.

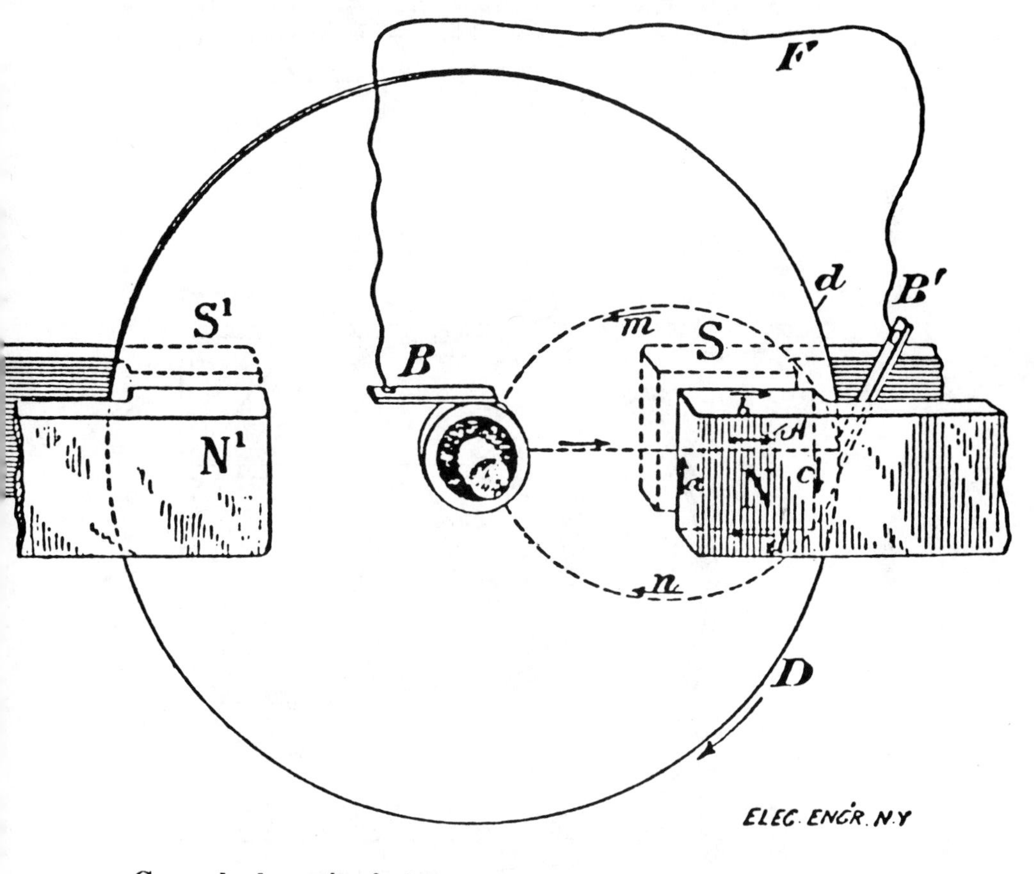

General schematic of a homopolar generator.

N. TESLA.
TURBINE.
APPLICATION FILED JAN. 17, 1911.

1,061,206. Patented May 6, 1913.

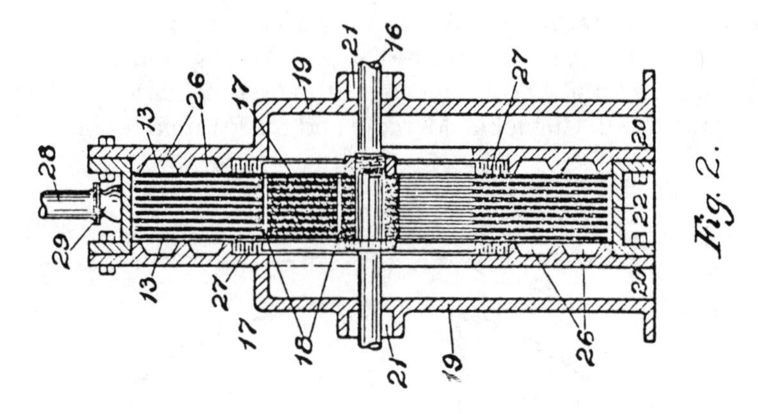

Fig. 2.

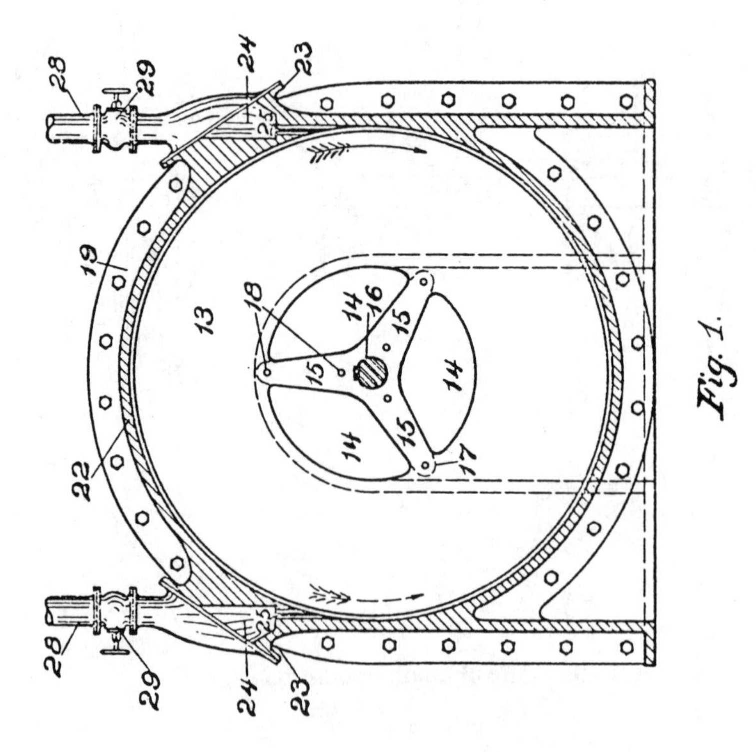

Fig. 1.

Witnesses:

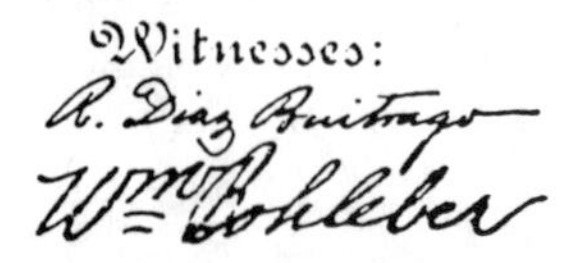

Nikola Tesla, Inventor

By his Attorneys
Kerr Page Cooper & Hayward

Chapter 8

THE LAST PATENTS (1913 TO 1928)

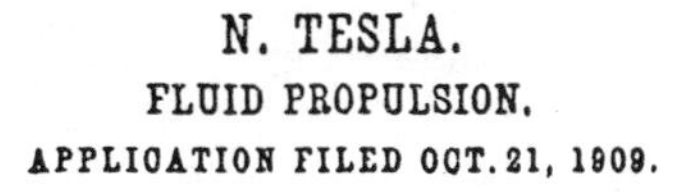

N. TESLA.
FLUID PROPULSION.
APPLICATION FILED OCT. 21, 1909.

1,061,142. Patented May 6, 1913.

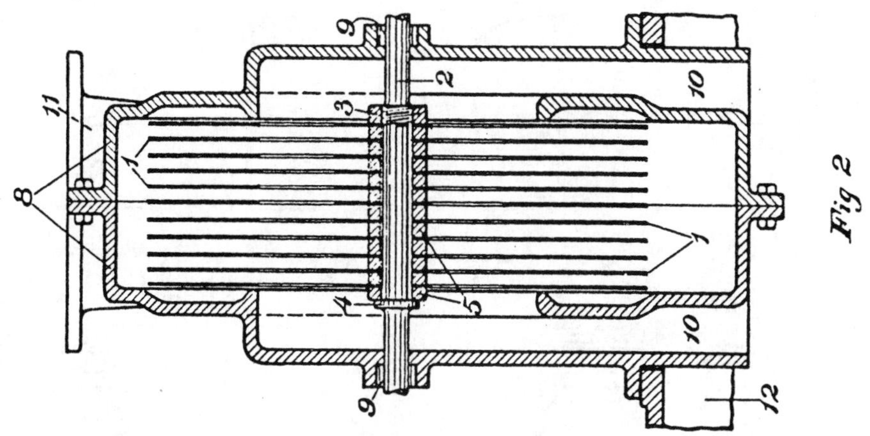

N. TESLA.
LIGHTNING PROTECTOR.
APPLICATION FILED MAY 6, 1916.

1,266,175. Patented May 14, 1918.

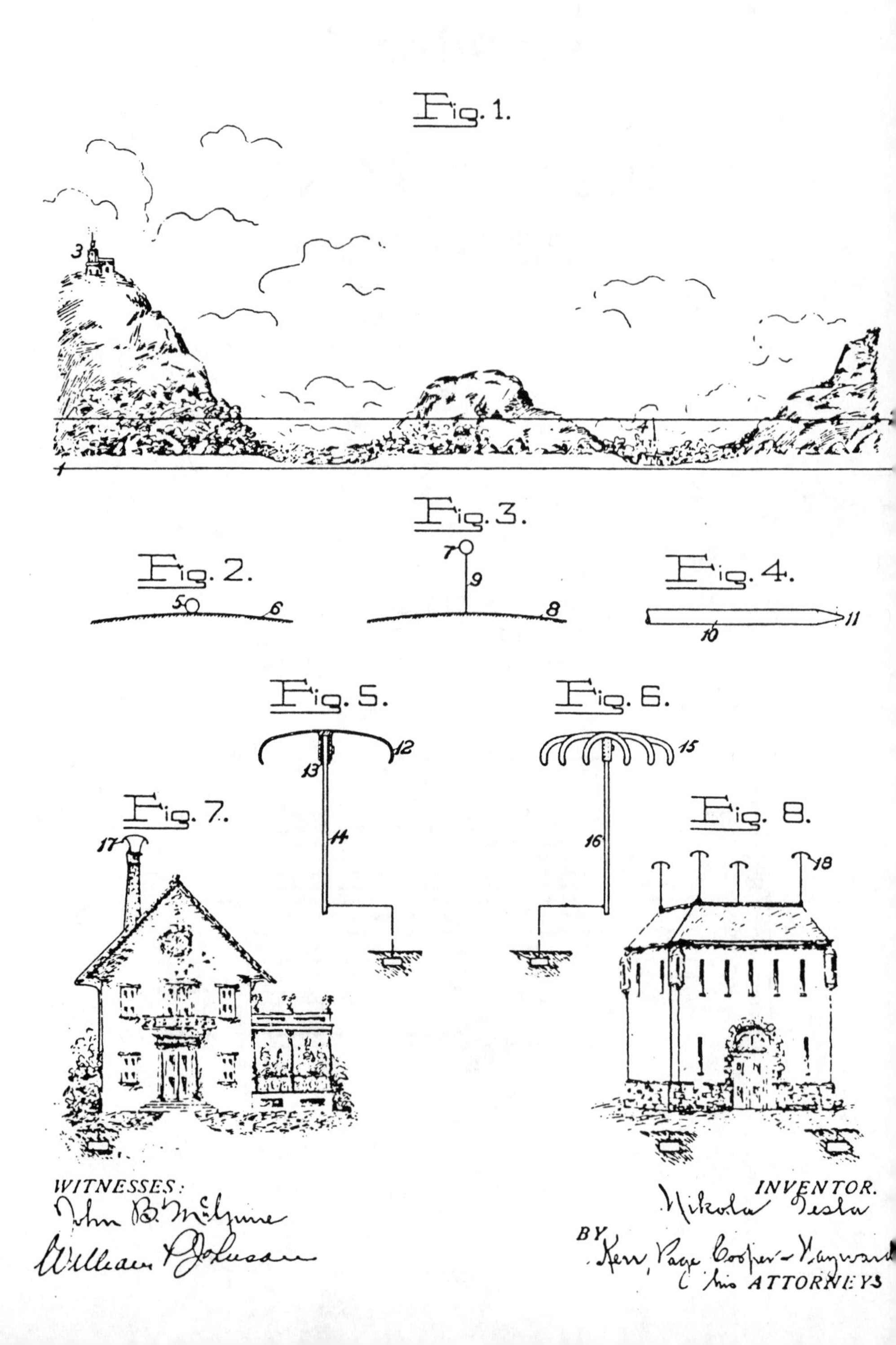

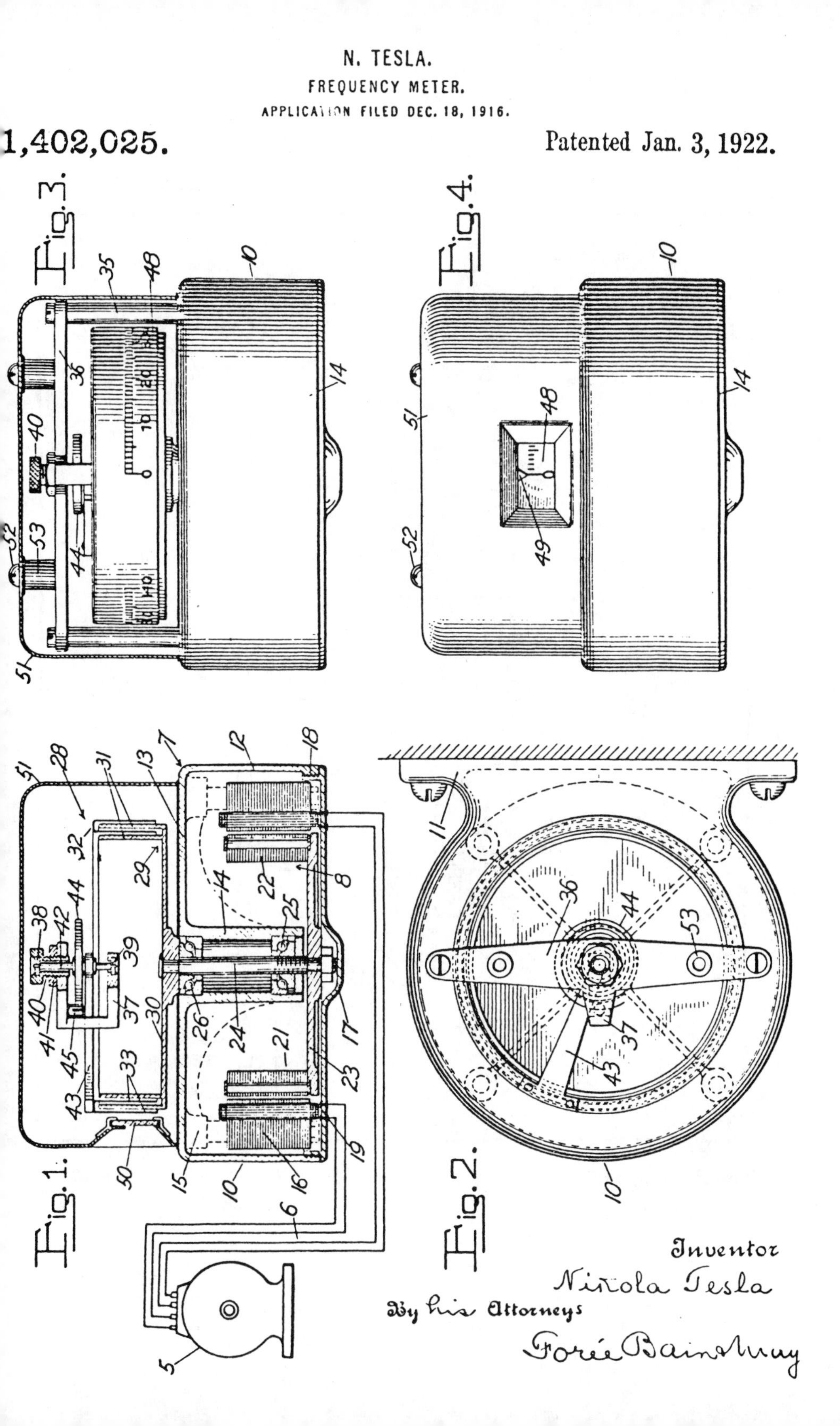
N. TESLA.
FREQUENCY METER.
APPLICATION FILED DEC. 18, 1916.
1,402,025.
Patented Jan. 3, 1922.
Fig. 3.
Fig. 4.
Fig. 1.
Fig. 2.
Inventor
Nikola Tesla
By his Attorneys
Fosée Bain & May

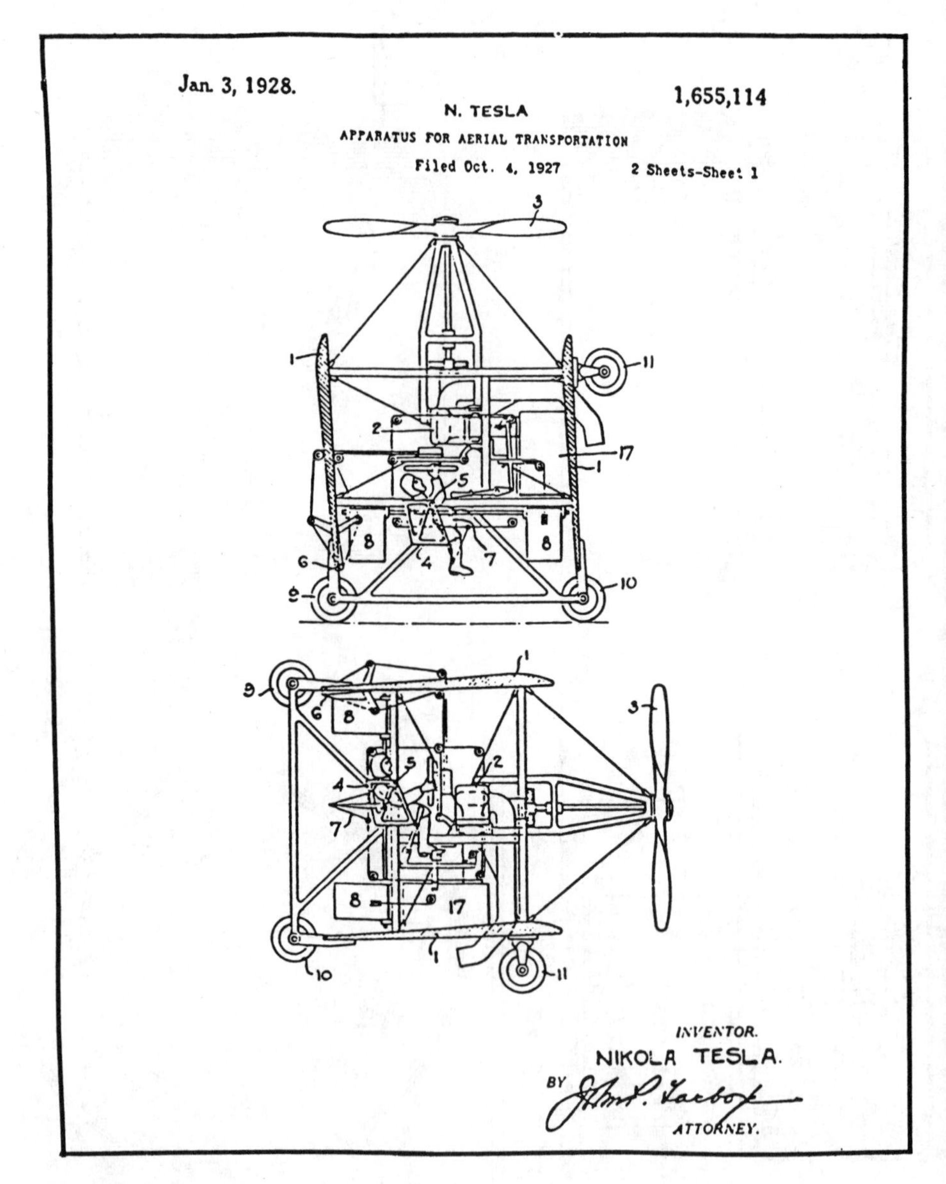

Tesla's design of a vertical takeoff aircraft.

Chapter 9

TESLA & THE PYRAMIDS OF MARS

The relationship between Tesla and Marconi is a fascinating study! While Tesla has become a popular figure to Revisionist Scientists in the last ten years, Marconi is still largely unknown and seen as an usurper of Tesla's inventions. Yet Guglielmo Marconi (1874-1937), was a brilliant scientist, and, in fact, Tesla's close friend.

In the esoteric writing of the Latin countries, Marconi has achieved a near legendary status, much as Tesla has recently in the United States. But most Tesla students are unaware that Marconi was supposed to have founded a secret high-tech city in the remote southern jungles of Venezuela.

The great Italian scientist Guglielmo Marconi was a former student of Tesla's. Marconi studied radio transmission theory with Tesla and made his first radio transmission in 1895. Marconi was fascinated by the transmission of power, and in 1896 received a British patent and sent a signal nine miles across the Bristol Channel. In 1899 he successfully set up a wireless station to communicate with a French station 31 miles across the English Channel.

It was thought that the curve of the earth's surface would limit radio transmission to 200 miles at the most. When, on December 11, 1901, Marconi transmitted a signal from Poldhu, Cornwall, to St. John's Newfoundland, 2000 miles away, he created a major sensation. For this Marconi replaced the wire receiver with a coherer, a glass tube filled with iron filings, which could conduct radio waves. At the time there was no scientific explanation for this phenomena of long-distance transmission, and it was postulated that there was a layer in the upper atmosphere—the ionosphere—which reflected back electromagnetic waves.

Marconi the Mysterious

Marconi was the son of a wealthy Italian landowner and an Irish mother. When interest in his first transmission in 1895 had not interested Italian authorities, he had gone to Britain. The Marconi Wireless Telegraph Company was formed in London in 1896 and Marconi made millions off his inventions.

Marconi and Tesla are both given credit for the invention of the radio. Marconi's historical radio transmission utilized a Heinrich Hertz spark arrester, a Popov antenna and an Edouard Bramely coherer for his simple device that was to go on to become the modern radio.

Marconi was given the Noble Prize for Physics in 1909 jointly with Karl Ferdinand Braun, who made important modification which considerably increased the range of the first Marconi transmitters.

Like Tesla, Marconi was a mysterious man in his later years, and was known to perform experiments, including anti-gravity experiments, aboard his yacht *Electra.* Marconi's yacht was a floating super-laboratory, from which he sent signals into space and lit lights in Australia in 1930. He did this with the aid of an Italian physicist named Landini by sending wave train signals through the earth, much as Tesla had done in Colorado Springs.

In June of 1936 Marconi demonstrated to Italian Fascist dictator Benito Mussolini a wave gun device that could be used as a defensive weapon. In the 1930's such devices were popularized as " death rays" as in a Boris Karloff film of the same name. Marconi demonstrated the ray on a busy highway north of Milan one afternoon. Mussolini had asked his wife Rachele to also be on the highway at precisely 3:30 in the afternoon. Marconi's device caused the electrical systems in all the cars, including Rachele's, to malfunction for half an hour, while her chauffeur and other motorists checked their fuel pumps and spark plugs. At 3:35 all the cars were able to start again. Rachele Mussolini later published this account in her autobiography.

Mussolini was quite pleased with Marconi's invention, however it is said that Pope Pius XI learned about the invention of the paralyzing rays and took steps to have Mussolini stop Marconi's research. According to Marconi's followers, Marconi then took his yacht to South America in 1937, after faking his own death.

The Secret City in South America

A number of European scientists were said to have gone with Marconi, including Landini. In the 1937, the enigmatic Italian physicist and alchemist Fulcanelli warned European physicists of the grave dangers of atomic weapons and then mysteriously vanished a few years later. He is believed to have joined Marconi's secret group in South America.

Ninety-eight scientists were said to have gone to South America where they built a city in an extinct volcanic cater in the southern jungles of Venezuela. In their secret city, financed by the great wealth they had created during their lives, they continued Marconi's work on solar energy, cosmic energy and anti-gravity. They worked secretly and apart from the world's nations, building free-energy motors and ultimately discoid aircraft with a form of gyroscopic anti-gravity. The community is said to be dedicated to universal peace and the common good of all mankind. Believing the rest of the world to be under the control of energy companies, multi-national bankers and the military-industrial complex, they have remained isolated from the rest of the world, working subversively to foster peace and a clean, ecological technology on the world.

We have information on this astonishing high-tech city from a number of sources. In the South America the story is a common subject among certain metaphysical groups. Says the French writer Robert Charroux in his book *The Mysteries of the Andes,* (1974, 1977, Avon Books) "...the Ciudad Subterranean de los Andes, is discussed in private from Caracas to Santiago." Charroux goes on to tell the story of Marconi and his secret city, plus the story of a Mexican journalist named Mario Rojas Avendaro who investigated the *Ciudad Subterranean de los Andes* (Underground City of the Andes) and concluded that it was a true story. Avendaro was contacted by man named Nacisso Genovese, who had been a student of Marconi's and was a physics teacher at a High School in Baja, Mexico.

Genovese was an Italian by origin and claimed to have lived for many years in the Ciudad Subterranean de los Andes. Sometime in the late 1950s he wrote an obscure book entitled "My Trip To Mars." Though the book was never published in English, it did appear in various Spanish, Portuguese and Italian editions.

Tesla Technology

Genovese claimed that the city had been built with large financial resources, was underground and had better research facilities than any other research facility in the world (at that time, at least). By 1946 the city already using a powerful collector of cosmic energy, the essential component of all matter, according to Marconi's theories, many of which he had derived from Tesla.

"In 1952," according to Genovese, "we travelled above all the seas and continents in a craft whose energy supply was continuous and practical inexhaustible. It reached a speed of half a million miles an hour and withstood enormous pressures, near the limit of resistance of the alloys that composed it. The problem was to slow it down at just the right time."

According to Genovese, the city is located at the bottom of a crater, is mostly underground and is entirely self-sufficient. The extinct volcano is covered in thick vegetation, is hundreds of miles from any roads, and is at thirteen thousand feet in the jungle mountains of the Amazon.

The French author Charroux expressed surprise and disbelief to the statement that the city was on a jungle covered mountain that was 13,000 feet high. Yet, the eastern side of Andean cordillera has many such mountains, from Venezuela to Bolivia, spanning thousands of miles. Several such cities, and mountains, could exist in this vast, unexplored, and perpetually cloud-covered region.

Yet, a secret city in a jungle crater was the least of the claims. Genovese claimed that flights to the Moon and Mars were made in their "flying saucers." He claimed that once the technology had been conquered, it was relatively simple to make the trip to the Moon (a few hours) or Mars (several days). Genovese does not mention pyramids or what they did on Mars. Perhaps they created a Martian base in one of the ancient, sand-blown pyramids of the Cydonia Region.

There have been many reports of UFOs in South America, especially along the edge of the mountainous jungles of the eastern Andes, from Bolivia to Venezuela. Is it possible that some of these UFOs are anti-gravity craft from the *Ciudad Subterranean de los Andes?*

In light of highly reliable sources who claim that a "Last Battalion" of German solders escaped via submarine in the last days of WWII to Antarctica and South America, it is possible that the Germans may have high tech super cities in the remote jungles of South America as well.

A number of military historians, such as Col. Howard Buechner, author of *Secrets of the Holy Lance* and *Hitler's Ashes,* maintain that the Germans had already created bases in Queen Maud Land, opposite South Africa during the war.

Afterwards, German U-Boats, in some reports as many as 100, took important scientists, aviators and politicians to the final fortress of Nazi Germany. Two of these U-boats surrendered in Argentina three months after the war. In 1947, the U.S. Navy invaded Antarctica, mainly Queen Maud Land with Admiral Byrd in command.

The Americans were defeated and several jets from the four aircraft carriers were said to have been shot down by discoid craft. The navy retreated and did not return until 1957.

According to the book, *Chronicle of Akakor,* a book first published in German by the journalist Karl Brugger, a German battalion had taken refuge in an underground city on the borders of Brazil and Peru. Brugger, a German journalist who lived in Manaus, was assassinated in the Rio de Janeiro suburb of Ipanema in 1981. His guide, Tatunca Nara, went on to become Jacques

Cousteau's guide on the upper Amazon. In fact, photographs of Tatunca Nara appear in Cousteau's large coffee-table book of color photographs called Cousteau's Amazon Journey. (For more information on Tatunca Nara, Karl Brugger, Underground Cities and Germans see *Lost Cities & Ancient Mysteries of South America.*)

While the secret cities of South America manufacturing flying saucers and battling the current powers of the world from their hidden jungle fortresses may sound too much like the plot of a James Bond movie, it appears to be based on fact!

Based upon the above scenario, it may not be totally fantastic to suggest, as some authors have, that Tesla was picked up during the late 1930s by a flying saucer. Yet, it would not have been a flying saucer from another planet, but one of Marconi's craft from the secret city in South America.

In the most incredible scenario so-far, and one that may well be true, Tesla was induced to fake his own death, just as Marconi and many of the other scientists had done, and was taken, by special discoid craft, to Marconi's high-tech super-city. Away from the outside world, the military governments, the oil companies, the arms and aircraft manufacturers, Marconi and Tesla, both supposedly dead, continued their experiments, in an atmosphere conducive to scientific achievement.

Who knows what they may have achieved? They were ten years ahead of the Germans and twenty years ahead of the Americans in their anti-gravity technology. Could they have developed discoid spacecraft in the early 1940s, and gone on to time travel machines and hyperspace drives? Perhaps Marconi and Tesla went into the future, and have already returned to the past!

Time Travel experiments, teleportation, pyramids on Mars, Armageddon and an eventual Golden Age on earth, may all have something to do with Tesla, Marconi and their suppressed inventions. While "UFO experts" and "former intelligence agents" tell us that flying saucers are extraterrestrial and are being currently retro-engineered by military scientists, Tesla, Marconi and their friends may be waiting for us at their space base at the pyramids and Face on Mars.

Our government, Hollywood and the media have trained us to certain beliefs and prejudices that amazing technology must be from extraterrestrials visiting our planet. To the scientist-philosopher who seeks knowledge... sometimes truth is stranger than fiction.

No. 676,332. Patented June 11, 1901.

G. MARCONI.

APPARATUS FOR WIRELESS TELEGRAPHY.

(Application filed Feb. 23, 1901.)

(No Model.) 2 Sheets—Sheet 2.

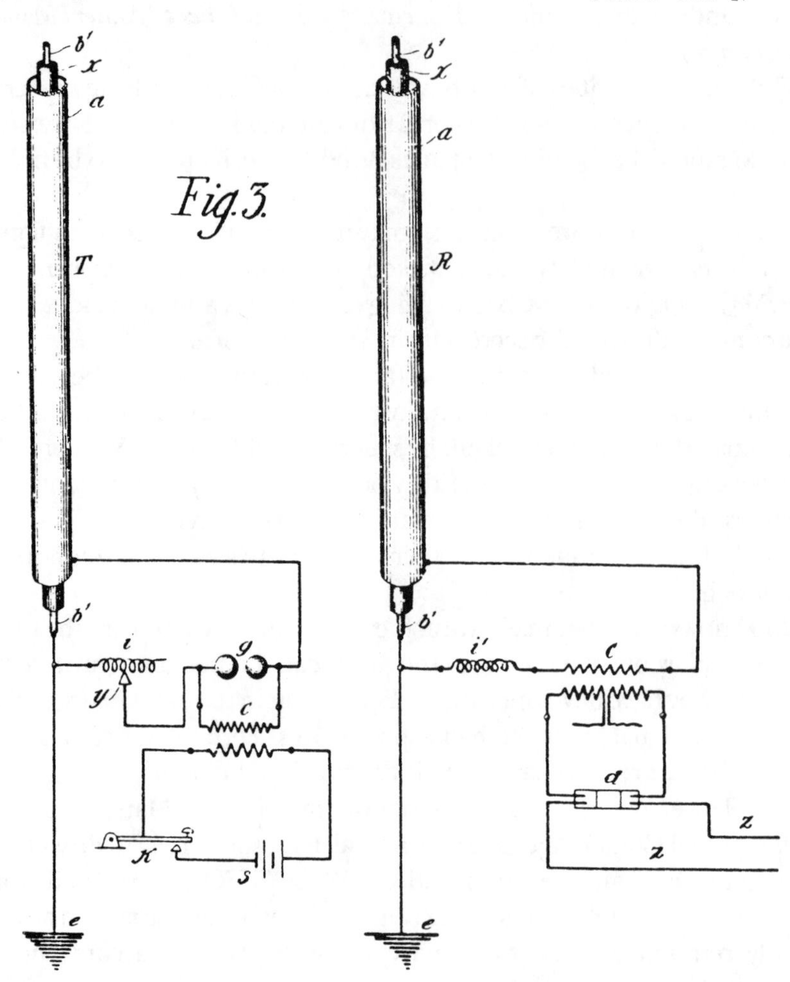

WITNESSES:

INVENTOR

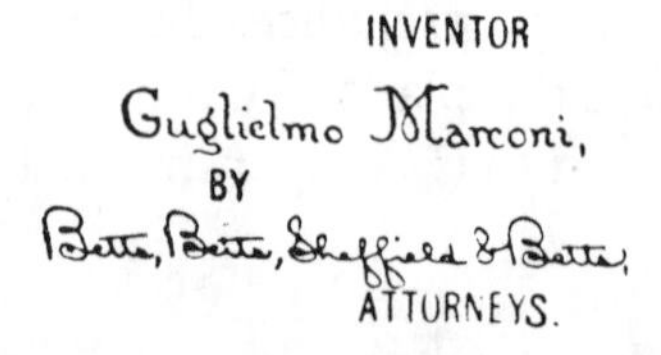

BY

ATTORNEYS.

A U.S. Patent for Guglielmo Marconi and his Wireless Telegraphy given on June 11, 1901. Marconi was as much of a genius as Tesla. When Marconi saw Tesla beaten by the powerful World Financiers, he hesitantly approached the Fascists of Italy with some of his inventions. After the Pope condemed his death-ray, Marconi faked his own death in 1936 and left with more than 100 scientists to South America aboard his yacht *Electra*.

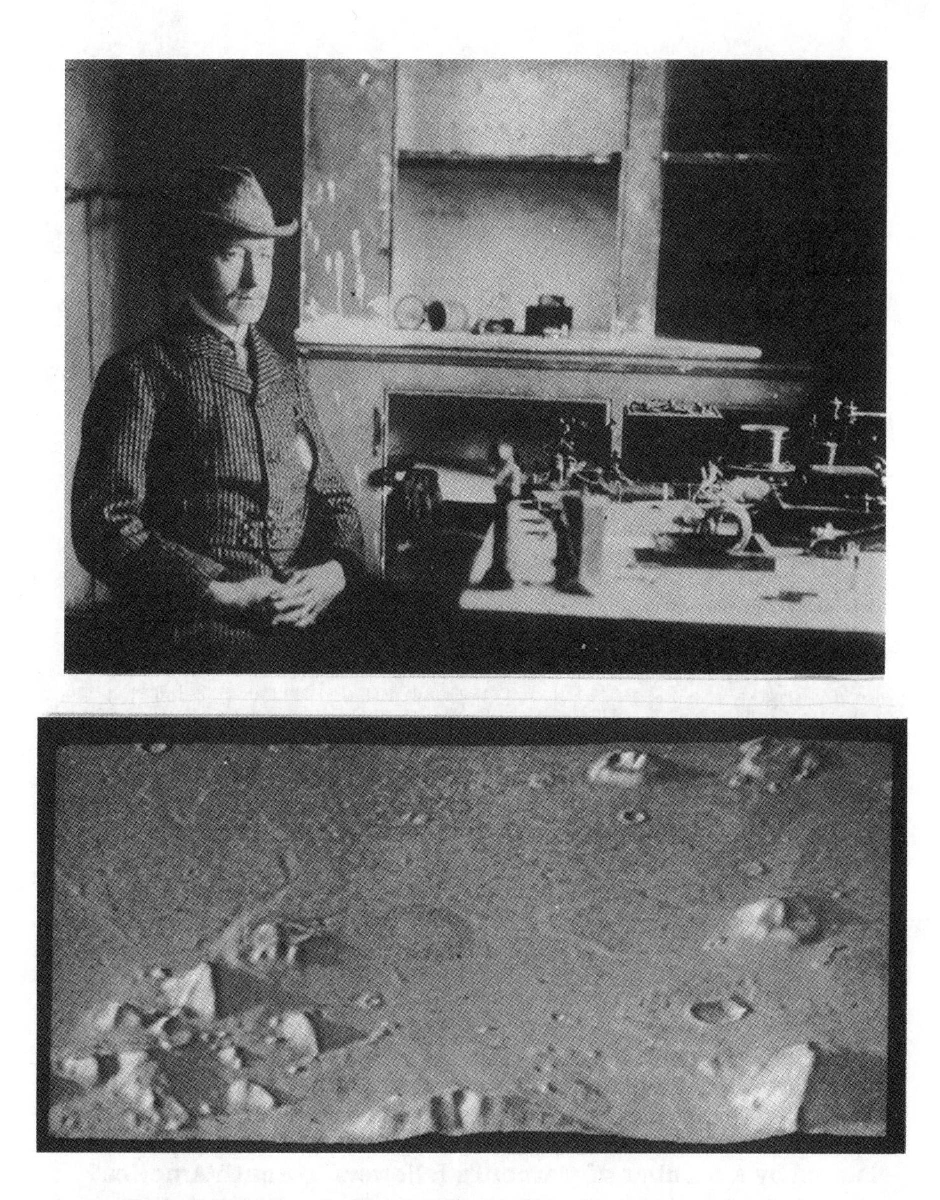

Above: Guglielmo Marconi, the Anglo-Italian inventor whose mentor was Nikola Tesla. In this photo, Marconi is seen at Signal Hill, Newfoundland, in 1901, with the instruments that he used to receive the first wireless signal across the Atlantic, sent to him from Polhu, Cornwall, England. Below: Mark Carlotto's computer illustration showing a perspective view of the Face on Mars and surrounding pyramids, looking from the west, and well above the Martian surface. Many experts on Marconi, Tesla and UFOs believe that Marconi and other scientists moved to a secret base in the Amazon jungle in 1937 and began making anti-gravity craft with which they reached Mars in the early 1950s or late 1940s.

NASA frame 35A72 of Mars, showing the famous "Face" (a) with its protective metal helmet, while the "city" (b) with its pyramids and the "D&M pyramid" (c) is farther to the lower right. Did Marconi and his scientists actually travel to Mars in the early 1950s as was claimed by a number of Marconi's followers in South America?

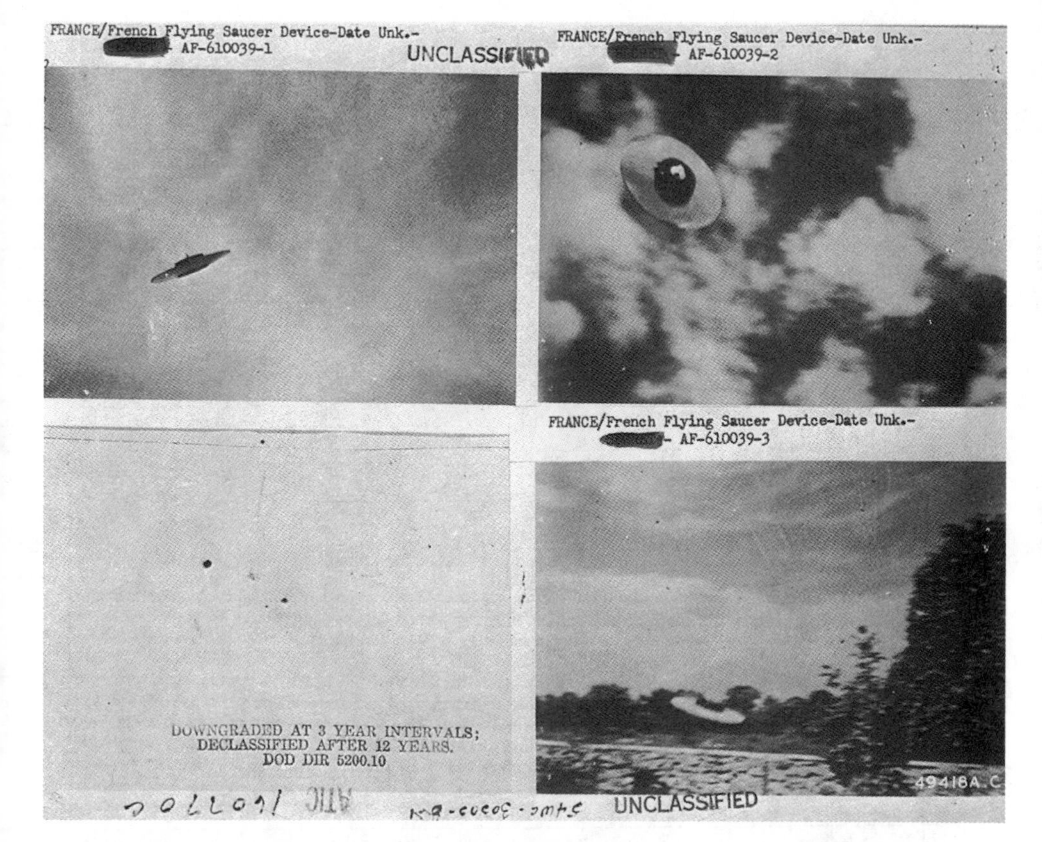

These four photos are from the U.S. Airforce Blue Book files and show a "French Flying Saucer Device" that is alleged to be one of the craft used by Marconi and built at the secret city in South America. The photos clearly show a man-made discoid craft hovering, tilting and landing. Although the margin notes state "Date Unk.—" the photos are known to have been taken in 1953. At first they were classified SECRET, but were later "downgraded" and finally marked "unclassified." The initials ATIC (upside down at bottom left) stand for Air Technical Intelligence Center, the parent military agency in the Air Force hierarchy sponsoring Project Blue Book.

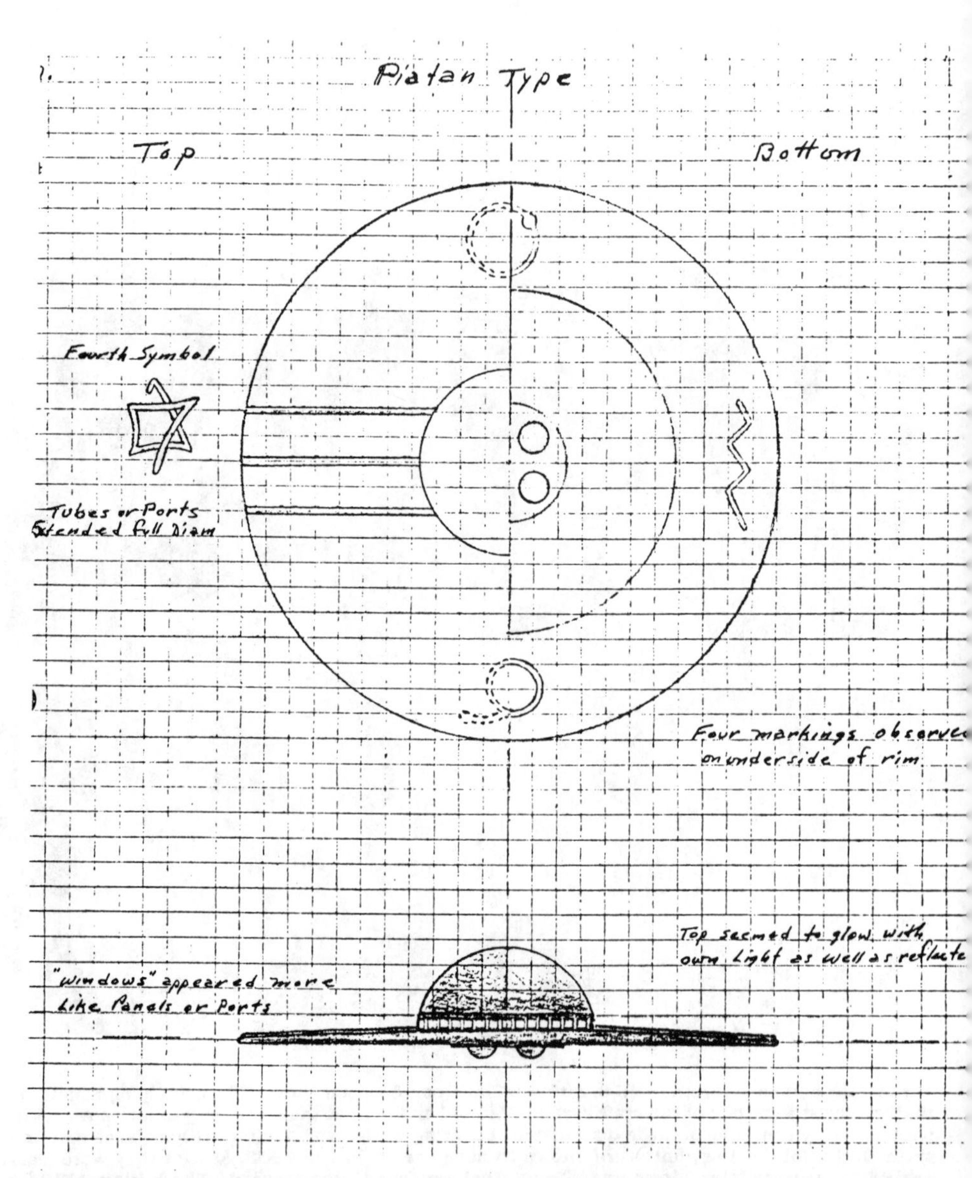

On April 24, 1959, Helio Augiar was driving along the Piata beach in Salvador, a beach in northern Brazil, when his engine suddenly stopped. He then witnessed a flying saucer flying along the beach and took four photos of it. It is interesting to note the symbols on the underside of the craft in this drawing. The craft had four hemispherical protrusions in the center of the craft on the bottom and three ribs or tubes on the top of the craft. This discoid vehicle is similar to the type of craft allegedly made at Marconi's secret city, and is also similar to the Schriever-Habermohl flying disc made by Germany at the BMW factory near Prague in 1944, and first flown on February 4, 1945 (*German Jet Genesis* by David Masters, 1982, Jane's Books, London. Page 135).

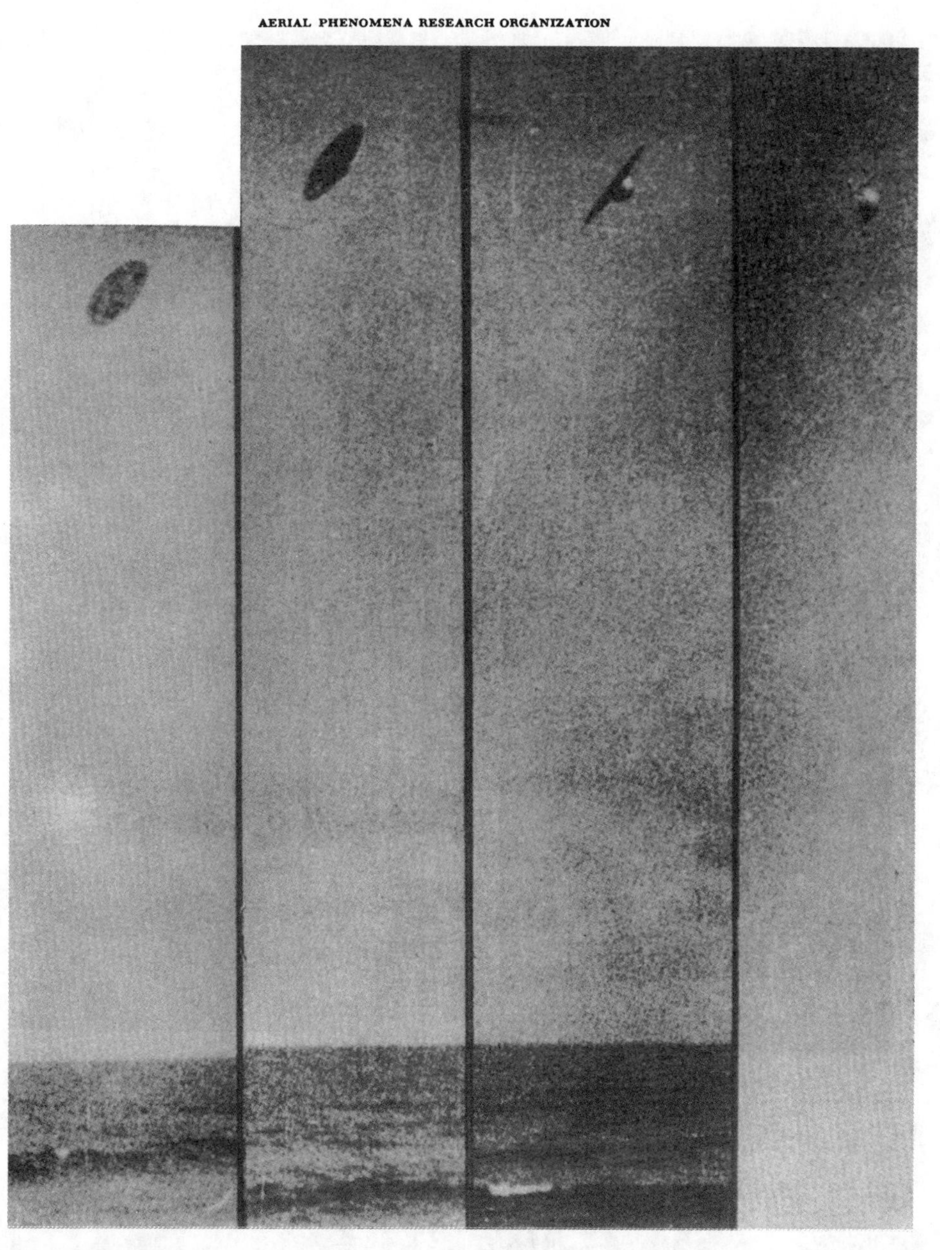

After taking these photos on Piata Beach, Salvador, Brazil, (April 24, 1959) Helio Aguiar lost conciousness. When he came to, he was clutching this message, written in his own hand: "ATOMIC EXPERIMENTS FOR WARLIKE PURPOSES SHALL BE DEFINITELY STOPPED ... THE EQUILIBRIUM OF THE UNIVERSE IS THREATENED. WE WILL REMAIN VIGILANT AND READY TO INTERFERE."

A Polaroid photo taken in Peru on October 19, 1973 by architect Hugo Vega. Vega, who happened to have a camera with him, was looking for a house of client about 34 miles east of Lima, the coastal capital of Peru. He was looking out over the Rimac River valley when a discoid craft with portholes between the domed top and main section, came into view. The craft flew along the valley and hovered for a few seconds against the jungle background, which is when Vega got his Polaroid shot. The old-fashioned design of the craft is noteworthy, with this 1973 saucer looking more like a 1950-type craft. Some UFO experts believe that this craft is one of Marconi's manufacture at the underground city.

Appendix

SUPREME COURT DOCUMENTS ON THE DISMANTLING OF THE WARDENCLYFFE TOWER

Tesla maintained a residence at the old Waldorf-Astoria Hotel in New York City from the spring of 1899 to 1915. He maintained by mortgaging the Wardenclyffe property and tower to the hotel and its owner, George C. Boldt. Since the anticipated income from the Wardenclyffe project of transmitting messages across the Atlantic was unrealized, Tesla was unable to repay the mortgages. Action was taken by the mortgagees in 1915 to foreclose on the property, and a bill of sale offered to Lester S. Holmes, a developer. The property would be sold to Holmes and the tower would be destroyed so that property could be developed.

Tesla appealed the foreclosure judgment, and the case was heard by the Supreme Court of the State of New York, Suffolk County. Tesla lost the case on appeal. Following the judgement on April 20, 1922, Tesla vacated his suite at the Waldorf-Astoria and took up residence at the Hotel St. Regis.

Though the 313 page transcript document largely contains the various lawyers arguing about the legal instruments of mortgaging and foreclosure actions, portions of the testimony are interesting. During Tesla's testimony, he give important information about the Wardenclyffe installation (transcript pages 163-181). Next is a portion of Ezra C. Bingham's testimony, chief engineer for the Waldorf-Astoria, in which he describes how the tower had been vandalized, and how poor the condition of the plant was (transcript pages 235-247). Tesla returns to the stand and gives more information on the purpose of the plant (transcript pages 269-275), and finally is Exhibit B, Tesla's inventory of the plant (transcript pages 309-312).

Supreme Court,

APPELLATE DIVISION—SECOND DEPARTMENT.

CLOVER BOLDT MILES AND GEORGE C. BOLDT, JR., AS EXECUTORS OF THE LAST WILL AND TESTAMENT OF GEORGE C. BOLDT, DECEASED,

Plaintiffs-Respondents,

against

NIKOLA TESLA,

Defendant-Appellant,

THOMAS G. SHEARMAN,

Defendant-Respondent,

et al.,

Case on Appeal.

WILLIAM RA[illegible]IN, JR.,
Attorney for Defendant-Appellant,
305 Broadway,
New York City.

BALDWIN & HUTCHINS,
Attorneys for Plaintiffs-Respondents.

WILLARD A. MITCHELL,
Attorney for Defendant-Respondent, THOMAS G. SHEARMAN.

A. Of course I had signed and the transaction was completed. 487

Q. And those papers were then in Mr. Hutchins' possession?

A. Yes those were almost his parting words.

Q. I think you said that conversation took place early in 1917 or late in 1916?

A. I think early in 1917, if I remember rightly, but my memory is a little—on account of the concentration——

Mr. Hawkins: I do not recall the date of that deed.

Mr. Fordham: Why don't you let your witness complete his answer about his memory? 488

Mr. Hawkins: I assumed he had.

By Mr. Fordham:

Q. What were you saying?

A. I answered all the questions to the best of my ability.

Q. No, counsel interrupted you intentionally in the middle of a sentence——

Mr. Hawkins: That is not true, that I interrupted him intentionally.

Mr. Fordham: Well, strike out the word intentionally. You interrupted him in the middle of the sentence. He can say what he started to say about his memory in connection with this transaction. The witness evidently thinks he does not need to pay any attention to what I say. Will your Honor kindly instruct the witness to complete his answer. 489

By the Referee:

Q. Had you completed your answer?

A. Yes, those were the parting words of Mr. Hutchins.

490 By Mr. Hawkins:

Q. As Mr. Fordham seems anxious to have you complete what you said, I heard what you said——

Mr. Fordham: He says he has completed.

A. Yes, in regard to the memory of the exact date, I say that I cannot exactly remember the dates on account of concentration on some other work that I am doing now, but I can easily ascertain all the dates from documents.

Q. Now at the time that you delivered that document to Mr. Hutchins, I refer now to the deed, will you please describe to the Court what there was upon the property?

491

A. Upon the property?

Q. Yes, described in the deed, which property is situated at Rocky Point.

Mr. Fordham: That is objected to on the ground it is incompetent, immaterial and irrelevant at this point what there was on the property.

The Referee: I will take it.

Mr. Fordham: We except.

The Referee: You mean structures, I suppose?

492

Mr. Hawkins: Yes, absolutely. Improvements, I had in mind, if there were any buildings there or structures.

Q. Tell the Court what there was there.

A. There was a brick building in which was located the power plant——

Q. Please describe the size of the brick building.

A. The building forms a square about one hundred by one hundred feet wide and it is one floor, rather high, with a roof covered with gravel, as they usually make them. This building was divided inside in four compartments, two of which were very large, one being the machine shop—— 493

Q. How large was that?

A. That was one hundred feet by about thirty-five feet, I should say.

Q. Now tell how big the other compartments were.

A. The other one was about one hundred by thirty-five and then these other two smaller ones where the engines were located on one side and the boilers on the other were about thirty by forty, thirty one way and forty the other.

Q. I think you said the building was one story high? 494

A. Yes.

Q. It had one floor, did it?

A. One floor, yes.

Q. Further describe the building, if there is any further description, and tell the Court whether there were any brick chimneys, outside chimneys?

A. Oh yes, right in the center of the building rose the chimney.

Q. How big was the chimney?

A. The chimney was four by four feet; it was calculated to give the proper speed to the products of combustion under the boilers. 495

Q. Of what was the chimney composed?

A. Brick.

Q. How high was the building?

A. The building might have been, I think the extent of the walls on one side, the lowest part of the roof might have been something like twenty-eight feet, I would say.

Q. Twenty-eight feet at the corners of the building?

A. Yes.

496 Q. And did it have a gable roof or a lantern roof?

A. Yes, as you call it in English—how is this roof called?

Q. I think it is a gable roof.

A. Gable roof. The building was resting on cement foundations and there were the usual modern conveniences and——

Q. Tell what you mean by the usual modern conveniences?

A. I mean the channels for leading off the waste, the rain drips and all that, and then attached to it was, of course, the water pump that 497 pumped the water for the building.

Mr. Fordham: I do not wish to interrupt counsel but what possible use can there be in a detailed description of the building on this property?

The Referee: I do not know at this time.

Mr. Fordham: Neither do I. It does not seem to me that we should burden the record indefinitely with these descriptive details.

The Referee: I will let him describe them.

Mr. Hawkins: It is a material part of the defense here.

498 The Referee: Go ahead. I will take it.

The Witness: I suppose what belongs to the buildings is the boiler plant, with two 300 horse power boilers on one side——

Q. That was two 300-horse power?

A. Two 300-horse power boilers, yes, and the pumps, injectors and other accessories, and then there were big water tanks that were placed around the chimney so as to utilize some of the waste heat. These tanks had a capacity of about 16,000 gallons, if I am correct.

Q. Of what were the tanks composed? 499

A. Of quarter inch thick sheet steel, galvanized.

Q. Those were all in one compartment, were they?

A. They were around the chimney under the roof, and for this purpose the room had an extension upward there. This could be shown on a photograph if his Honor wishes to see the photograph.

Q. Just a moment please. Now describe the other three compartments of the building.

A. Well, I have described the boiler plant. Now right opposite to the boiler plant lengthwise was a corresponding compartment and therein were located the engines. 500 Of these engines there was one 400-horse power Westinghouse reciprocating engine, driving a directly connected dynamo which was specially made for my purposes. Then there was a 35-kilowatt Westinghouse outfit also driving the dynamo, which was for the purpose of lighting and other work, a permanent attachment to the building to furnish all conveniences. There was then a high pressure compressor which also formed an essential part of the equipment. And then there was a low pressure compressor or blower. Then there was a high pressure pump and a reciprocating low pressure pump. That was all—— 501

Q. Water pumps?

A. Water pumps, yes. Those were all in that compartment, and of course this compartment also contained the switches and the switchboard and all that which goes with the equipment of the plant. Then there was a gallery on the top on which certain parts were placed and arranged that were needed daily in the operation.

Q. Those were parts of what?

A. Well they were the tools, you know, that were needed in the plant.

502 Q. Please describe another compartment.

A. The compartment that was towards the railroad, that was the machine shop.

Q. Which part of the building was that, the north, south, east or west?

A. I cannot locate it——

The Referee: The north side.

The Witness: Towards the road, facing the road. That compartment was one hundred by thirty-five feet with a door in the middle and it contained I think eight lathes.

The Referee: You are speaking now, when 503 you said facing the road, that is on the south side, the travel road or——

The Witness: Facing the railroad. It is just close on the railroad track, your Honor, this building. That contained I think eight lathes ranging in swing from eight inches to thirty-two, I believe. Then there was a milling machine and there was a planer, and shaper, a spliner, a vertical machine for splining. Then there were three drills, one very large, another medium and a third quite small one. Then there were four motors which operated the machinery. Also a grinder and an ordinary grind- 504 stone, a forge——

Q. Blacksmith's forge?

A. Yes, a blacksmith's forge. Then a special high temperature stove and the blower for the forge. Of course the shop was full of counter shafting and there were a few special tools which suited certain purposes which I contemplated there. I cannot at present recall them exactly, but there were five or six of them.

Q. Were those stationary tools or hand tools?

A. No, some of them were attachments to the

ordinary lathes or milling machines, suitable for certain work and others were of course portable. 505

Q. Now have you described the four compartments of the building?

A. No. Now the compartment opposite, that is facing further away from the railroad, which also was one hundred feet, the whole length of the building, by about thirty-five, there is where the real expensive apparatus was located. That contained also the desks and the office accessories. Shall I describe now this one?

Q. Yes, describe any stationary fixtures there were in this other compartment.

506

A. Well, is machinery a stationary fixture?

Q. Yes we call that a stationary fixture.

A. Right along the back wall that separated this compartment from the rest of the building there were two special glass cases in which I kept the historical apparatus which was exhibited and described in my lectures and scientific articles. There were probably at least a thousand bulbs and tubes each of which represented a certain phase of scientific development. Then close, beginning with these two glass cases, there were five large tanks. Four of those contained special transformers according to my design, made by the Westinghouse Electric Manufacturing Company. These were to transform the energy for the plant. They were about, I should say, seven feet high and about five by five feet each, and were filled with special oil which we call transformer oil, to stand an electric tension of 60,000 volts. Then besides these four tanks there was another similar tank which was for special purposes, containing a transformer. Then there were two doors, one door that led to the other compartment and the other one led in the closets, and between those two doors there was a space on which 507

508 was placed my electric generating apparatus. This apparatus I used in my laboratory demonstrations in two laboratories before, and I had also used it in the Colorado experiments where I erected a wireless plant in 1889. That apparatus was precious because it could flash a message across the Atlantic, and yet it was built in 1894 or 1895. That is a complicated and very expensive apparatus.

Then beyond the door there were again four tanks, big tanks almost the same size as those I described. These four tanks were to contain the condensers, what we call electric condensers, which store the energy and then discharge and make it go around the world. These condensers, some of them, were in an advanced state of construction, two, I think, and the others were not. They were according to a principle of discovery. Then there was a very expensive piece of apparatus that the Westinghouse Company furnished me; only two of this kind of apparatus were made by the Westinghouse Company, one for me and one for themselves. It was developed together by myself and their engineers. That was a steel tank which contained a very elaborate assemblage of coils, an elaborate regulating apparatus, and it was intended to give every imaginable regulation that I wanted in my measurements and control of energy. Then on the last side, where I had described the first four big tanks there was a special 100-horse power motor and this motor was equipped with elaborate devices for rectifying the alternating currents and then sending them into the condensers. On this apparatus alone I spent thousands of dollars. The 100-horse power motor was specially constructed for me by the Westinghouse Company, but the other parts were all made by myself and that took a considerable portion of space there and it was a

509

510

wonderful piece of apparatus. I have photographs of these which will make this description very clear. 511

Then along the center of the room, I had a very precious piece of apparatus. That was a boat which was illustrative of my discovery of teletaumatics; that is a boat which was controlled without wire, which would do anything you wanted, but there was no connection. This boat was exhibited by me on many occasions.

Q. The boat was not stationary, was it?

A. It was stationary, yes, on the supports. It was stationary on the supports but as I say that boat was my wireless boat; that is a boat that you commanded it and it would perform as many evolutions as you wanted, by just commanding it. 512

Q. Was that about all there was, generally speaking?

A. Oh, no, nowhere near. Then there were on each side long specially made, how do you call them, not desks or shelves, but closets, I might say, which were specially made to contain the apparatus, because I had accumulated for years hundreds of different kinds of appliances which stand for a certain principle, and this apparatus was stored in there, and on top of these I had again all full of apparatus, each representing a different phase. 513 And then on one side there were the desks and then on the other side there were the drawing implements and tools. And then in the corner, when you looked at the railroad side, on the right side in the corner there was my testing room and that contained—there were two precious instruments among these that Lord Kelvin made especially for me. He was a great friend of mine. A device for measurement invented by him; it is called a breach; and another a voltmeter of his. Both of these things were given to me and prepared for me by his

514 special instructions. There were a lot of other instruments, voltmeters, wattmeters, ampere meters; in that small space there was a fortune in there.

Mr. Fordham: The last, that there was a fortune in there, calls for a conclusion as to the salable value of the stuff and I think it should go out.

The Referee: Yes, strike it out.

Q. I think you said this building was constructed of brick, did you not?

A. Yes.

Q. How thick were the walls of it?

515 A. That I cannot tell now exactly, but I should say about twelve inches.

Q. It was more than one brick thick, at any rate?

A. Oh, I should say so. I paid something——

Q. I presume this building had windows in it?

A. Oh yes there were large windows which were divided into panels.

Q. And what were the window sash made of, metal or wood?

A. Wood sashes.

Q. I show you a document, Defendant's Exhibit C, and call your attention to the signature on that

516 document, and ask you if that is your signature?

A. Yes sir, that is my signature.

Q. Do you recognize the instrument?

A. Yes sir, that was one of the——

Q. That is the deed which you delivered, is it not?

A. Yes.

Q. I call your attention to the date of the deed, March 30, 1915.

A. 1915?

Q. Yes.

A. Well that was—1915? 517

Q. Yes.

A. I was under the impression it was a little later.

Q. Well that is the only deed which you delivered in the transaction to Mr. Holmes, is it not?

A. Hutchins?

Q. Hutchins, yes.

A. So far as I know.

Q. Then would you like to change your testimony when you said it was in 1917? The date of this in March 30, 1915.

A. I have stated that I was not sure about the dates, but I could ascertain it exactly by looking at the documents. 518

Q. Well there is the document.

A. Well it must be so because it is there.

Q. It is 1915 then instead of 1917?

A. Yes, but my impression was that this was another attorney who had it first and it was made to Mr. Hutchins later.

Q. I do not know what you mean by saying it was made to Mr. Hutchins. The grantee in the deed is Lester S. Holmes.

A. Yes, Lester S. Holmes.

519

By the Referee:

Q. The transaction you had was with Mr. Hutchins?

A. Yes that is all.

The Referee: I do not think there is any dispute about that.

Mr. Fordham: There was only one, I understand. The witness does not claim there were two, one in 1915 and another in 1917.

The Witness: No.

520 By Mr. Hawkins:

Q. No there was only one and Mr. Holmes was the grantee in the deed which you gave to Mr. Hutchins, was he not?

A. Yes, and I recall the transaction with Mr. Hutchins.

Q. Were there any other structures upon the property aside from the building?

By the Referee:

Q. Did you read that paper at the time you executed it?

521 A. Yes at the request of Mr. Hutchins.

By Mr. Hawkins:

Q. Were there any other structures upon the premises other than that brick factory or laboratory which you have just described?

A. Yes sir, there was the structure which in a certain sense was the most important structure, because the power plant was only an accessory to it. That was the tower.

Q. Please describe the tower as to dimensions and material and method of construction and kind of construction?

522

Mr. Fordham: We renew our objection, if the Court please. This is entirely immaterial, irrelevant and incompetent until after they have succeeded in establishing their contention that the deed is a mortgage.

The Referee: I will take it.

Mr. Fordham: Exception.

A. The tower was 187 feet high from the base to the top. It was built of special timber and it was built in such a way that every stick could be taken out at any time and replaced if it was necessary. The design of the tower was a matter of considerable difficulty. It was made in the shape of an octagon and pyramidal form for strength and was supporting what I have termed in my scientific articles a terminal. 523

By the Referee:

Q. There was sort of a globe at the top?

A. Yes. That, your Honor, was only the carrying out of a discovery I made that any amount of electricity within reason could be stored provided you make it of a certain shape. Electricians even today do not appreciate that yet. But that construction enabled me to produce with this small plant many times the effect that could be produced by an ordinary plant of a hundred times the size. And this globe, the framework, was all specially shaped, that is the girders had to be bent in shape and it weighed about fifty-five tons. 524

By Mr. Hawkins:

Q. Of what was it constructed?

A. Of steel, all the girders being specially bent into shape.

Q. Was the tower that supported it entirely constructed of wood or partly steel? 525

A. That part alone on top was of steel. The tower was all timbers and of course the timbers were held together by specially shaped steel plates.

The Referee: Braces?

The Witness: Yes, steel plates. I had to construct it this way for technical reasons.

The Referee: We are not interested in that.

Q. Was the tower enclosed or open?

A. The tower, at the time of the execution of this

526 deed, was open, but I have photographs to show how it looked exactly and how it would have looked finished.

Q. After you delivered the deed was the tower ever enclosed?

A. No, it was just open.

Q. Now the dome or the terminal at the top, was that enclosed?

A. No sir.

Q. Never enclosed?

A. Never enclosed, no.

Q. Had that structure ever been completed?

A. The structure so far, if I understand the terms
527 right, yes, the structure was all completed but the accessories were not placed on it yet. For instance that globe there was to be covered with specially pressed plates. These plates——

Q. That had not been done, had it?

A. That had not been done, although I had it all prepared. I had prepared everything, I had designed and prepared everything, but it was not done.

Q. Was the structure of the tower in any manner connected with the brick building or power plant?

A. The tower was separate.

528 Q. I understand, but was there any connection between them?

A. There were of course two channels. One was for communicating, for bringing into the tower compressed air and water and such things as I might have needed for operations, and the other one was to bring in the electric mains.

By the Referee:

Q. In order to do that there was, as a matter of fact, was there not, a well-like shaft going down right in the middle of the tower into the ground some fifty or sixty feet?

A. Yes. You see the underground work is one of the most expensive parts of the tower. In this system that I have invented it is necessary for the machine to get a grip of the earth, otherwise it cannot shake the earth. It has to have a grip on the earth so that the whole of this globe can quiver, and to do that it is necessary to carry out a very expensive construction. I had in fact invented special machines. But I want to say this underground work belongs to the tower. 529

By Mr. Hawkins:

Q. Anything that was there, tell us about. 530

A. There was, as your Honor states, a big shaft about ten by twelve feet goes down about one hundred and twenty feet and this was first covered with timber and the inside with steel and in the center of this there was a winding stairs going down and in the center of the stairs there was a big shaft again through which the current was to pass, and this shaft was so figured in order to tell exactly where the nodal point is, so that I could calculate every point of distance. For instance I could calculate exactly the size of the earth or the diameter of the earth and measure it exactly within four feet with that machine.

Q. And that was a necessary appurtenance to your tower? 531

A. Absolutely necessary. And then the real expensive work was to connect that central part with the earth, and there I had special machines rigged up which would push the iron pipes, one length after another, and I pushed these iron pipes, I think sixteen of them, three hundred feet, and then the current through these pipes takes hold of the earth. Now that was a very expensive part of the work, but it does not show on the tower, but it belongs to the tower.

532 By Mr. Fordham:

Q. Was the hole really one hundred and twenty feet deep, did you say?

A. Yes, you see the ground water on that place is about one hundred and twenty feet. We are above the ground water about one hundred and twenty feet. In the well we struck water at about eighty feet.

By the Referee:

Q. What you call the main water table?

A. Yes the main well we struck at eighty feet, 533 but there we had to go deeper.

By Mr. Hawkins:

Q. Tell the Court generally, not in detail, the purpose of that tower and the equipment which you have described connected with it?

Mr. Fordham: How is that material?

The Referee: I will take it.

Mr. Fordham: We except.

A. Well, the primary purpose of the tower, your Honor, was to telephone, to send the human voice and likeness around the globe.

534

By the Referee:

Q. Through the instrumentality of the earth.

A. Through the instrumentality of the earth. That was my discovery that I announced in 1893, and now all the wireless plants are doing that. There is no other system being used. And the idea was to reproduce this apparatus and then connect it just with a central station and telephone office, so that you may pick up your telephone and if you wanted to talk to a telephone subscriber in Australia you would simply call up that plant and the plant would connect immediately with that subscriber, no matter where in the world, and you could talk to him. And I had contemplated to have press messages, stock quotations, pictures for the press and these reproductions of signatures, checks and everything transmitted from there throughout the world, but—— 535

By Mr. Hawkins:

Q. The purpose then briefly was for wireless communication to various parts of the world?

A. Yes and the tower was so designed that I could apply to it any amount of power and I was planning to give a demonstration in the transmission of power which I have so perfected that power can be transmitted clear across the globe with a loss of not more than five per cent, and that plant was to serve as a practical demonstration. And then I was going to interest people in a larger project and the Niagara people had given me 10,000-horse power—— 536

Q. What do you mean by power, energy?

A. Yes, power in any amount.

Q. Were there any other structures upon the premises? 537

A. No, just these two big structures.

Q. I call your attention, Mr. Tesla, to Defendants Exhibit A which I characterize as a bill of sale and ask you to notice the signature there.

A. That is my signature, sir.

Q. Now the date of this document is the 30th day of March, 1915?

A. Yes sir.

Q. Is that the bill of sale that was delivered the same time the deed was delivered?

A. Yes.

538 Q. I do not wish to repeat this, but when you stated that that was also on or about the early part of 1917 you had in mind this document which you delivered in March, 1915?

A. Yes, but what stands out in my mind strongest is the construction of the tower, and that is the reason I have that in mind, the construction of the tower.

Q. Do you recall the testimony of Mr. Hutchins, that the Waldorf entered possession of the property?

A. Of Hutchins?

Q. Do you recall the testimony of Mr. Hutchins?

539 A. Yes, I recall something of that which he stated.

Q. And when was that done, in 1917, before or subsequent to the destruction of the tower?

A. It was done some time before the actual destruction of the tower.

Q. Do you recall when the tower was destroyed?

A. It was about in 1917, as near as I can recall, but I can ascertain——

Q. When was the tower erected?

A. The tower was erected from 1901 to 1902.

Q. What had you done to it to preserve it?

540 A. I spent considerable money on it by painting all the metal parts over three times, I think, each time at a cost of about a thousand dollars.

Q. Was there anything done to preserve the wooden portion of the structure?

A. Oh yes, we carefully watched everything, and——

Q. I know, but did you apply anything to it?

A. No not to the wood.

Q. Did not paint it?

A. No, not the wood.

Q. Had the wood been treated in any manner prior to being put in the construction, to preserve it? 541

Mr. Fordham: How is this material, your Honor? All this detail of preliminary work?

The Referee: I want to give counsel much latitude, but I suggest to be just as brief as you can about it.

Mr. Hawkins: Yes. My idea is this, if the wood had been creosoted or treated in any way to preserve it that was part of its value.

Mr. Fordham: Not unless it could have been sold for more money. It is absolutely immaterial. 542

The Referee: I will let him state if it had been treated.

A. No, but it was the finest timber.

Q. What was the timber?

A. Pine.

Q. What kind of pine?

A. I cannot tell you, there are so many kinds of pine in America.

The Referee: I think it was yellow pine.

The Witness: I could ascertain exactly.

The Referee: Timbers of that sort generally are. 543

Q. Now prior to the time when the tower was taken down did you have a conversation with Mr. Hutchins concerning that?

A. Concerning the tower?

Q. Concerning the destruction of the tower?

A. Concerning the destruction of the tower?

Q. Yes.

A. No, certainly not. He gave me a friendly assurance that nothing would be done in an unfriendly way.

The Referee: Overruled. 703

Mr. Hawkins: Exception.

Mr. Fordham: You may proceed, Mr. Bingham.

A. What is the question again, please?

The question was read as follows: Have you had any experience, and if so what, in the purchase and sale of machinery?

A. Well, I have not had any in a great many years no, sir, either direct purchase or selling.

Q. Well, have you had any experience so that you are qualified to speak as to the value of machinery?

A. Only partially, I think. 704

Q. I call your attention to the testimony of the defendant Tesla, which appears on pages 88 to 161 of the record here inclusive, at the hearing on January 26, 1922, and ask you if you have read that testimony?

A. Yes, sir, I read that whole paragraph through.

Q. Are you acquainted with the premises referred to in the complaint in this action and the deed which is in evidence of the premises of the defendant Tesla?

Mr. Hawkins: That is objected to as immaterial, irrelevant and incompetent. 705

The Referee: Overruled.

Mr. Hawkins: Exception.

A. Yes, sir.

Q. When did you first visit those premises?

Mr. Hawkins: Same objection.

The Referee: Same ruling.

Mr. Hawkins: Exception.

A. Well, it is hard for me to state just when I first——

706 Q. (Interrupting.) Well, as near as you recall?

A. I would say about 1913.

Q. What was the occasion of your visit then?

Mr. Hawkins: Same objection.

The Referee: Same ruling.

Mr. Hawkins: Exception.

A. Mr. Boldt held a mortgage on this property and he sent me down there to check up and go over the condition of it and see what condition it was in.

Q. Yes; and how many times were you there?

Mr. Hawkins: Same objection.

The Referee: Same ruling.

707 Mr. Hawkins: Exception.

A. Oh, up until the time that I received that notice from Mr. Hutchins, I presume twenty times.

Q. The notice to which you refer is the letter of July 20, 1915, of which I show you a copy?

Mr. Hawkins: Same objection.

The Referee: Same ruling.

Mr. Hawkins: Exception.

A. Yes, sir.

Q. You haven't the original letter in your possession, have you?

708 A. No, sir; I have not.

Q. This is a correct copy?

A. Yes, sir.

The Referee: May I see that, Mr. Fordham? I do not just bear it in mind.

Mr. Fordham: Yes. I offer this letter in evidence.

Mr. Hawkins: Objected to as incompetent irrelevant and immaterial and further on the ground that it is a self-serving declaration: and I further object to it because it is not the original document.

The Referee: I will take it. 709

Mr. Hawkins: Exception.

(Letter marked Plaintiffs' Exhibit No. 8.)

Q. Mr. Bingham, between the time you first went there to the property in 1913 and July 20, 1915, so far as you recall, how many visits did you make to the property?

Mr. Hawkins: Same objection.

The Referee: Same ruling.

Mr. Hawkins: Exception.

A. I could not say. I used to go there on an average of once or twice a month.

710

Q. During that period?

A. During that period.

Q. Will you please tell the Court what you found on the property during those visits?

Mr. Hawkins: Objected to as incompetent, immaterial and irrelevant and certainly can have no bearing upon the question as to whether these instruments were delivered as absolute conveyances or as security.

The Referee: Overruled.

Mr. Hawkins: Exception.

A. The first time I went down there I found the windows—well, I should say there was half a dozen of them that were wide open, and in the big room such as was used for experimenting and things of that kind there was probably a dozen or fifteen desks in there and a great many wardrobes, that is closets and things of that kind, and among them was a—what you would call a model submarine. Well, this place had practically been wrecked. 711

Mr. Hawkins: I ask that be stricken out.

The Referee: Yes, strike out the "practically been wrecked." Just describe its condition.

712 A. There had been a desk that the drawers had been opened, pulled out and thrown on the floor and all the tops of the desks—they were roll-tops desks—they had been ripped off and thrown on the floor, the doors were ripped off the closets and the books and stuff that was in there, I would say there was four truckloads of that thrown all over this big room, and I came back and made a report to Mr. Boldt of the condition we found things.

Mr. Hawkins: I ask that be stricken out.

The Referee: No, the fact that he made a report, let it stand.

713 Mr. Hawkins: Exception.

A. (Continuing.) And in two or three days I took a couple of carpenters and we went down there and nailed up the windows.

Mr. Hawkins: I ask that be stricken out.

Q. (Interrupting.) Well, Mr. Bingham, bear in mind the question I was asking you; the first time you went there in 1913 you did not go down and nail up windows, did you?

A. I did in two or three days.

Mr. Hawkins: Is the last part of that answer stricken out on my motion?

714 The Referee: Yes.

A. (Continuing.) And put in some light pieces of board, such as "Compo" board where the glass was gone out, so as to kind of protect the place, as at his suggestion he thought I better do that.

Mr. Hawkins: I ask that be stricken out.

The Referee: Strike out the last part.

Mr. Hawkins: And also that they nailed up boards.

The Referee: No, I will let that stand.

Mr. Hawkins: Exception.

Q. Proceed. What did you then discover? 715

A. About two weeks later I went down again and I found these things all ripped open again and the doors open, and I came back and locked them up the best I could and went over to see the station agent and they didn't know anything about what had happened or anything of that kind.

Mr. Hawkins: I ask that be stricken out.

The Referee: Granted.

A. (Continuing.) And in the meantime there had been some of these desks that was in there that was completely smashed up and taken away, I should say there was about half of them gone. 716

Mr. Hawkins: I ask that be stricken out.

The Referee: Denied.

Mr. Hawkins: Exception.

A. (Continuing.) And I went down, I would not say just how soon again, but probably within a month because I had to go there that often, Mr. Boldt insisted on my going down there and keeping a check on it.

Mr. Hawkins: I ask that be stricken out.

The Referee: Strike that out.

Q. You may say what you did? 717

A. I continually kept going down there up until the time I received this notice from Mr. Hutchins. Up until that time——

The Referee (interrupting): That is the exhibit that has just been offered.

Mr. Fordham: Yes, Exhibit 8, dated July 20, 1915.

Q. Proceed.

A. Up until that time they had practically stripped the place of everything; they had stolen off all the railings and everything that might per-

718 tain to brass of any description, even the boiler feed pumps they had taken the tops off and taken the valves and valve seats out; all the toilets, they had taken off the toilets and taken all the lead pipe back of the toilet and everything that could be possibly sold that could be drawn in any kind of a wagon had been taken away, I suppose for junk, that is the only thing they could possibly use it for.

Mr. Hawkins: I ask that be stricken out.

The Referee: The supposition strike out.

A. (Continuing.) The boilers were there, simply the headers and tubes; everything that pertained to 719 them were gone, they had stolen and dragged away; the dynamos were still there, the main part.

Mr. Hawkins: I ask the part of the answer in effect had been stolen be stricken out.

The Referee: Yes, the characterization stolen we will strike out. The fact that they were not there we will let stand.

Q. Proceed.

A. Well, that had been taken away.

The Referee: Well, they were gone?

The Witness: They were not taken for ornaments; they were gone.

720

The Referee: Yes.

The Witness: The engines, the main part of the engines were there, that is the foundation and the fly wheels, because they could not take them away; and some of the big part of the machinery, the different lathes and milling machines and the main drill-press; all small lathes and motors and everything of that kind were gone. What had become of them I could not say, but I would say they were stolen. And when I got this notice from Mr. Hutchins I went down then and got the notice the same as today. I made the signs up and went down tomorrow and put up the signs, and in about a week or ten days from then I took a couple of trucks and went down there and brought the big machinery away

721

Q Just what did you bring away?

A. I brought away a large drill-press, milling machine, planer and two lathes.

Q. Do you know the value of those articles which you brought away?

A. I do not know exactly the value of those things. I have everything yet at the Waldorf, with the exception of the milling machine.

722

Q. Well, was the value a few hundred dollars or was it a great many thousand?

Mr. Hawkins: That is objected to on the ground the witness is not qualified.

The Referee: Objection sustained.

Mr. Fordham: We except.

Q. You have the things now, with the exception of the milling machine?

A. I have, with the exception of the milling machine, yes, sir.

Q. Do you know what became of that?

723

Mr. Hawkins: That is objected to as immaterial.

The Referee: Overruled.

Mr. Hawkins: Exception.

A. That was sold; I don't know just now who it was.

Q. You don't know?

A. No, but I could find out, I could look the book up and find out who did buy it, but I don't remember.

724 Q. Do you know what was received for it?

A. No, I do not.

Q. I show you Defendants' Exhibit A, a certain bill of sale, and call your attention to the schedules setting forth the various items purporting to have been conveyed by that bill of sale, and ask you to look over those items and to tell the Court what, if any of them, were on the property on July 20, 1915?

Mr. Hawkins: That is objected to as incompetent, irrelevant and immaterial.

The Referee: Overruled.

725 Mr. Hawkins: Exception.

A. You want me to start at the top of it?

Q. Yes, and go right through it. if you please?

A. No. 1 Westinghouse Compound Engine was there.

Q. What was its condition? Describe its condition.

Mr. Hawkins: That is objected to as immaterial.

The Referee: I will take it.

Mr. Hawkins: Exception.

A. The compound engine was there without any 726 trimming, as I have stated before, everything was stripped.

Q. Yes, but——

A. (Interrupting.) There wasn't anything left on it.

Q. Yes, but repeat your statement as it applies to each of those items?

A. Both numbers, 1, the Westinghouse Alternating and the Westinghouse Compound Engine were there. that is the bodies of the engine; and the direct connected double current generator was there, the 25 k.w.; the 15 horsepower motor, and

No. 1 item here, 16235, was not there; the transformers were not there; the tank was not there; the truck was not there; Fairbanks Scale was not there; Laidlaw-Dunn-Gordon Pump was there, but the inside was out of it; Westinghouse Electric Motor was not there; milling machine was not there; lathe No. 1 was not there, there was no tools of any description left there; the work benches were still there, but nothing on them; vises were gone; Westinghouse Type C 2-horsepower motor was gone; Westinghouse Type C inducting motor was gone; Westinghouse Type C 5-horsepower motor was gone; Westinghouse Motor about one-quarter horsepower was gone; the three lathes that he mentions here, only two could have been there at most at that time, the two that I have; I don't know the names of them. 727 728

Mr. Hawkins: I ask that be stricken out. only two could have been there.

The Referee: Well, you only got two, is that what you mean?

The Witness: Yes, sir.

Q. How many were there?

A. I don't know how many there was, quite a good many the first time I looked in there, but I know at the time we took possession from the time I went there, they were all carted away, some truck came in there for some place around there one day and I asked the agent there, and he said Mr. Tesla told this fellow—he runs a garage over there—that he could have them and he took a lot of stuff of that class. 729

Q. When was that?

A. I think that was along about a year before I got that notice.

Q. Proceed, please, with the other items.

730 A. Planer made by the Headley people, I see no planer there; planer made by Pedrick, no planers at all; no drill-press; that was gone; one large drill-press that I have; 36 lockers, they were all ripped to pieces; one testing fan motor——

Mr. Hawkins (interrupting): I ask that be stricken out.

The Referee: Denied.

Mr. Hawkins: Exception.

A. (Continuing.) That was gone; telephone and bell wire gone; quantity of lead cable gone; 4 radiators, they were gone; drills, bits, reamers, taps 731 and all tools for milling machines and lathes at present time in storeroom located inside workshop, that was all gone; oil tanks, they were ripped up and they evidently had torn them apart because they wanted to get something inside of them, either lead or copper, I don't know which.

Mr. Hawkins: That is objected to.

The Referee: Strike it out.

Q. Leave out your conclusions about why they did it; what was the condition of them?

A. Just ripped to pieces; all the meters and starting boxes and switches had all been stripped 732 off, only the bare slates left there; 2 Babcock & Wilcox Boilers, everything but the shell and tubes were gone, and feed pumps, just the housing was left; one hand——

The Referee (interrupting): What do you mean by that?

The Witness: Well, it is made out of cast iron, and the insides are brass, that is the valve seats and valves, they are always brass. They had been taken out. One hand blacksmith's forge was gone; toilets, urinals, wash basins, all ripped to pieces; 7 rheostats, desks, safes, 3 meters, all those things were gone; one set of storage batteries, tanks, submarine boat, Westinghouse Motor 28292, Westinghouse Motor Type C 5-horsepower No. 62320, Westinghouse Motor Type C 5-horsepower No. 22070; 4 high-tension transformers in tanks and switchboards, wiring drums, drafting boards and tools all gone. Chairs, there was two or three old chairs left there, was all; clocks, no clocks; radiators, no radiators at all. 733

Q. What did you find the condition of the tower to be?

Mr. Hawkins: I ask that the entire testimony of this witness concerning the items mentioned in the Defendants' Exhibit A, I think it was, be stricken from the record as incompetent, irrelevant and immaterial, especially because that testimony bears no weight upon the question as to whether that bill of sale was delivered as a security or as an absolute conveyance. 734

The Referee: Denied:

Mr. Hawkins: Exception.

Mr. Fordham: Read the last question.

The question was read by the stenographer.

Mr. Hawkins: That is objected to as incompetent, irrelevant and immaterial. 735

The Referee: I will let him describe what he found. Overruled.

Mr. Hawkins: Exception.

A. The tower was badly rotted, the main supports going up where the stairs were, the great big timbers were rotted out, they were half gone and it is a wonder they could stand up.

Mr. Hawkins: I move to strike that out.

The Referee: Yes, the wonder they ever stood, strike that out.

736 Mr. Hawkins: Yes.

The Witness: The stairs leading up to the top of the ball were half rotted away so that we could not get up to the ball. I wanted to see what the ball was made out of and I took a man down there, a rigger, and he went up about two-thirds of the way, climbing up over it, and he was so afraid he came back.

Mr. Hawkins: I ask that be stricken out.

The Referee: That he was afraid and came back, yes. I will let the fact stand that he did not go on up.

737 Mr. Hawkins: I ask the other be stricken out, that the rigger was sent up there.

The Referee: Denied.

Mr. Hawkins: Exception. And I also move that it be stricken out on the ground that it is in no sense rebuttal.

The Referee: Motion denied.

Mr. Hawkins: Exception.

Q. You may tell what the rigger did?

A. This big ball on top of the tower, you could not tell what it was made out of, whether it was brass or steel, as the ends of the wires where it had been grounded had rusted out and blown away, 738 and there was a thousand and one little wires sticking out in every direction, so you could not see what it was made up of.

The Referee: You could not get up?

The Witness: You could not get up. You could get up so you could see the fibres of everything up there, you could see it plain enough, but the tower was rotted in no end of places, it had never been taken care of, nothing had ever been done to it.

Mr. Hawkins: I ask that that be stricken out.

The Referee: Yes, granted as to the last part. 739

Mr. Fordham: What is granted?

The Referee: He said nothing had been done to it.

Mr. Hawkins: And I ask also that it be stricken out that the tower had not been taken care of.

The Referee: Granted. Describe its condition.

Q. Yes, you may tell the condition of the tower, Mr. Bingham. Was the condition secure or insecure?

A. Insecure. There was none of the woodwork that had ever been painted, all that held it together 740 was the big steel plates on the sides of it.

Q. As I understood, you say the woodwork was badly rotted out?

A. Rotted away, yes, sir.

Q. So that the tower in that condition was a menace to anybody passing near it, in view of its insecurity?

Mr. Hawkins: That is objected to as calling for a conclusion and a speculation.

The Referee: I think so.

Mr. Fordham: Not a speculation. The man is an expert in his own line.

The Referee: Objection sustained. 741

Mr. Hawkins: I ask that it be stricken out.

The Referee: Stricken out.

Q. Tell the Court from your own knowledge of structural materials, as an engineer, whether the tower was safe or unsafe, as you found it at that time?

Mr. Hawkins: That is objected to.

A. Absolutely unsafe.

The Referee: Have you sufficiently qualified him in that regard to testify at this time?

Q. (Interrupting.) Yes or no. 805

Mr. Fordham: Just a moment. If the Court please, I object to this on the ground that no foundation has been laid to qualify this witness.

The Referee: I will let him answer that question. Overruled.

Mr. Fordham: Exception.

A. Yes.

Q. At that time, to what use could the property be put?

Mr. Fordham: That is objected to, if the Court please, on the ground that it calls for the conclusion of an expert witness, and that there has no foundation been laid to qualify Mr. Tesla as an expert on real property value. 806

Mr. Hawkins: No, I have not asked him the value in that respect.

The Referee: I will take it.

Mr. Fordham: We except.

(The question was read.)

A. The property was expressly built for the transmission of wireless impulses.

The Referee: I do not think you understand the question, do you, Doctor? The question was to what use it was fitted, is that right? 807

Mr. Hawkins: Yes.

Mr. Fordham: I move to strike out this answer.

The Referee: Yes, strike it out.

Mr. Hawkins: I will formally except.

The Referee: I thought he misunderstood it.

A. The use it was built——

Q. (Interrupting.) No. Tell to what use it could be put at the time that deed was made?

808 A. At the time that deed was made it could have been used as a receiving wireless station.

The Referee: Yes.

The Witness: Pardon me for adding, it could also have been made use of as a transmitting station, but not to the extent that it could in the fully developed plant.

Q. But although it was not fully developed or permanently equipped, it could at that time have been used as a transmitting station?

A. Yes, sir.

Q. And as a receiving station?

809 A. As a receiving station.

Q. And there is now a large station of a similar kind very near your premises at Rocky Point, is there not?

Mr. Fordham: That is objected to, if the Court please, as immaterial and irrelevant.

The Referee: Overruled.

Mr. Fordham: Exception.

A. Yes, sir, there is, but it is of incomparably smaller power than mine.

Q. Are you familiar with the equipment and structures for the purpose of the receiving and

810 transmission of wireless messages?

A. Yes, sir, I am.

Q. What has been your experience in that line, Doctor?

A. I have worked for thirty years on the art and have given all of the fundamental principles to it; and during at least twenty years I have been making apparatus of that kind and experimenting with it.

Q. Have you been making aparatus of that kind for sale?

A. Yes and no. I did attempt to start manufacturing several times, but could not find sufficient encouragement, because at the time that I started the art was not sufficiently developed for the general public to have faith in it. I was ahead of the time, and that was the only reason why it was impracticable to start manufacturing. 811

Q. Have you invented and put on the market electrical apparatus for use in connection with wireless operations?

A. Yes, sir.

Q. At the time the deed was given, what was the value of the premises at Rocky Point in their condition at that time for the purpose of wireless receiving and transmitting uses? 812

Mr. Fordham: I object to that, if the Court please, on the ground there is no proper foundation laid to qualify this witness to speak of the commercial or financial value of the property at that time or at any other time. As to the scientific value or possibilities of it, he has already been interrogated. And it is objected to, if that is the point of the question, that is objected to on the ground it is needless repetition.

Mr. Hawkins: I submit, if your Honor please, the witness is qualified to answer this specific question, and has shown it by his testimony here. 813

The Referee: I don't understand, Mr. Hawkins. Are you asking him to testify as to the value of the land?

Mr. Hawkins: No, sir; I am asking him to testify as to the value of the entire premises, including the land and the buildings, but particularly the buildings.

Mr. Fordham: Well, commercially and

814 financially what is their value? And he knows nothing about it.

The Referee: I will overrule your objection and take it.

Mr. Fordham: Well, we except, if the Court please. I particularly call your Honor's attention to the fact that the testimony shows that the witness could not have known, because he had not been there for months before.

The Referee: I will take it for what it is worth.

Mr. Fordham: We except.

815 The Referee: Answer the question, Mr. Tesla, if you can.

A. At the time the deed was given a fair estimate of the value of the property would have been something like $350,000, because the income——

Q. Never mind all that, you have answered my question.

The Referee: You mean by that, taking in the land and your scientific development on it?

The Witness: No; I estimate it on the basis of earning power as a transmitting and receiving plant for the purpose for which it was made.

816 The Referee: Had it ever earned anything at that time?

The Witness: Yes, but because I was carrying on the plan which would ultimately have yielded $25,000 a day income, but at that time——

Q. (Interrupting.) Never mind, don't go on with that.

Mr. Fordham: I move to strike out the answer on the ground that the witness' explanation shows he is not qualified to make an estimate, and that his estimate as made is not based on any sound financial or legal or other ground. 817

The Referee: I am inclined to agree with you, but—I don't see, Mr. Hawkins, that that is admissible.

Mr. Hawkins: I submit that that is admissible. The man shows he has worked in that line of business for many years and knows the value of that equipment for that purpose.

The Referee: If you want it to stand, I will let it stand.

Mr. Hawkins: Yes, sir I do.

818

Mr. Fordham: We except.

The Referee: The objection is overruled.

Mr. Fordham: Our motion is denied to strike out?

The Referee: Yes, motion denied.

Mr. Fordham: We except.

Q. Did that condition which you have just described, and those values, obtain at the time the bill of sale was given?

A. Oh, at the time the bill of sale was given the property was very much more valuable, it was worth—it could have earned at least five times as much as the Tuckerton plant on Long Island, and they had an income of something like forty thousand or fifty thousand dollars a year.

819

The Referee: Well, it could have earned if it had been completed. Now, was it in that position to earn?

The Witness: I must explain it. If it had been completed, it could have earned $25,000 a day, but in that time in the state it was, if it had not been for my pushing the plant to com-

820 one hundred thousand or one hundred and twenty-five thousand dollars a year.

The Referee: What was it earning at that time?

The Witness: It was earning nothing.

Mr. Fordham: If the Court please, I move to strike out this last answer on the ground that it is contradictory to former testimony, because it appears that the deed and bill of sale were both given the same day, and it is impossible there could have been a wide difference in value between the few minutes when the deed was given and the bill of sale was given.

821

The Referee: I will let it stand.

The Witness: May I explain?

Q. Yes, explain.

A. Pardon me then, I did not understand the question. When I was asked when the deed was given, I had in mind when I first placed the property with Mr. Boldt, that was the valuation at that time.

Q. That was the first mortgage, wasn't it?

A. Oh, at the time the deed was given, now I understand better. Yes, that was 1915, the property was worth very much more because the art had been developed, the power stations had multiplied, the receivers had multiplied and where I would have had a hundred customers, then I would have thousands.

822

Q. Doctor, when you speak of the value at the time the bill of sale was given, do you mean the value at the time you first made a mortgage to Mr. Boldt?

A. No, sir, I mean at the time that the deed

was given, the property was worth more than $350,000. 823

Q. Yes, but what did you have in mind as the value when you spoke of the value as of the time the bill of sale was given?

A. I had in mind the value at the time I gave the mortgage to Mr. Boldt.

Q. Yes, the first mortgage?

A. Yes, the first mortgage.

Q. Upon the property to Mr. Boldt?

A. Yes, sir.

Q. Doctor, the property was developed for the purpose and use of a commercial wireless station, was it not? 824

A. Yes, sir.

Q. And so far as you know, it had no particular value for any other purpose, did it?

A. Yes, it might have been used for an electrical power plant for distribution.

Q. Yes.

A. In fact, the proposition was made to me at one time for that purpose.

Q. But looking at the situation from the local real estate market, it had no particular market value for any other purpose than that of wireless telegraphy, did it? 825

A. It might have as a factory building.

Q. But you are predicating your statement of values upon its uses for the purposes of wireless telegraphy, are you not?

A. For the purposes of the wireless art, yes.

Q. Wireless art?

A. Yes, in all its numerous applications.

Q. Do you remember Mr. Bingham saying that he went down to the property?

A. Yes, sir, I remember.

Defendants' Exhibit B.

925

KNOW ALL MEN BY THESE PRESENTS, That I, WILLIAM N. HALLOCK, of the City, County and State of New York, party of the first part, for and in consideration of the sum of One Hundred and more dollars, lawful money of the United States, to me in hand paid, at or before the ensealing and delivery of these presents, by Waldorf-Astoria Hotel Company, party of the second part, the receipt whereof is hereby acknowledged, have bargained and sold, and by these presents do grant and convey, unto the said party of the second part, its successors and assigns, all and several the chattels located in the brick factory building near Skeleton Tower on premises owned or heretofore owned by Nikola Tesla, immediately adjoining on the southerly side the railroad tracks of the Long Island Railroad at Shoreham Station, Long Island, in the Town of Brookhaven, Suffolk County, New York, including but not limited to the chattels specifically set out on the Schedule hereto annexed.

926

TO HAVE AND TO HOLD the same unto the said party of the second part, its successors and assigns forever. And I do for my heirs, executors and administrators, covenant and agree to and with the said party of the second part, to warrant and defend the sale of the said chattels hereby sold unto the said party of the second part, its successors and assigns against all and every person and persons whomsoever.

927

IN WITNESS WHEREOF, I have hereunto set my hand and seal the eighth day of April in the year one thousand nine hundred and fifteen.

WILLIAM N. HALLOCK.

[L. S.]

928 SCHEDULE OF FOREGOING BILL OF SALE.

IN THE GENERATING ROOM.

1 Westinghouse auto compound No. 1497, size 16 by 27 by 16,

1 direct connecting Westinghouse alternating current generator 200 Kw., Serial No. 155407, complete with lubricator, gauge, Rheostadt, switchboard and switches,

1 Westinghouse engine, No. 4750, size 8½ by 8, with direct connected double current generator, 25 kw., Serial No. 168362, complete with lubricator, gauge, Rheostats, switchboard and

929 switches,

1 15 H. P. Westinghouse motor, No. 162315,

4 Westinghouse transformers, 15 kw. type O. D.,

1 tank manufactured by Stoutenborough,

1 truck,

1 Fairbank's scale

1 Laidlaw Dunn-Gordon pump, No. 16473.

IN THE WORKSHOP.

1 Westinghouse electric motor, used for power to drive machine shop, type C, induction motor, 6 H. P., No. 162319,

930 1 Milling machine with tools complete, made by Brown & Sharp Manufacturing Company,

1 lathe made by Pond Machine Tool Company, No. P-3040, with tools, belting and shafting,

11 work benches,

4 vises,

1 Westinghouse, type C, 2 H. P. induction motor, No. 162278,

1 Westinghouse, type C, induction motor, 2 H. P. Serial No. 162272

1 Westinghouse, type C, induction motor, 5 H. P., No. L-74487

Defendants' Exhibit B.

---

1 Westinghouse motor, about 1/4 H. P., No. 22190 931
3 lathes made by F. E. Reed of Worcester, Mass. with shafting, belting and tools,
1 plainer made by Hendey Machine Co., with shafting, belting and tools.
1 plainer made by Pedrick & Ayr, with shafting, belting and tools.
1 F. E. Reed, hand drill press, shafting, belting and tools,
1 large drill press by Prentice Brothers, with shafting, belting and tools
36 lockers containing miscellaneous supply of valves, joints, lubricators, fittings, scales, switches, single and double pole, socket, wrenches, fuses and plugs, 932
1 testing fan motor,
A quantity of telephone and bell wire,
A quantity of lead cable material,
4 radiators,
A quantity of drills, rose bits, reamers, taps, and all tools for milling machine and lathes, at present time in store room located in said workshop,
2 oil tanks,
1 testing motion by Crocker Wheeler, 1/2 H.P. with Rheostat, No. 1000, 933
1 submarine boat,
1 clock
All of the aforesaid motors with starting boxes and switches.

BOILER ROOM.

2 Babcock & Wilcox boilers with steam gauges and water columns and with Metropolitan injector and Worthington feed pump,

Defendants' Exhibit B.

934 1 other feed pump

1 hand blacksmith and forge

7 toilets,
1 urinal,
6 wash basins, } all adjoining boiler room.

Testing or Laboratory Room.

7 Rheostats,
4 desks,
2 safes,
3 motors,
1 set of storage batteries and tanks
1 submarine boat,
1 Westinghouse motor, No. 28292
935
1 Westinghouse motor, type C, 5 H.P. No. 62320
1 Westinghouse motor, type C, 5 H.P. No. 22070,
4 high-tension transformers in tanks; and switchboards
Wiring drums
Drafting boards and tools,
24 chairs
2 clocks
14 radiators

State of New York,
County of New York } ss

936 On this eight day of April in the year of our Lord one thousand nine hundred and fifteen before me the undersigned personally came and appeared William N. Hallock to me known and known to me to be the individual described in and who executed the foregoing instrument, and he acknowledged to me that he executed the same.

Isidor W. Muller
Notary Public No. 45, Bronx County
Certificate filed New York County No. 85
Register's No. 6216
Commission expires March 30th, 1916

[notarial seal]

"The Supreme Court Documents on the Dismantling of the Wardenclyffe Tower" is a private legal document and under copyright.

Reproduced with permission of the publisher from the book *Nikola Tesla On His Work With Alternating Currents and Their Application to Wireless Telegraphy, Telephony, and Transmission of Power.*

NIKOLA TESLA: COMPLETE BIBLIOGRAPHY

Anderson, Leland I. Bibliography: Dr. Nikola Tesla (1856-1943) 2d enlarged edition. Minneapolis, Tesla Society [1956].

Belgrad. Muzej Nikole Tesle. Centenary of the birth of Nikola Tesla,1856-1956. Beograd, 1959.

Cheney, Margaret. Tesla, Man Out of Time :
Englewood Cliffs, N.J. : Prentice-Hall, 1981.

Hunt, Inez. Lightning in his Hand: The Life Story of Nikola Tesla. Hawthorne, CA: Omni Publications,1964.

Muzej Nikole Tesle. Tribute to Nikola Tesla. Presented in articles, letters, documents. Beograd, Nikola Tesla Museum, 1961.

Nikola Tesla—Covek i Pronalazac. <Dusan Nedeljkovic: Predgovor>. Beograd, Univerzitet—Elektrotehnicki fakultet, 1968.

O'Neill, John J. (John Joseph), 1889- Prodigal Genius: the Life ofNikola Tesla Angriff Press, 1981.

O'Neill, John Joseph, 1889- Nenadmasni genije, zwot Nikole Tesla.Predgovor Sava N. Kosanovic. Beograd, Prosveta, 1951.

Popovich, Vojislav, Nikola Tesla. Beograd, Tehnicka knjiga, 1951.

Ratzlaff, John T. Dr. Nikola Tesla bibliography : Ragusan Press, Palo Alto, Calif. 1979

Tesla Centennial Symposium (1984 : Colorado College) Proceedings of theTesla Centennial Symposium, held at Colorado College, Colorado Springs,Colorado, United States of America, August 9-12, 1984 editors, ElizabethAnn Rauscher and Toby Grotz. Colorado Springs, Colo. : International TeslaSociety, 1985.

12. Tesla, Nikola. The Problem of Increasing Human Energy. High Energy Enterprises, 1989.

13. Proceedings of the 1990 International Tesla Symposium, 1990, International Tesla Society, Colorado Springs.

14. Tesla, Nikola, 1856-1943. Moji pronalasci = My inventions NikolaTesla; preveli Tomo Bosanac, Vanja Aljinovic ; pogovor napisao TomoBosanac; pogovor preveo na engleski Janko Paravic ; urednik BranimiraValic. Zagreb : Školska knjiga ; [New York : distributed by W. S.Heinman], 1977.

15. Tesla, Nikola, 1856-1943. My Inventions : The Autobiography of Nikola Tesla edited, with an introduction, by Ben Johnston. Hart Bros., 1982.

16. Walters, Helen B. Nikola Tesla, Giant of Electricity. Illustrated by Leonard Everett Fisher. New York, Crowell, 1961.

17. Bearden, T, E. Fer-De-Lance: A Briefing on Soviet Scaler Electromagnetic Weapons. Tesla Book Co., 1986.

18. 43.Tesla, Nikola. Inventions, Researches, & Writings. Angriff Press

19. Corum, James F.: A Personal Computer Analysis of Spark Gap Tesla Coils. Corum & Associates, Inc,. 1988

20. Johnston, Benjamin H. And in Creating Live : The Early Life of Nikola Tesla. Hart Brothers Publishing, 1990.

21. Martin, T. C. Inventions, Reaserches & Writings of Nikola Tesla. Gordon Press Publishers, 1986

22. Martin, Thomas Commerford, 1856-1924. The inventions, researches, and writing of Nikola Tesla, with special reference to his work in polyphase currents and high potential lighting. by Thomas Commerford Martin. Omni Publications, Hawthorne, CA,1977.

23. Commander X, Nikola Tesla, Free Energy and the White Dove, 1992, Abelard.

24. Proceedings of the 1988 International Tesla Symposium, 1988, International Tesla Society, Colorado Springs.

25. Ratzlaff, John T., ed. Tesla: Complete Patents. Gordon Press Publishers,

26. Ratzlaff, John T. Tesla Said. Tesla Book Co., 1984.

27. Ratzlaff, John T. ed. Dr. Nikola Tesla-Selected Patent Wrappers from the National Archives. Tesla Book Co., 1981.

28. Ratzlaff, John T. Dr. Nikola Tesla Bibliography. Ragusan Press, 1979.

29. Ratzlaff, John T. Dr. Nikola Tesla—Complete Patents. 2nd ed. Tesla Book Co.,

30. Michael X, Tesla, Man Of Mystery, 1992, Inner Light.

31. Tesla, Nikola. The Tesla Coil. Revisionist Press, 1991.

32. Tesla, Nikola. Catalogue of Patents: A. Radmila. Vanous, Arthur, Co., 1988.

33. Tesla, Nikola. Museum Catalogue - Museum. Vanous, Arthur, Co., 1987.

34. Tesla, Nikola. Expirements with Alternating Currents. Gordon Press Publishers, 1986

35. Tesla, Nikola. Nikola Tesla: Colorado Springs Notes 1899-1900. Gordon Press Publishers, 1986

36. Tesla, Nikola. My Inventions: The Autobiography of Nikola Tesla. Hart Bothers Publishing. 1982.

37. Bearden, T E. Solutions to Tesla's Secrets & the Soviet Tesla Weapons with Reference Articles for Solutions to Tesla's Secrets. Tesla Book Co., 1982.

38. Norman, Ruth E. Tesla Speaks. Unarius Publications, 1973.

39. Tesla, Nikola. Tribute to: Museum. Vanous, Arthur, Co., 1961.

40. Hayes, Jeffery A. Boundary Layer Breakthrough: The Bladeless Tesla Turbine. 1990, High Energy Enterprises.

41. Tesla, Nikola. Colorado Springs Notes 1899 - 1900. Angriff Press.

42. Tesla, Nikola. Experiments with A. C. & Transmission of Electric Energy Without Wires. Angriff Press.

43. Anderson, Leland I., Nikola Tesla On His Work With Alternating Currents and Their Application to Wireless Telegraphy, Telephony, and Transmission of Power. Twenty-First Century Books, Colorado, 1993, 2002. www.tfcbooks.com

The Adventures Unlimited Catalog

Visit us online at:
www.adventuresunlimitedpress.com

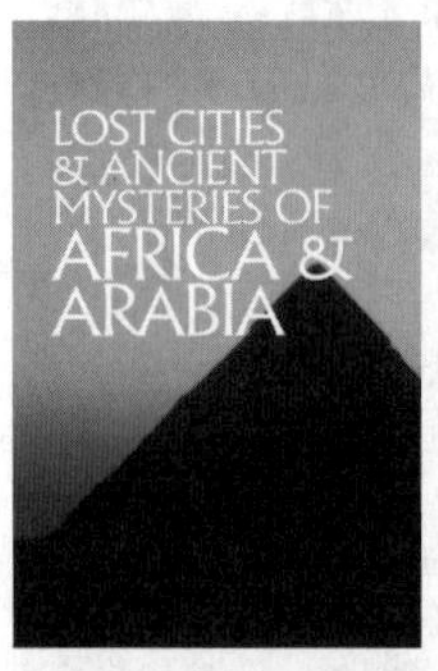

LOST CITIES & ANCIENT MYSTERIES OF AFRICA & ARABIA

by David Hatcher Childress

Childress continues his world-wide quest for lost cities and ancient mysteries. Join him as he discovers forbidden cities in the Empty Quarter of Arabia; "Atlantean" ruins in Egypt and the Kalahari desert; a mysterious, ancient empire in the Sahara; and more. This is the tale of an extraordinary life on the road: across war-torn countries, Childress searches for King Solomon's Mines, living dinosaurs, the Ark of the Covenant and the solutions to some of the fantastic mysteries of the past.

423 pages. 6x9 Paperback. Illustrated. $14.95. code: AFA

LOST CITIES OF ATLANTIS, ANCIENT EUROPE & THE MEDITERRANEAN

by David Hatcher Childress

Childress takes the reader in search of sunken cities in the Mediterranean; across the Atlas Mountains in search of Atlantean ruins; to remote islands in search of megalithic ruins; to meet living legends and secret societies. From Ireland to Turkey, Morocco to Eastern Europe, and around the remote islands of the Mediterranean and Atlantic, Childress takes the reader on an astonishing quest for mankind's past. Ancient technology, cataclysms, megalithic construction, lost civilizations and devastating wars of the past are all explored in this book.

524 pages. 6x9 Paperback. Illustrated. $16.95. code: MED

LOST CITIES OF CHINA, CENTRAL ASIA & INDIA

by David Hatcher Childress

Like a real life "Indiana Jones," maverick archaeologist David Childress takes the reader on an incredible adventure across some of the world's oldest and most remote countries in search of lost cities and ancient mysteries. Discover ancient cities in the Gobi Desert; hear fantastic tales of lost continents, vanished civilizations and secret societies bent on ruling the world; visit forgotten monasteries in forbidding snow-capped mountains with strange tunnels to mysterious subterranean cities! A unique combination of far-out exploration and practical travel advice, it will astound and delight the experienced traveler or the armchair voyager.

429 pages. 6x9 Paperback. Illustrated. Footnotes & Bibliography. $14.95. Code: CHI

LOST CITIES OF ANCIENT LEMURIA & THE PACIFIC

by David Hatcher Childress

Was there once a continent in the Pacific? Called Lemuria or Pacifica by geologists, Mu or Pan by the mystics, there is now ample mythological, geological and archaeological evidence to "prove" that an advanced and ancient civilization once lived in the central Pacific. Maverick archaeologist and explorer David Hatcher Childress combs the Indian Ocean, Australia and the Pacific in search of the surprising truth about mankind's past. Contains photos of the underwater city on Pohnpei; explanations on how the statues were levitated around Easter Island in a clockwise vortex movement; tales of disappearing islands; Egyptians in Australia; and more.

379 pages. 6x9 Paperback. Illustrated. Footnotes & Bibliography. $14.95. Code: LEM

A HITCHHIKER'S GUIDE TO ARMAGEDDON

by David Hatcher Childress

With wit and humor, popular Lost Cities author David Hatcher Childress takes us around the world and back in his trippy finalé to the Lost Cities series. He's off on an adventure in search of the apocalypse and end times. Childress hits the road from the fortress of Megiddo, the legendary citadel in northern Israel where Armageddon is prophesied to start. Hitchhiking around the world, Childress takes us from one adventure to another, to ancient cities in the deserts and the legends of worlds before our own. In the meantime, he becomes a cargo cult god on a remote island off New Guinea, gets dragged into the Kennedy Assassination by one of the "conspirators," investigates a strange power operating out of the Altai Mountains of Mongolia, and discovers how the Knights Templar and their off-shoots have driven the world toward an epic battle centered around Jerusalem and the Middle East.

320 PAGES. 6x9 PAPERBACK. ILLUSTRATED. BIBLIOGRAPHY. INDEX. $16.95. CODE: HGA

TECHNOLOGY OF THE GODS

The Incredible Sciences of the Ancients

by David Hatcher Childress

Childress looks at the technology that was allegedly used in Atlantis and the theory that the Great Pyramid of Egypt was originally a gigantic power station. He examines tales of ancient flight and the technology that it involved; how the ancients used electricity; megalithic building techniques; the use of crystal lenses and the fire from the gods; evidence of various high tech weapons in the past, including atomic weapons; ancient metallurgy and heavy machinery; the role of modern inventors such as Nikola Tesla in bringing ancient technology back into modern use; impossible artifacts; and more.

356 PAGES. 6x9 PAPERBACK. ILLUSTRATED. BIBLIOGRAPHY. $16.95. CODE: TGOD

VIMANA AIRCRAFT OF ANCIENT INDIA & ATLANTIS

by David Hatcher Childress, introduction by Ivan T. Sanderson

In this incredible volume on ancient India, authentic Indian texts such as the *Ramayana* and the *Mahabharata* are used to prove that ancient aircraft were in use more than four thousand years ago. Included in this book is the entire Fourth Century BC manuscript *Vimaanika Shastra* by the ancient author Maharishi Bharadwaaja. Also included are chapters on Atlantean technology, the incredible Rama Empire of India and the devastating wars that destroyed it.

334 PAGES. 6x9 PAPERBACK. ILLUSTRATED. $15.95. CODE: VAA

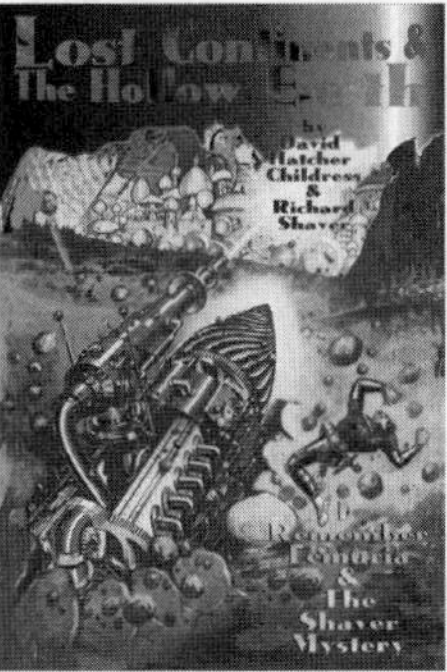

LOST CONTINENTS & THE HOLLOW EARTH

I Remember Lemuria and the Shaver Mystery

by David Hatcher Childress & Richard Shaver

Shaver's rare 1948 book *I Remember Lemuria* is reprinted in its entirety, and the book is packed with illustrations from Ray Palmer's *Amazing Stories* magazine of the 1940s. Palmer and Shaver told of tunnels running through the earth—tunnels inhabited by the Deros and Teros, humanoids from an ancient spacefaring race that had inhabited the earth, eventually going underground, hundreds of thousands of years ago. Childress discusses the famous hollow earth books and delves deep into whatever reality may be behind the stories of tunnels in the earth. Operation High Jump to Antarctica in 1947 and Admiral Byrd's bizarre statements, tunnel systems in South America and Tibet, the underground world of Agartha, the belief of UFOs coming from the South Pole, more.

344 PAGES. 6x9 PAPERBACK. ILLUSTRATED. $16.95. CODE: LCHE

ATLANTIS & THE POWER SYSTEM OF THE GODS

by David Hatcher Childress and Bill Clendenon

Childress' fascinating analysis of Nikola Tesla's broadcast system in light of Edgar Cayce's "Terrible Crystal" and the obelisks of ancient Egypt and Ethiopia. Includes: Atlantis and its crystal power towers that broadcast energy; how these incredible power stations may still exist today; inventor Nikola Tesla's nearly identical system of power transmission; Mercury Proton Gyros and mercury vortex propulsion; more. Richly illustrated, and packed with evidence that Atlantis not only existed—it had a world-wide energy system more sophisticated than ours today.

246 PAGES. 6X9 PAPERBACK. ILLUSTRATED. $15.95. CODE: APSG

THE ANTI-GRAVITY HANDBOOK

edited by David Hatcher Childress

The new expanded compilation of material on Anti-Gravity, Free Energy, Flying Saucer Propulsion, UFOs, Suppressed Technology, NASA Cover-ups and more. Highly illustrated with patents, technical illustrations and photos. This revised and expanded edition has more material, including photos of Area 51, Nevada, the government's secret testing facility. This classic on weird science is back in a new format!

230 PAGES. 7X10 PAPERBACK. ILLUSTRATED. $16.95. CODE: AGH

ANTI–GRAVITY & THE WORLD GRID

Is the earth surrounded by an intricate electromagnetic grid network offering free energy? This compilation of material on ley lines and world power points contains chapters on the geography, mathematics, and light harmonics of the earth grid. Learn the purpose of ley lines and ancient megalithic structures located on the grid. Discover how the grid made the Philadelphia Experiment possible. Explore the Coral Castle and many other mysteries, including acoustic levitation, Tesla Shields and scalar wave weaponry. Browse through the section on anti-gravity patents, and research resources.

274 PAGES. 7X10 PAPERBACK. ILLUSTRATED. $14.95. CODE: AGW

ANTI–GRAVITY & THE UNIFIED FIELD

edited by David Hatcher Childress

Is Einstein's Unified Field Theory the answer to all of our energy problems? Explored in this compilation of material is how gravity, electricity and magnetism manifest from a unified field around us. Why artificial gravity is possible; secrets of UFO propulsion; free energy; Nikola Tesla and anti-gravity airships of the 20s and 30s; flying saucers as superconducting whirls of plasma; anti-mass generators; vortex propulsion; suppressed technology; government cover-ups; gravitational pulse drive; spacecraft & more.

240 PAGES. 7X10 PAPERBACK. ILLUSTRATED. $14.95. CODE: AGU

THE TIME TRAVEL HANDBOOK

A Manual of Practical Teleportation & Time Travel

edited by David Hatcher Childress

The Time Travel Handbook takes the reader beyond the government experiments and deep into the uncharted territory of early time travellers such as Nikola Tesla and Guglielmo Marconi and their alleged time travel experiments, as well as the Wilson Brothers of EMI and their connection to the Philadelphia Experiment—the U.S. Navy's forays into invisibility, time travel, and teleportation. Childress looks into the claims of time travelling individuals, and investigates the unusual claim that the pyramids on Mars were built in the future and sent back in time. A highly visual, large format book, with patents, photos and schematics. Be the first on your block to build your own time travel device!

316 PAGES. 7X10 PAPERBACK. ILLUSTRATED. $16.95. CODE: TTH

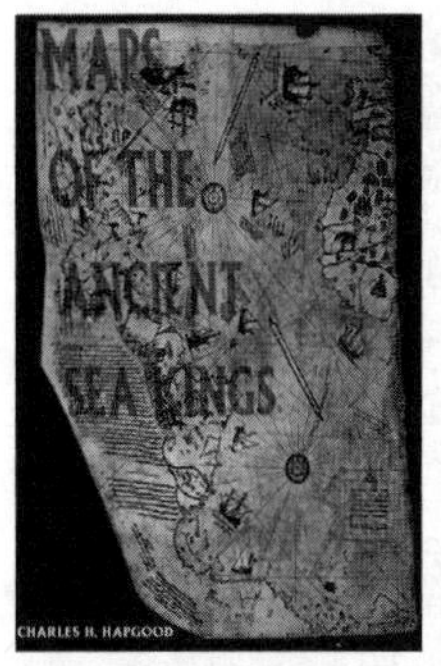

MAPS OF THE ANCIENT SEA KINGS

Evidence of Advanced Civilization in the Ice Age

by Charles H. Hapgood

Charles Hapgood has found the evidence in the Piri Reis Map that shows Antarctica, the Hadji Ahmed map, the Oronteus Finaeus and other amazing maps. Hapgood concluded that these maps were made from more ancient maps from the various ancient archives around the world, now lost. Not only were these unknown people more advanced in mapmaking than any people prior to the 18th century, it appears they mapped all the continents. The Americas were mapped thousands of years before Columbus. Antarctica was mapped when its coasts were free of ice!

316 PAGES. 7x10 PAPERBACK. ILLUSTRATED. BIBLIOGRAPHY & INDEX. $19.95. CODE: MASK

PATH OF THE POLE

Cataclysmic Pole Shift Geology

by Charles H. Hapgood

Maps of the Ancient Sea Kings author Hapgood's classic book *Path of the Pole* is back in print! Hapgood researched Antarctica, ancient maps and the geological record to conclude that the Earth's crust has slipped on the inner core many times in the past, changing the position of the pole. *Path of the Pole* discusses the various "pole shifts" in Earth's past, giving evidence for each one, and moves on to possible future pole shifts.

356 PAGES. 6x9 PAPERBACK. ILLUSTRATED. $16.95. CODE: POP

SECRETS OF THE HOLY LANCE

The Spear of Destiny in History & Legend

by Jerry E. Smith

Secrets of the Holy Lance traces the Spear from its possession by Constantine, Rome's first Christian Caesar, to Charlemagne's claim that with it he ruled the Holy Roman Empire by Divine Right, and on through two thousand years of kings and emperors, until it came within Hitler's grasp—and beyond! Did it rest for a while in Antarctic ice? Is it now hidden in Europe, awaiting the next person to claim its awesome power? Neither debunking nor worshiping, *Secrets of the Holy Lance* seeks to pierce the veil of myth and mystery around the Spear. Mere belief that it was infused with magic by virtue of its shedding the Savior's blood has made men kings. But what if it's more? What are "the powers it serves"?

312 PAGES. 6x9 PAPERBACK. ILLUSTRATED. BIBLIOGRAPHY. $16.95. CODE: SOHL

THE FANTASTIC INVENTIONS OF NIKOLA TESLA

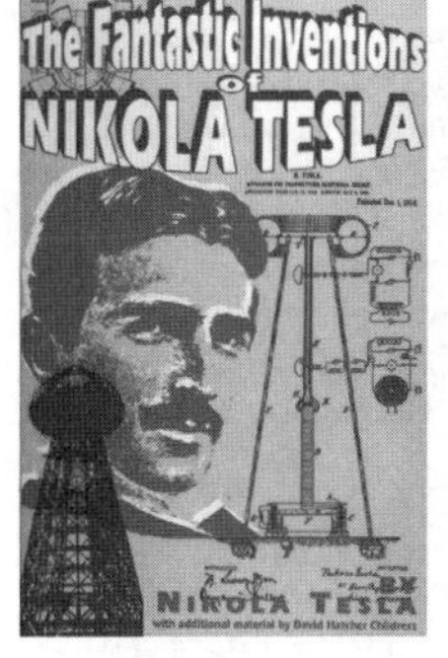

by Nikola Tesla with additional material by David Hatcher Childress

This book is a readable compendium of patents, diagrams, photos and explanations of the many incredible inventions of the originator of the modern era of electrification. In Tesla's own words are such topics as wireless transmission of power, death rays, and radio-controlled airships. In addition, rare material on a secret city built at a remote jungle site in South America by one of Tesla's students, Guglielmo Marconi. Marconi's secret group claims to have built flying saucers in the 1940s and to have gone to Mars in the early 1950s! Incredible photos of these Tesla craft are included. •His plan to transmit free electricity into the atmosphere. •How electrical devices would work using only small antennas. •Why unlimited power could be utilized anywhere on earth. •How radio and radar technology can be used as death-ray weapons in Star Wars.

342 PAGES. 6x9 PAPERBACK. ILLUSTRATED. $16.95. CODE: FINT

REICH OF THE BLACK SUN

Nazi Secret Weapons & the Cold War Allied Legend

by Joseph P. Farrell

Why were the Allies worried about an atom bomb attack by the Germans in 1944? Why did the Soviets threaten to use poison gas against the Germans? Why did Hitler in 1945 insist that holding Prague could win the war for the Third Reich? Why did US General George Patton's Third Army race for the Skoda works at Pilsen in Czechoslovakia instead of Berlin? Why did the US Army not test the uranium atom bomb it dropped on Hiroshima? Why did the Luftwaffe fly a non-stop round trip mission to within twenty miles of New York City in 1944? *Reich of the Black Sun* takes the reader on a scientific-historical journey in order to answer these questions. Arguing that Nazi Germany actually won the race for the atom bomb in late 1944,

352 PAGES. 6X9 PAPERBACK. ILLUSTRATED. BIBLIOGRAPHY. $16.95. CODE: ROBS

THE GIZA DEATH STAR

The Paleophysics of the Great Pyramid & the Military Complex at Giza

by Joseph P. Farrell

Was the Giza complex part of a military installation over 10,000 years ago? Chapters include: An Archaeology of Mass Destruction, Thoth and Theories; The Machine Hypothesis; Pythagoras, Plato, Planck, and the Pyramid; The Weapon Hypothesis; Encoded Harmonics of the Planck Units in the Great Pyramid; High Fregquency Direct Current "Impulse" Technology; The Grand Gallery and its Crystals: Gravito-acoustic Resonators; The Other Two Large Pyramids; the "Causeways," and the "Temples"; A Phase Conjugate Howitzer; Evidence of the Use of Weapons of Mass Destruction in Ancient Times; more.

290 PAGES. 6X9 PAPERBACK. ILLUSTRATED. $16.95. CODE: GDS

THE GIZA DEATH STAR DEPLOYED

The Physics & Engineering of the Great Pyramid

by Joseph P. Farrell

Farrell expands on his thesis that the Great Pyramid was a maser, designed as a weapon and eventually deployed—with disastrous results to the solar system. Includes: Exploding Planets: A Brief History of the Exoteric and Esoteric Investigations of the Great Pyramid; No Machines, Please!; The Stargate Conspiracy; The Scalar Weapons; Message or Machine?; A Tesla Analysis of the Putative Physics and Engineering of the Giza Death Star; Cohering the Zero Point, Vacuum Energy, Flux: Feedback Loops and Tetrahedral Physics; and more.

290 PAGES. 6X9 PAPERBACK. ILLUSTRATED. $16.95. CODE: GDSD

THE GIZA DEATH STAR DESTROYED

The Ancient War For Future Science

by Joseph P. Farrell

Farrell moves on to events of the final days of the Giza Death Star and its awesome power. These final events, eventually leading up to the destruction of this giant machine, are dissected one by one, leading us to the eventual abandonment of the Giza Military Complex—an event that hurled civilization back into the Stone Age. Chapters include: The Mars-Earth Connection; The Lost "Root Races" and the Moral Reasons for the Flood; The Destruction of Krypton: The Electrodynamic Solar System, Exploding Planets and Ancient Wars; Turning the Stream of the Flood: the Origin of Secret Societies and Esoteric Traditions; The Quest to Recover Ancient Mega-Technology; Non-Equilibrium Paleophysics; Monatomic Paleophysics; Frequencies, Vortices and Mass Particles; "Acoustic" Intensity of Fields; The Pyramid of Crystals; tons more.

292 pages. 6x9 paperback. Illustrated. $16.95. Code: GDES

THE TESLA PAPERS

Nikola Tesla on Free Energy & Wireless Transmission of Power

by Nikola Tesla, edited by David Hatcher Childress

David Hatcher Childress takes us into the incredible world of Nikola Tesla and his amazing inventions. Tesla's fantastic vision of the future, including wireless power, anti-gravity, free energy and highly advanced solar power. Also included are some of the papers, patents and material collected on Tesla at the Colorado Springs Tesla Symposiums, including papers on: •The Secret History of Wireless Transmission •Tesla and the Magnifying Transmitter •Design and Construction of a Half-Wave Tesla Coil •Electrostatics: A Key to Free Energy •Progress in Zero-Point Energy Research •Electromagnetic Energy from Antennas to Atoms •Tesla's Particle Beam Technology •Fundamental Excitatory Modes of the Earth-Ionosphere Cavity

325 PAGES. 8x10 PAPERBACK. ILLUSTRATED. $16.95. CODE: TTP

UFOS AND ANTI-GRAVITY

Piece For A Jig-Saw

by Leonard G. Cramp

Leonard G. Cramp's 1966 classic book on flying saucer propulsion and suppressed technology is a highly technical look at the UFO phenomena by a trained scientist. Cramp first introduces the idea of 'anti-gravity' and introduces us to the various theories of gravitation. He then examines the technology necessary to build a flying saucer and examines in great detail the technical aspects of such a craft. Cramp's book is a wealth of material and diagrams on flying saucers, anti-gravity, suppressed technology, G-fields and UFOs. Chapters include Crossroads of Aerodymanics, Aerodynamic Saucers, Limitations of Rocketry, Gravitation and the Ether, Gravitational Spaceships, G-Field Lift Effects, The Bi-Field Theory, VTOL and Hovercraft, Analysis of UFO photos, more.

388 PAGES. 6x9 PAPERBACK. ILLUSTRATED. $16.95. CODE: UAG

THE COSMIC MATRIX

Piece for a Jig-Saw, Part Two

by Leonard G. Cramp

Cramp examines anti-gravity effects and theorizes that this super-science used by the craft—described in detail in the book—can lift mankind into a new level of technology, transportation and understanding of the universe. The book takes a close look at gravity control, time travel, and the interlocking web of energy between all planets in our solar system with Leonard's unique technical diagrams. A fantastic voyage into the present and future!

364 PAGES. 6x9 PAPERBACK. ILLUSTRATED. BIBLIOGRAPHY. $16.00. CODE: CMX

THE A.T. FACTOR

A Scientists Encounter with UFOs

by Leonard Cramp

British aerospace engineer Cramp began much of the scientific anti-gravity and UFO propulsion analysis back in 1955 with his landmark book *Space, Gravity & the Flying Saucer* (out-of-print and rare). In this final book, Cramp brings to a close his detailed and controversial study of UFOs and Anti-Gravity.

324 PAGES. 6x9 PAPERBACK. ILLUSTRATED. BIBLIOGRAPHY. INDEX. $16.95. CODE: ATF

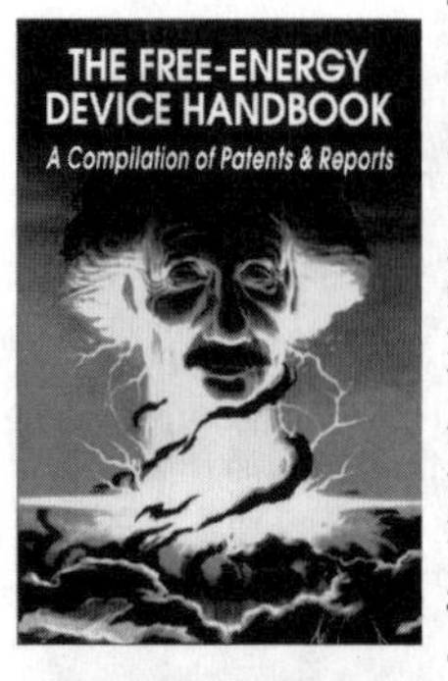

THE FREE-ENERGY DEVICE HANDBOOK
A Compilation of Patents and Reports
by David Hatcher Childress

A large-format compilation of various patents, papers, descriptions and diagrams concerning free-energy devices and systems. *The Free-Energy Device Handbook* is a visual tool for experimenters and researchers into magnetic motors and other "over-unity" devices. With chapters on the Adams Motor, the Hans Coler Generator, cold fusion, superconductors, "N" machines, space-energy generators, Nikola Tesla, T. Townsend Brown, and the latest in free-energy devices. Packed with photos, technical diagrams, patents and fascinating information, this book belongs on every science shelf.

292 PAGES. 8X10 PAPERBACK. ILLUSTRATED. $16.95. CODE: FEH

THE ENERGY GRID
Harmonic 695, The Pulse of the Universe
by Captain Bruce Cathie

This is the breakthrough book that explores the incredible potential of the Energy Grid and the Earth's Unified Field all around us. Cathie's first book, *Harmonic 33*, was published in 1968 when he was a commercial pilot in New Zealand. Since then, Captain Bruce Cathie has been the premier investigator into the amazing potential of the infinite energy that surrounds our planet every microsecond. Cathie investigates the Harmonics of Light and how the Energy Grid is created. In this amazing book are chapters on UFO Propulsion, Nikola Tesla, Unified Equations, the Mysterious Aerials, Pythagoras & the Grid, Nuclear Detonation and the Grid, Maps of the Ancients, an Australian Stonehenge examined, more.

255 PAGES. 6X9 TRADEPAPER. ILLUSTRATED. $15.95. CODE: TEG

THE BRIDGE TO INFINITY
Harmonic 371244
by Captain Bruce Cathie

Cathie has popularized the concept that the earth is crisscrossed by an electromagnetic grid system that can be used for anti-gravity, free energy, levitation and more. The book includes a new analysis of the harmonic nature of reality, acoustic levitation, pyramid power, harmonic receiver towers and UFO propulsion. It concludes that today's scientists have at their command a fantastic store of knowledge with which to advance the welfare of the human race.

204 PAGES. 6X9 TRADEPAPER. ILLUSTRATED. $14.95. CODE: BTF

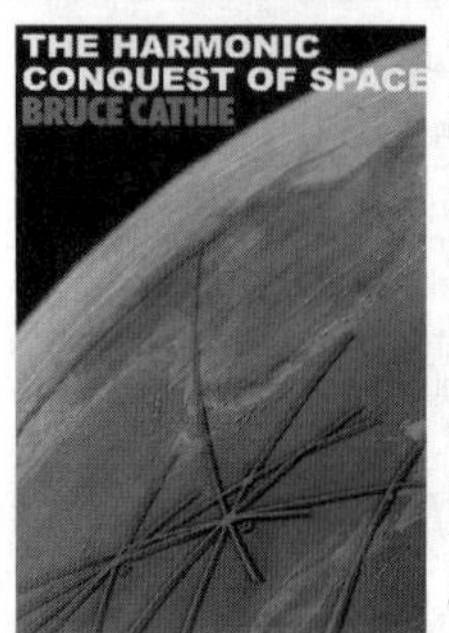

THE HARMONIC CONQUEST OF SPACE
by Captain Bruce Cathie

Chapters include: Mathematics of the World Grid; the Harmonics of Hiroshima and Nagasaki; Harmonic Transmission and Receiving; the Link Between Human Brain Waves; the Cavity Resonance between the Earth; the Ionosphere and Gravity; Edgar Cayce—the Harmonics of the Subconscious; Stonehenge; the Harmonics of the Moon; the Pyramids of Mars; Nikola Tesla's Electric Car; the Robert Adams Pulsed Electric Motor Generator; Harmonic Clues to the Unified Field; and more. Also included are tables showing the harmonic relations between the earth's magnetic field, the speed of light, and anti-gravity/gravity acceleration at different points on the earth's surface. New chapters in this edition on the giant stone spheres of Costa Rica, Atomic Tests and Volcanic Activity, and a chapter on Ayers Rock analysed with Stone Mountain, Georgia.

248 PAGES. 6X9. PAPERBACK. ILLUSTRATED. BIBLIOGRAPHY. $16.95. CODE: HCS

GRAVITATIONAL MANIPULATION OF DOMED CRAFT
UFO Propulsion Dynamics
by Paul E. Potter

Potter's precise and lavish illustrations allow the reader to enter directly into the realm of the advanced technological engineer and to understand, quite straightforwardly, the aliens' methods of energy manipulation: their methods of electrical power generation; how they purposely designed their craft to employ the kinds of energy dynamics that are exclusive to space (discoverable in our astrophysics) in order that their craft may generate both attractive and repulsive gravitational forces; their control over the mass-density matrix surrounding their craft enabling them to alter their physical dimensions and even manufacture their own frame of reference in respect to time. Includes a 16-page color insert.

624 pages. 7x10 Paperback. Illustrated. References. $24.00. Code: GMDC

TAPPING THE ZERO POINT ENERGY
Free Energy & Anti-Gravity in Today's Physics
by Moray B. King

King explains how free energy and anti-gravity are possible. The theories of the zero point energy maintain there are tremendous fluctuations of electrical field energy imbedded within the fabric of space. This book tells how, in the 1930s, inventor T. Henry Moray could produce a fifty kilowatt "free energy" machine; how an electrified plasma vortex creates anti-gravity; how the Pons/Fleischmann "cold fusion" experiment could produce tremendous heat without fusion; and how certain experiments might produce a gravitational anomaly.

180 PAGES. 5x8 PAPERBACK. ILLUSTRATED. $12.95. CODE: TAP

QUEST FOR ZERO-POINT ENERGY
Engineering Principles for "Free Energy"
by Moray B. King

King expands, with diagrams, on how free energy and anti-gravity are possible. The theories of zero point energy maintain there are tremendous fluctuations of electrical field energy embedded within the fabric of space. King explains the following topics: TFundamentals of a Zero-Point Energy Technology; Vacuum Energy Vortices; The Super Tube; Charge Clusters: The Basis of Zero-Point Energy Inventions; Vortex Filaments, Torsion Fields and the Zero-Point Energy; Transforming the Planet with a Zero-Point Energy Experiment; Dual Vortex Forms: The Key to a Large Zero-Point Energy Coherence. Packed with diagrams, patents and photos.

224 PAGES. 6x9 PAPERBACK. ILLUSTRATED. $14.95. CODE: QZPE

DARK MOON
Apollo and the Whistleblowers
by Mary Bennett and David Percy

Did you know a second craft was going to the Moon at the same time as Apollo 11? Do you know that potentially lethal radiation is prevalent throughout deep space? Do you know there are serious discrepancies in the account of the Apollo 13 'accident'? Did you know that 'live' color TV from the Moon was not actually live at all? Did you know that the Lunar Surface Camera had no viewfinder? Do you know that lighting was used in the Apollo photographs—yet no lighting equipment was taken to the Moon? All these questions, and more, are discussed in great detail by British researchers Bennett and Percy in *Dark Moon,* the definitive book (nearly 600 pages) on the possible faking of the Apollo Moon missions. Tons of NASA photos analyzed for possible deceptions.

568 PAGES. 6x9 PAPERBACK. ILLUSTRATED. BIBLIOGRAPHY. INDEX. $25.00. CODE: DMO

THE MYSTERY OF THE OLMECS
by David Hatcher Childress

The Olmecs were not acknowledged to have existed as a civilization until an international archeological meeting in Mexico City in 1942. Now, the Olmecs are slowly being recognized as the Mother Culture of Mesoamerica, having invented writing, the ball game and the "Mayan" Calendar. But who were the Olmecs? Where did they come from? What happened to them? How sophisticated was their culture? Why are many Olmec statues and figurines seemingly of foreign peoples such as Africans, Europeans and Chinese? Is there a link with Atlantis? In this heavily illustrated book, join Childress in search of the lost cities of the Olmecs! Chapters include: The Mystery of Quizuo; The Mystery of Transoceanic Trade; The Mystery of Cranial Deformation; more.

296 Pages. 6x9 Paperback. Illustrated. Bibliography. Color Section. $20.00. Code: MOLM

THE LAND OF OSIRIS
An Introduction to Khemitology
by Stephen S. Mehler

Was there an advanced prehistoric civilization in ancient Egypt who built the great pyramids and carved the Great Sphinx? Did the pyramids serve as energy devices and not as tombs for kings? Mehler has uncovered an indigenous oral tradition that still exists in Egypt, and has been fortunate to have studied with a living master of this tradition, Abd'El Hakim Awyan. Mehler has also been given permission to present these teachings to the Western world, teachings that unfold a whole new understanding of ancient Egypt . Chapters include: Egyptology and Its Paradigms; Asgat Nefer—The Harmony of Water; Khemit and the Myth of Atlantis; The Extraterrestrial Question; more.

272 pages. 6x9 Paperback. Illustrated. Color Section. Bibliography. $18.00 Code: LOOS

ABOMINABLE SNOWMEN: LEGEND COME TO LIFE
The Story of Sub-Humans on Six Continents from the Early Ice Age Until Today
by Ivan T. Sanderson

Do "Abominable Snowmen" exist? Prepare yourself for a shock. In the opinion of one of the world's leading naturalists, not one, but possibly four kinds, still walk the earth! Do they really live on the fringes of the towering Himalayas and the edge of myth-haunted Tibet? From how many areas in the world have factual reports of wild, strange, hairy men emanated? Reports of strange apemen have come in from every continent, except Antarctica.

525 pages. 6x9 Paperback. Illustrated. Bibliography. Index. $16.95. Code: ABML

INVISIBLE RESIDENTS
The Reality of Underwater UFOS
by Ivan T. Sanderson

In this book, Sanderson, a renowned zoologist with a keen interest in the paranormal, puts forward the curious theory that "OINTS"—Other Intelligences—live under the Earth's oceans. This underwater, parallel, civilization may be twice as old as Homo sapiens, he proposes, and may have "developed what we call space flight." Sanderson postulates that the OINTS are behind many UFO sightings as well as the mysterious disappearances of aircraft and ships in the Bermuda Triangle. What better place to have an impenetrable base than deep within the oceans of the planet? Sanderson offers here an exhaustive study of USOs (Unidentified Submarine Objects) observed in nearly every part of the world.

298 pages. 6x9 Paperback. Illustrated. Bibliography. Index. $16.95. Code: INVS

PIRATES & THE LOST TEMPLAR FLEET
The Secret Naval War Between the Templars & the Vatican
by David Hatcher Childress

Childress takes us into the fascinating world of maverick sea captains who were Knights Templar (and later Scottish Rite Free Masons) who battled the ships that sailed for the Pope. The lost Templar fleet was originally based at La Rochelle in southern France, but fled to the deep fiords of Scotland upon the dissolution of the Order by King Phillip. This banned fleet of ships was later commanded by the St. Clair family of Rosslyn Chapel (birthplace of Free Masonry). St. Clair and his Templars made a voyage to Canada in the year 1298 AD, nearly 100 years before Columbus! Later, this fleet of ships and new ones to come, flew the Skull and Crossbones, the symbol of the Knights Templar.

320 pages. 6x9 Paperback. Illustrated. Bibliography. $16.95. code: PLTF

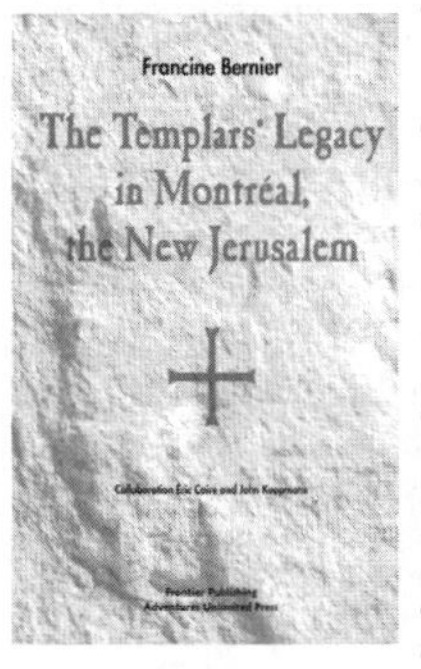

TEMPLARS' LEGACY IN MONTREAL
The New Jerusalem
by Francine Bernier

The book reveals the links between Montreal and: John the Baptist as patron saint; Melchizedek, the first king-priest and a father figure to the Templars and the Essenes; Stella Maris, the Star of the Sea from Mount Carmel; the Phrygian goddess Cybele as the androgynous Mother of the Church; St. Blaise, the Armenian healer or "Therapeut"- the patron saint of the stonemasons and a major figure to the Benedictine Order and the Templars; the presence of two Black Virgins; an intriguing family coat of arms with twelve blue apples; and more.

352 pages. 6x9 Paperback. Illustrated. Bibliography. $21.95. Code: TLIM

THE HISTORY OF THE KNIGHTS TEMPLARS
by Charles G. Addison, introduction by David Hatcher Childress

Chapters on the origin of the Templars, their popularity in Europe and their rivalry with the Knights of St. John, later to be known as the Knights of Malta. Detailed information on the activities of the Templars in the Holy Land, and the 1312 AD suppression of the Templars in France and other countries, which culminated in the execution of Jacques de Molay and the continuation of the Knights Templars in England and Scotland; the formation of the society of Knights Templars in London; and the rebuilding of the Temple in 1816. Plus a lengthy intro about the lost Templar fleet and its North American sea routes.

395 pages. 6x9 Paperback. Illustrated. $16.95. code: HKT

OTTO RAHN AND THE QUEST FOR THE HOLY GRAIL
The Amazing Life of the Real "Indiana Jones"
by Nigel Graddon

Otto Rahn led a life of incredible adventure in southern France in the early 1930s. The Hessian language scholar is said to have found runic Grail tablets in the Pyrenean grottoes, and decoded hidden messages within the medieval Grail masterwork *Parsifal*. The fabulous artifacts identified by Rahn were believed by Himmler to include the Grail Cup, the Spear of Destiny, the Tablets of Moses, the Ark of the Covenant, the Sword and Harp of David, the Sacred Candelabra and the Golden Urn of Manna. Some believe that Rahn was a Nazi guru who wielded immense influence on his elders and "betters" within the Hitler regime, persuading them that the Grail was the Sacred Book of the Aryans, which, once obtained, would justify their extreme political theories and revivify the ancient Germanic myths. But things are never as they seem, and as new facts emerge about Otto Rahn a far more extraordinary story unfolds.

450 pages. 6x9 Paperback. Illustrated. Appendix. Index. $18.95. Code: ORQG

EYE OF THE PHOENIX
Mysterious Visions and Secrets of the American Southwest
by Gary David

GaryDavid explores enigmas and anomalies in the vast American Southwest. Contents includes: The Great Pyramids of Arizona; Meteor Crater—Arizona's First Bonanza?; Chaco Canyon—Ancient City of the Dog Star; Phoenix—Masonic Metropolis in the Valley of the Sun; Along the 33rd Parallel—A Global Mystery Circle; The Flying Shields of the Hopi Katsinam; Is the Starchild a Hopi God?; The Ant People of Orion—Ancient Star Beings of the Hopi; Serpent Knights of the Round Temple; The Nagas—Origin of the Hopi Snake Clan?; The Tau (or T-shaped) Cross—Hopi/Maya/Egyptian Connections; The Hopi Stone Tablets of Techqua Ikachi; The Four Arms of Destiny—Swastikas in the Hopi World of the End Times; and more.

348 pages. 6x9 Paperback. Illustrated. Bibliography. $16.95. Code: EOPX

THE ORION PROPHECY
Egyptian and Mayan Prophecies on the Cataclysm of 2012
by Patrick Geryl and Gino Ratinckx

In the year 2012 the Earth awaits a super catastrophe: its magnetic field will reverse in one go. Phenomenal earthquakes and tidal waves will completely destroy our civilization. These dire predictions stem from the Mayans and Egyptians—descendants of the legendary Atlantis. The Atlanteans were able to exactly predict the previous world-wide flood in 9792 BC. They built tens of thousands of boats and escaped to South America and Egypt. In the year 2012 Venus, Orion and several others stars will take the same 'code-positions' as in 9792 BC!

324 PAGES. 6x9 PAPERBACK. ILLUSTRATED. $16.95. CODE: ORP

PRODIGAL GENIUS
The Life of Nikola Tesla
by John J. O'Neill

This special edition of O'Neill's book has many rare photographs of Tesla and his most advanced inventions. Tesla's eccentric personality gives his life story a strange romantic quality. He made his first million before he was forty, yet gave up his royalties in a gesture of friendship, and died almost in poverty. Tesla could see an invention in 3-D, from every angle, within his mind, before it was built; how he refused to accept the Nobel Prize; his friendships with Mark Twain, George Westinghouse and competition with Thomas Edison. Tesla is revealed as a figure of genius whose influence on the world reaches into the far future. Deluxe, illustrated edition.

408 pages. 6x9 Paperback. Illustrated. Bibliography. $18.95. Code: PRG

NOSTRADAMUS AND THE LOST TEMPLAR LEGACY
by Rudy Cambier

An analysis of the verses of Nostradamus' "prophecies" has shown that the language spoken in the verses belongs to the medieval times of the 14th Century, and the Belgian borders. The documents known as Nostradamus' prophecies were not written ca. 1550 by the French "visionary" Michel de Nostradame. Instead, they were composed between 1323 and 1328 by a Cistercian monk, Yves de Lessines, prior of the abbey of Cambron, on the border between France and Belgium. According to the author, these documents reveal the location of a Templar treasure.

204 PAGES. 6x9 PAPERBACK. ILLUSTRATED. BIBLIOGRAPHY. $17.95. CODE: NLTL

THE CRYSTAL SKULLS
Astonishing Portals to Man's Past
by David Hatcher Childress and Stephen S. Mehler

Childress introduces the technology and lore of crystals, and then plunges into the turbulent times of the Mexican Revolution form the backdrop for the rollicking adventures of Ambrose Bierce, the renowned journalist who went missing in the jungles in 1913, and F.A. Mitchell-Hedges, the notorious adventurer who emerged from the jungles with the most famous of the crystal skulls. Mehler shares his extensive knowledge of and experience with crystal skulls. Having been involved in the field since the 1980s, he has personally examined many of the most influential skulls, and has worked with the leaders in crystal skull research, including the inimitable Nick Nocerino, who developed a meticulous methodology for the purpose of examining the skulls.

294 pages. 6x9 Paperback. Illustrated. Bibliography. $18.95. Code: CRSK

THE INCREDIBLE LIGHT BEINGS OF THE COSMOS
Are Orbs Intelligent Light Beings from the Cosmos?
by Antonia Scott-Clark

Scott-Clark has experienced orbs for many years, but started photographing them in earnest in the year 2000 when the "Light Beings" entered her life. She took these very seriously and set about privately researching orb occurrences. The incredible results of her findings are presented here, along with many of her spectacular photographs. With her friend, GoGos lead singer Belinda Carlisle, Antonia tells of her many adventures with orbs. Find the answers to questions such as: Can you see orbs with the naked eye?; Are orbs intelligent?; What are the Black Villages?; What is the connection between orbs and crop circles? Antonia gives detailed instruction on how to photograph orbs, and how to communicate with these Light Beings of the Cosmos.

334 pages. 6x9 Paperback. Illustrated. References. $19.95. Code: ILBC

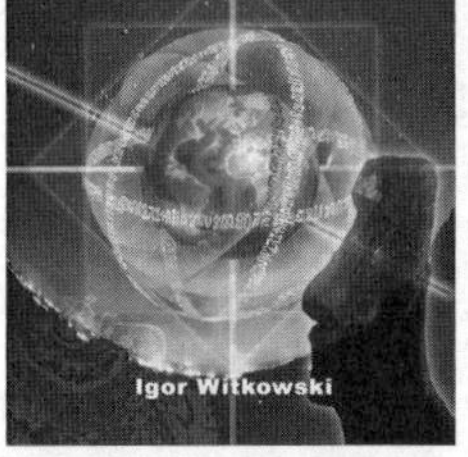

AXIS OF THE WORLD
The Search for the Oldest American Civilization
by Igor Witkowski

Polish author Witkowski's research reveals remnants of a high civilization that was able to exert its influence on almost the entire planet, and did so with full consciousness. Sites around South America show that this was not just one of the places influenced by this culture, but a place where they built their crowning achievements. Easter Island, in the southeastern Pacific, constitutes one of them. The Rongo-Rongo language that developed there points westward to the Indus Valley. Taken together, the facts presented by Witkowski provide a fresh, new proof that an antediluvian, great civilization flourished several millennia ago.

220 pages. 6x9 Paperback. Illustrated. References. $18.95. Code: AXOW

LEY LINE & EARTH ENERGIES
An Extraordinary Journey into the Earth's Natural Energy System
by David Cowan & Chris Arnold

The mysterious standing stones, burial grounds and stone circles that lace Europe, the British Isles and other areas have intrigued scientists, writers, artists and travellers through the centuries. How do ley lines work? How did our ancestors use Earth energy to map their sacred sites and burial grounds? How do ghosts and poltergeists interact with Earth energy? How can Earth spirals and black spots affect our health? This exploration shows how natural forces affect our behavior, how they can be used to enhance our health and well being.

368 PAGES. 6x9 PAPERBACK. ILLUSTRATED. $18.95. CODE: LLEE

SECRETS OF THE UNIFIED FIELD
The Philadelphia Experiment, the Nazi Bell, and the Discarded Theory
by Joseph P. Farrell

Farrell examines the now discarded Unified Field Theory. American and German wartime scientists and engineers determined that, while the theory was incomplete, it could nevertheless be engineered. Chapters include: The Meanings of "Torsion"; Wringing an Aluminum Can; The Mistake in Unified Field Theories and Their Discarding by Contemporary Physics; Three Routes to the Doomsday Weapon: Quantum Potential, Torsion, and Vortices; Tesla's Meeting with FDR; Arnold Sommerfeld and Electromagnetic Radar Stealth; Electromagnetic Phase Conjugations, Phase Conjugate Mirrors, and Templates; The Unified Field Theory, the Torsion Tensor, and Igor Witkowski's Idea of the Plasma Focus; tons more.

340 pages. 6x9 Paperback. Illustrated. Bibliography. Index. $18.95. Code: SOUF

NAZI INTERNATIONAL
The Nazi's Postwar Plan to Control Finance, Conflict, Physics and Space
by Joseph P. Farrell

Beginning with prewar corporate partnerships in the USA, including some with the Bush family, he moves on to the surrender of Nazi Germany, and evacuation plans of the Germans. He then covers the vast, and still-little-known recreation of Nazi Germany in South America with help of Juan Peron, I.G. Farben and Martin Bormann. Farrell then covers Nazi Germany's penetration of the Muslim world including Wilhelm Voss and Otto Skorzeny in Gamel Abdul Nasser's Egypt before moving on to the development and control of new energy technologies including the Bariloche Fusion Project, Dr. Philo Farnsworth's Plasmator, and the work of Dr. Nikolai Kozyrev. Finally, Farrell discusses the Nazi desire to control space, and examines their connection with NASA, the esoteric meaning of NASA Mission Patches.

412 pages. 6x9 Paperback. Illustrated. References. $19.95. Code: NZIN

ARKTOS
The Polar Myth in Science, Symbolism & Nazi Survival
by Joscelyn Godwin

Explored are the many tales of an ancient race said to have lived in the Arctic regions, such as Thule and Hyperborea. Progressing onward, he looks at modern polar legends: including the survival of Hitler, German bases in Antarctica, UFOs, the hollow earth, and the hidden kingdoms of Agartha and Shambala. Chapters include: Prologue in Hyperborea; The Golden Age; The Northern Lights; The Arctic Homeland; The Aryan Myth; The Thule Society; The Black Order; The Hidden Lands; Agartha and the Polaires; Shambhala; The Hole at the Pole; Antarctica; more.

220 Pages. 6x9 Paperback. Illustrated. Bib. Index. $16.95. Code: ARK

MIND CONTROL, WORLD CONTROL
The Encyclopedia of Mind Control
by Jim Keith

Keith uncovers a surprising amount of information on the technology, experimentation and implementation of Mind Control technology. Various chapters in this shocking book are on early C.I.A. experiments such as Project Artichoke and Project RIC-EDOM, the methodology and technology of implants, Mind Control Assassins and Couriers, various famous "Mind Control" victims such as Sirhan Sirhan and Candy Jones. Also featured in this book are chapters on how Mind Control technology may be linked to some UFO activity and "UFO abductions.

256 Pages. 6x9 Paperback. Illustrated. References. $14.95. Code: MCWC

HAARP
The Ultimate Weapon of the Conspiracy
by Jerry Smith

The HAARP project in Alaska is one of the most controversial projects ever undertaken by the U.S. Government. Jerry Smith gives us the history of the HAARP project and explains how works, in technically correct yet easy to understand language. At at worst, HAARP could be the most dangerous device ever created, a futuristic technology that is everything from super-beam weapon to world-wide mind control device. Topics include Over-the-Horizon Radar and HAARP, Mind Control, ELF and HAARP, The Telsa Connection, The Russian Woodpecker, GWEN & HAARP, Earth Penetrating Tomography, Weather Modification, Secret Science of the Conspiracy, more. Includes the complete 1987 Eastlund patent for his pulsed super-weapon that he claims was stolen by the HAARP Project.

256 pages. 6x9 Paperback. Illustrated. Bib. $14.95. Code: HARP

WEATHER WARFARE
The Military's Plan to Draft Mother Nature
by Jerry E. Smith

Weather modification in the form of cloud seeding to increase snow packs in the Sierras or suppress hail over Kansas is now an everyday affair. Underground nuclear tests in Nevada have set off earthquakes. A Russian company has been offering to sell typhoons (hurricanes) on demand since the 1990s. Scientists have been searching for ways to move hurricanes for over fifty. years. In the same amount of time we went from the Wright Brothers to Neil Armstrong. Hundreds of environmental and weather modifying technologies have been patented in the United States alone – and hundreds more are being developed in civilian, academic, military and quasi-military laboratories around the world *at this moment!* Numerous ongoing military programs do inject aerosols at high altitude for communications and surveillance operations.

304 Pages. 6x9 Paperback. Illustrated. Bib. $18.95. Code: WWAR

FREE ENERGY PIONEER
John Worrell Keely
by Theo Paijmans, foreword by John A. Keel

Over a century ago, a man in Philadelphia made the most important discovery of all time: a mysterious source of free, unlimited energy. He experimented with the substance for years, building a staggering 2,000 machines and devices that ran on his esoteric force. His eccentric vision led him to experiment with anti-gravity and the disintegration of solid matter. Lots of illustrations; Keely's mysterious source of unlimited, everlasting energy; Keely's many inventions; more.

416 pages. 6x9 Paperback. Illustrated. Bib. Index. $19.95. code: FEP

THE BOOK OF ENOCH
The Prophet
translated by Richard Laurence

This is a reprint of the Apocryphal *Book of Enoch the Prophet* which was first discovered in Abyssinia in the year 1773 by a Scottish explorer named James Bruce. One of the main influences from the book is its explanation of evil coming into the world with the arrival of the "fallen angels." Enoch acts as a scribe, writing up a petition on behalf of these fallen angels, or fallen ones, to be given to a higher power for ultimate judgment. Christianity adopted some ideas from Enoch, including the Final Judgment, the concept of demons, the origins of evil and the fallen angels, and the coming of a Messiah and ultimately, a Messianic kingdom.

224 PAGES. 6X9 PAPERBACK. ILLUSTRATED. INDEX. $16.95. CODE: BOE